Multi-faceted Deep Learning

Jenny Benois-Pineau • Akka Zemmari
Editors

Multi-faceted Deep Learning

Models and Data

 Springer

Editors
Jenny Benois-Pineau
LaBRI UMR 5800
University of Bordeaux
Talence Cedex, France

Akka Zemmari
LaBRI UMR 5800
University of Bordeaux
Talence Cedex, France

ISBN 978-3-030-74480-9   ISBN 978-3-030-74478-6   (eBook)
https://doi.org/10.1007/978-3-030-74478-6

This Springer imprint is published by the registered company Springer Nature Switzerland AG
The registered company address is: Gewerbestrasse 11, 6330 Cham, Switzerland

"And now, if you will set us to our task,
We will serve you four and twenty hours a
day . . ."
  Rudyard Kipling, The secret of machines

We dedicate this book to our students.
Be curious, be inventive, persevere and
serve the Dame Science!

Preface

Today, artificial intelligence approaches penetrate all areas of societal activity. One of its main branches, artificial neural networks, got a new life with the drastic augmentation of computational capacities due to graphical processing units and cloud computing. Neural networks have become deep. Deep learning is now a winner in all supervised machine learning approaches which have been ever used for data mining and decision-making.

These tools are specifically interesting in the field which has been traditionally called "multimedia." Indeed, this field supplies a huge amount of heterogeneous data: images, video, audio and music, text, and multimodal signals. Furthermore, these data have a spatio-temporal grid structure which is convenient for one of the varieties of deep learning networks, such as convolutional neural networks.

Hence, in this book, we tried to provide a snapshot of methods, models, and data which are being developed or used in this research community. This book is a collective work of selected researchers at the French National Network GDR-ISIS and ACM Special Interest Group on Multimedia. We hope this book will be interesting for young researchers, student, and professionals who are employing existing models and designing new ones in the framework of deep learning.

Bordeaux, France
December 2020

Jenny Benois-Pineau
Akka Zemmari

Acknowledgments

The editors of this book acknowledge the French National Research Network CNRS GDR-ISIS and also ACM-SIGMM which enabled the authors of this book to collaborate together.

Contents

Chapter 1
Introduction

Jenny Benois-Pineau

Artificial Intelligence (AI), and particularly Deep Learning, is changing our daily life. In last few years, this domain gained much interest both from theoretical and practical point of views. AI based solutions are deployed in many applications and are used in many fields. This includes financial applications, security, trading, autonomous vehicles, etc.

Deep Learning gained very high interest with the huge amount of available data and the new computation capabilities. It benefits also from a long research activities on neural networks started by the first works of W. McCulloch and W. Pitts on 1943 who first defined the formal neuron and crowned by the award of the Turing prize to Y. LeCun, Y. Bengio and G. Hinton in 2018.

In this book, we present most popular Artificial Intelligence methods for data mining of nowadays form multifaceted perspective: in problems, methods and data. The book gives a rich overview of ongoing research in the community and is written by the first rank international researchers.

The book starts by introducing the design and implementation of various architectures for Deep Learning, together with optimization algorithms. It discusses the most state-of-the-art networks, such as Artificial Neural Networks, Convolutional Neural Networks and Recurrent Networks. Then it presents some other models like Generative Neural Networks, Autoencoders and Siamese CNNs.

As a first application of Deep Learning methods, we consider its use for semantic segmentation. A chapter reviews the image semantic segmentation task and recent advanced strategies to face typical training issues (few training samples, specific data, strong target imbalance, . . .) in a variety of application domains. Another chapter considers image and video captioning using deep learning. It aims at giving

J. Benois-Pineau (✉)
LaBRI UMR 5800, University of Bordeaux, Talence Cedex, France
e-mail: jenny.benois-pineau@u-bordeaux.fr

© Springer Nature Switzerland AG 2021
J. Benois-Pineau, A. Zemmari (eds.), *Multi-faceted Deep Learning*,
https://doi.org/10.1007/978-3-030-74478-6_1

insights on how to generate descriptive sentences from images and videos. A third application investigates the use of the 3D Convolutional Neural Networks for action recognition with application to sport gesture recognition.

As mentioned above, the impressive success of Deep Learning is due to the huge amount of available data. However, for supervised learning, this data have to be labeled. In a dedicated chapter, we present solutions based on three families of methods for learning with less expensive labelling. We detail an approach based on these methods applied for using incomplete annotations for medical image segmentation.

Another interesting problem is related to similarity metric learning. It aims to model the general semantic similarities and distances between classes of objects (e.g. related to persons) in order to recognize them. As opposed to supervised learning, the class labels are usually not known in advance and not explicitly trained. An overview of the principle methods and models used for similarity metric learning with neural networks is given and several recent applications on person re-identification and face verification in images are presented.

Zero-shot learning (ZSL) deals with the ability to recognize objects without any visual training samples. To counterbalance this lack of visual data, each class to recognize is associated to a semantic prototype that reflects the essential features of the object. The general approach is to learn a mapping from visual data to semantic prototypes, then use it at inference to classify visual samples from the class prototypes only. A chapter presents a review of the approaches based on deep neural networks to tackle the ZSL problem. We also propose several contributions to improve the performances of usual approaches to address the most interesting practical use cases.

Video compression is one of the interesting problems in multimedia research. A chapter of this book explores the use of Deep Learning to improve video coding with the primary purpose of improving video compression rates while retaining same video quality. The chapter provides an overview of state-of-the-art research work, providing examples of few prominent published papers that illustrate and further explain the different highlighted topics in the field of using Deep Learning for video compression.

A chapter provides an overview of how Deep Learning techniques can be used for audio signals. It reviews the three main classes of applications: audio recognition, audio transformation and audio synthesis for three main classes of audio content: speech, music and environmental sounds.

The recent focus of AI and Pattern Recognition communities on the supervised learning approaches, and particularly on Deep Learning/AI, resulted in considerable increase of performance of AI systems, but also raised the question of the trustfulness and applicability of their predictions for decision-making. A chapter discusses how Explainable AI affords such issues in medical imaging. It presents the most part of machine learning approaches developed for breast cancer diagnosis.

A chapter covers deep learning methodologies that can be employed to recover image and video quality. Most of the covered approaches will be based on conditional Generative Adversarial Networks (GAN) which have the benefit to produce images which look more natural.

We invite our readers to make an excursion into this diversified and rich panorama of Deep Learning approaches.

Chapter 2
Deep Neural Networks: Models and Methods

Akka Zemmari and Jenny Benois-Pineau

2.1 Artificial Neural Networks

Artificial neural networks consist of distributed information processing units. In this section, we define the components of such networks. We will first introduce the elementary unit: the formal neuron introduced by McCulloch and Pitts in [MP43]. Further we will explain how such units can be assembled to design simple neural networks.

2.1.1 Formal Neuron

The neuron is the elementary unit of the neural network. It receives input signals $(x_1, x_2, \cdots, x_p)$, applies an activation function f to a linear combination z of the signals. This combination is determined by a vector of weights $(w_1, w_2, \cdots, w_p)$ and a bias b. More formally, the output neuron value y is defined as follows:

$$y = f(z) = f\left(\sum_{i=1}^{p} w_i x_i + b\right).$$

Figure 2.1a sums up the definition of a formal neuron. A compact representation of the same is given in Fig. 2.1b.

A. Zemmari (✉) · J. Benois-Pineau
LaBRI UMR 5800, University of Bordeaux, Talence Cedex, France
e-mail: akka.zemmari@u-bordeaux.fr; jenny.benois-pineau@u-bordeaux.fr

© Springer Nature Switzerland AG 2021
J. Benois-Pineau, A. Zemmari (eds.), *Multi-faceted Deep Learning*,
https://doi.org/10.1007/978-3-030-74478-6_2

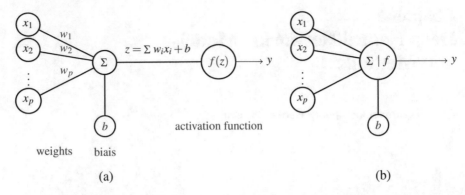

Fig. 2.1 Two representations of a formal neuron: A detailed representation (**a**) and a compact one (**b**). Note that when it is a part of a neural network, the neuron is simply represented by a vertex

2.1.1.1 Activation Functions

Different activation functions are commonly encountered in neural networks:

The Step Function ξ_c

$$\xi_c(x) = \begin{cases} 1 \text{ if } x > c \\ 0 \text{ otherwise.} \end{cases} \tag{2.1}$$

This simplistic function was the first activation function considered. Its main problem is that it can activate different labels to 1, and the problem of classification is not solved. As a consequence, the use of smooth functions is preferred, as it gives analog activations rather than binary ones, thus the risk of having several labels scored 1 is widely reduced for smooth activation functions.

The Sigmoid Function σ

$$\sigma(x) = \frac{1}{1 + e^{-x}}. \tag{2.2}$$

It is one of the most popular activation functions. It maps $\mathbb{R}$ onto the interval $[0, 1]$. This function is a smooth approximation of the step function, see Fig. 2.2a. It has many interesting properties.

The continuity of the function enables to properly train networks for non-binary classification tasks. Its differentiability is a good property in theory because of the

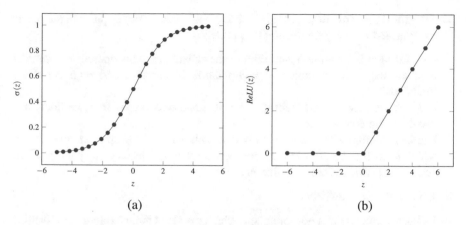

Fig. 2.2 Two particular activation functions: (**a**) the sigmoid function and (**b**) the Rectified Linear Unit (Relu) function

way neural networks "learn". Furthermore, it has a steep gradient around 0, which means that this function has a tendency to bring the y values to either end of the curve: this is a good behaviour for classification, as it makes clear distinctions between predictions. Another good property of this activation function is that it is bounded: this prevents divergence of the activations.

The biggest problem of the sigmoid is that it has very small gradient when the argument is distant from 0. This is responsible for a phenomenon called the vanishing of the gradient: the learning is drastically slowed down, if not stopped.

The Tanh Function

$$\tanh(x) = \frac{2}{1 + e^{-2x}} - 1 = 2\sigma(2x) - 1. \tag{2.3}$$

As the above equation suggests, this function is in fact a scaled and vertically shifted sigmoid function that maps $\mathbb{R}$ onto the interval $[-1, 1]$, thus it shares the same pros and cons. The main difference between tanh and sigmoid lies in the strength of the gradient: tanh tends to have stronger gradient values. Just like the sigmoid, tanh is also a very popular activation function.

The ReLU Function

$$\mathrm{ReLU}(x) = \max(0, x). \tag{2.4}$$

The Rectified Linear Unit (ReLU) (Fig. 2.2b) has become very popular in the recent years. It indeed has very interesting properties:

- This function is a simple tresholding of the activations. This operation is simpler than the expensive exponential computation in sigmoid and tanh activation functions.
- ReLU tends to accelerate training a lot. It supposedly comes from its linear and non-bounded component.
- Unlike in sigmoid-like activations where each neuron fires up in an analog way which is costly, the 0-horizontal component of ReLU leads to a sparsity of the activations, which is computationally efficient.

It also has its own drawbacks:

- Its linear component is non-bounded, which may lead to an exploding activation.
- The sparsity of activations can become detrimental to the training of a network: when a neuron activates in the 0-horizontal component of ReLU, the gradient vanishes and training stops for this neuron. The neuron "dies". ReLU can potentially make a substantial part of the network passive.
- It is non-differentiable in 0: problem with the computation of the gradient near 0.

Many other activation functions exist. They all have in common the non-linear property, which is essential: if they were linear, the whole neural network would be linear (a linear combination is fed to a linear activation, which is fed to a linear combination etc...), but in this case no matter how many layers we have, those linear layers would be equivalent to a unique linear layer and we would then loose the multi-layer architecture characteristic to neural networks: the final layer would simply become a linear transformation applied to the input of the first layer.

Softmax or How to Transform Outputs into Probabilities

Given a vector $\mathbf{y} = (y_1, y_2, \cdots, y_k)$ with positive real-valued coordinates, the softmax function aims to transform the values of y to a vector $\mathbf{s} = (p_1, p_2 \cdots, p_k)$ of real values in the range $(0, 1)$ that sums to 1. More precisely, it is defined for each $i \in \{1, 2, \cdots, k\}$ by:

$$p_i = \frac{e^{y_i}}{\sum_{j=1}^{k} e^{y_j}}.$$

The softmax function is used in the last layer of multi-layer neural networks which are trained under a cross-entropy (we will define this function in next paragraphs) regime. When used for image recognition, the softmax computes the estimated probabilities, for each input data, of being in a class from a given taxonomy.

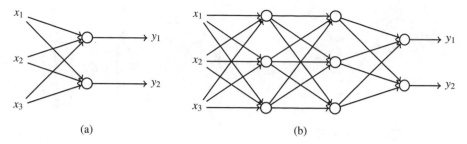

Fig. 2.3 Two neural networks: both have three inputs x_1, x_2 and x_3, and two outputs y_1 and y_2. Network (**a**) is a single-layer neural network and (**b**) a multilayers network with two hidden layers

2.1.2 Artificial Neural Networks and Deep Neural Networks

An artificial Neural Network (ANN) incorporates the biological geometry and behaviour into its architecture in a simplified way: a neural network consists of a set of layers disposed linearly, and each layer is a set of (artificial) neurons. The model is simplified so that signals can only circulate from the first layer to the last one.

Each neuron in the ith layer of the neural network is connected to all the neurons in the $(i-1)$th layer, and the neurons in a given layer are all independent from each other. Figure 2.3 gives two examples of simple artificial neural networks. It is "feedforward network" as the information flows from the input **x** to the output **y** only in one direction. It is also important to note that *all* the neurons of the same layer have the *same* activation function.

A network with all the inputs connected directly to the outputs is a single-layer network. Figure 2.3a shows an example of such a network. In [MP43], McCulloch and Pitts prove that a network with a single unit can represent the basic Boolean functions AND and OR. However, it is also easy to prove that a network with a single layer can not be used to represent XOR function.

All the neurons of the same layer l have he same activation function $f^{(l)}$. That is, given a neural network with $L > 0$ layers, if we denote by $\mathbf{y}^{(l)}$ the vector with the outputs of layer l, for any $1 < l < L$, then, one can write:

$$\mathbf{y}^{(l+1)} = f^{(l+1)}\left(w^{(l)^T}\mathbf{y}^{(l)} + \mathbf{b}^{(l+1)} \right). \tag{2.5}$$

Expression (2.5) explains how the values given as input to the network are forwarded to compute the value of the output $\hat{\mathbf{y}} = \mathbf{y}^{(L)}$ (Fig. 2.4).

Any neural network must contain at least two layers: one for input and one for output. As discussed, some functions can not be implemented using a neural network without any hidden layer (XOR gate for example). However, Cybenko, in [Cyb89], proved that any continuous function can be approximated with any desired accuracy, in terms of the uniform norm, with a network of one hidden layer whose

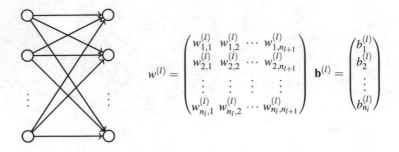

$$w^{(l)} = \begin{pmatrix} w_{1,1}^{(l)} & w_{1,2}^{(l)} & \cdots & w_{1,n_{l+1}}^{(l)} \\ w_{2,1}^{(l)} & w_{2,2}^{(l)} & \cdots & w_{2,n_{l+1}}^{(l)} \\ \vdots & \vdots & \vdots & \vdots \\ w_{n_l,1}^{(l)} & w_{n_l,2}^{(l)} & \cdots & w_{n_l,n_{l+1}}^{(l)} \end{pmatrix} \quad \mathbf{b}^{(l)} = \begin{pmatrix} b_1^{(l)} \\ b_2^{(l)} \\ \vdots \\ b_{n_i}^{(l)} \end{pmatrix}$$

layer l layer $l+1$
with n_l units.   with n_{l+1} units.

Fig. 2.4 Two layers, the weights matrix between the two layers l and $l+1$ and the bias vector of layer l

activation functions are sigmoids. A *deep* neural network is an artificial neural network with *at least one* hidden layer.

2.2 Convolutional Neural Networks

We introduce here convolutional neural networks which were originally designed for image analysis. First of all we will expose some general principles, then go into detail layer-by-layer and finally briefly overview most popular convolutional neural networks architectures.

2.2.1 General Principles

Regular neural networks do not scale well for visual information mining. Their architecture, for a given neuron in a certain layer, connects all the activations from the previous layer to that neuron. In the context of Computer Vision, data takes the form of images (or videos), that is at least three dimensional objects. Let's imagine that we want to connect a neuron of the first layer to each pixel of a color image of rather small size $200 \times 200 \times 3$. This would give no less than 120,000 weights, just for one neuron in the first layer. Even shallow neural networks would become unmanageable regarding the number of weights (recall that each weight should be learned during the training).

On the other hand, it is crucial in image processing to consider the spatially-local correlation in images. For instance, if we wish to train an edge detector, it does not make sense to consider the whole image all at once for each neuron, we should rather

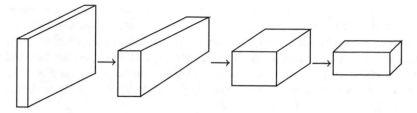

Fig. 2.5 In a CNN, the dimensions decrease layers after layers

decompose the image into different windows and apply the filter on each window, in order to make it react to windows with strong edges.

Convolutional neural networks (abbreviated CNNs) were specifically designed for Computer Vision tasks and their architecture presents differencies compared to regular neural networks. The most obvious difference is that the layers of a CNN have their neurons arranged in three dimensions (height, width and depth) so as to match the geometric shape of the data (images can be seen as cuboids of pixels).

In Fig. 2.5, one can notice that the spatial dimensions decrease layers after layers, whereas the depth dimension increases. In most cases, the first layers of a CNN (sometimes called bottom layers) are locally-connected, that is their spatial support (also called receptive field) is limited which helps obtaining spatially-local correlation information. Furthermore, those first layers reduce the spatial dimensions by pooling operation. Once the spatial dimensions are small enough (when the top layers are reached), fully-connected layers are often used just like in a regular neural network, and the last layer is a fully-connected layer that computes the class scores.

The layers used in the bottom layers of CNNs are obviously characteristic of this particular type of neural network, due to their locally-connected property. Convolutional neural networks are indeed biologically-inspired: Hubel and Wiesel showed in [HW68] from the observation of animal's visual cortex that the latter consists of complex arrangements of cells, and that those cells are sensitive only to limited parts of the global visual field.

2.2.2   Layers of a CNN

2.2.2.1   Convolutional Layers

Convolutional layers are at the core of the CNN architecture: they react to spatially-local correlation in input images. Those layers consist of a set of filters of limited spatial size, and those filters have their weights trainable. The name convolutional layer comes from the way this layer type behaves: it makes its filters slide across the whole image exactly as we compute convolution of an image with filter. On the contrary to the image processing operations the shift of filter mask when sliding along the input map can be greater than 1. This shift is called "stride" parameter and is a part of general settings or "hyper-parameters" of a Deep Convolutional Neural Network. At each position taken by a filter, a dot product is performed between the filter and the corresponding region in the image. Once all the filters are done sliding over the whole image, an activation map is obtained for each of those filters. Those activation maps are stored along the depth, which explains why the depth dimension tends to increase in CNNs layer after layer.

In practice, each filter learned by the CNN will activate when they encounter a specific visual feature. The filters tend to become more and more abstract when the layers go up. Filters from the bottom layers tend to react to simple objects such as edges, specific shapes or colors, whereas filters from the upper layers can react to more complex objects, such as buildings, animals, etc...depending on the dataset they were trained on.

In the sequel, we will give the intuition on how the use of CNN reduces the number of parameters (weights) of the network. For the sake of simplicity, consider that each image is encoded using a matrix whose elements are 0 or 1 as explained in Fig. 2.6.

Given a matrix (corresponding to an image), the convolution layers learn a set of filters. Figure 2.7 gives an example of the result of such operation when applied to the matrix of Fig. 2.6 with a filter of size 3×3 using a stride equal to 1.

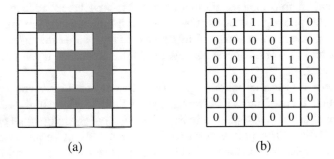

(a)   (b)

Fig. 2.6 Image encoding: (**a**) a handwritten digit and (**b**) its encoding

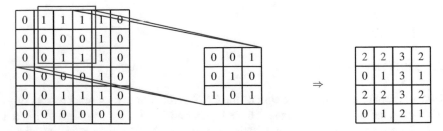

Fig. 2.7 The convolution of a matrix (6×6) corresponding to the image in Fig. 2.6 and a filter (3×3) and the resulting matrix

Fig. 2.8 The first two layers of the CNN for the previous example: the two units (red and blue) are connected to a subset of the input layer units and not to all of them and the two units share the same weights for some of the previous units

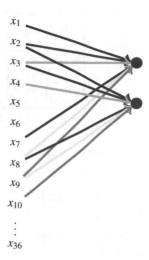

Figure 2.7 explains how the convolution of an image (6×6) with a filter (3×3) and a stride equal to 1 is computed. We can observe that the resulting image is of smaller size (4×4). Filters correspond to the weights to learn and induce the connection between the input entries (of a layer) and the units of the convolution layer. This is illustrated in Fig. 2.8.

The key points are the following:

- the units of the convolution layer are not connected to all the units in the previous layer.
- The units share the same weights.

This is illustrated in Fig. 2.8. It is clear also that the number of connections, and hence the number of parameters (weights) to learn is reduced compared to fully connected networks.

2.2.2.2 Max-Pooling layers

Pooling reduces the computational complexity for the upper layers and summarizes the outputs of neighboring groups of neurons from the same kernel map. It reduces the size of each input feature map by the acquisition of a value for each receptive field of neurons of the next layer. Once again, this reduces the number of parameters to learn. Pooling is a general operation and many variants exist. However, we consider here the Max-Pooling operation, meaning that, for each rectangle, the maximum value is kept and the others are discarded as shown in Fig. 2.9.

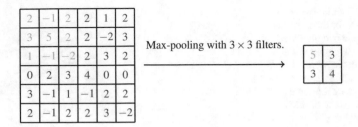

Fig. 2.9 An illustration of a max-pooling operation: the size of the image is reduced

This layer can be seen as a form of non-linear sub-sampling. It presents two interests:

- It reduces the spatial dimensions, which reduces the computational cost for the upper layers.
- It provides a form of translation invariance. To see this, consider the case of a max-pooling on regions of size 2×2 followed by a convolutional layer. Regions can be translated of one pixel in eight directions. Among the eight configurations, three will produce the exact same output at the convolutional layer.

One should pay attention to the excessive use of pooling: indeed, reducing the spatial dimensions results in a loss of information which can affect the training. Note that there exists different types of pooling layers, such as average pooling and L^2-norm pooling, but the max pooling version has proven to be the most effective most of the time.

2.2.2.3 Dropout

One of the bottlenecks of supervised learning approach is the so-called "overfitting phenomenon". This means that the classifier after a training process classifies training data with a small error, but is not able to generalize well on unseen data. Overfitting can be avoided using model combination. This involves averaging the outputs of many separately trained neural networks which is extremely expensive, especially for deep neural networks.

Moreover, such type of averaging assumes (for effectiveness) that the averaged models are very different: they should either have different architectures or be trained on different data. The former is difficult to achieve because properly tuning an architecture is complex, and tuning many of them is even more. The latter is difficult to achieve for large models: deep neural networks indeed require more data to be properly trained, which means that different neural networks would probably exceed the available amount of training data.

The dropout method addresses those problems by sampling many thinned versions of the original neural network. Those versions are obtained by randomly discarding units and their connections from the neural network. More details can be

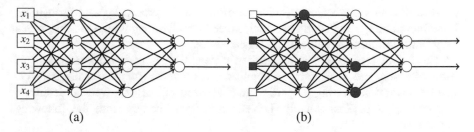

Fig. 2.10 Illustration of the dropout regularization: (**a**) the neural network and (**b**) the dropout application. At the beginning of any iteration in the training phase, each unit have probability p to be removed, together with its connections making the network becomes thinner. During the test phase, if the dropout probability is p, then each weight times $1 - p$

found in [SHh+14]. The dropout regularization is illustrated in Fig. 2.10. We note that in CNNs Dropout is performed on Fully Connected layer and indeed increases network performances.

2.2.3   Some Well-Known CNNs Architectures

It is extremely difficult to be exhaustive in the overview of popular architectures. We will limit ourselves to those which serve as a basis for quite a large set of applications.

2.2.3.1   LeNet Architecture and MNIST Dataset

LeNet is an extremely popular architecture, which was first introduced in 1998 by LeCun et al. [LBBH98]. This CNN architecture was designed to solve the problem of digit recognition in a very robust manner. It has a straightforward and rather shallow architecture (five hidden layers only), making it a very popular CNN to understand the basics of Deep Learning.

The core principle of LeNet family architectures is the repetition of the pattern convolutional layer + activation + max-pooling layer in the lower layers. Resulting spatial reduction enables the use of cascading fully-connected layers for the upper layers, and a softmax classifier gives the score of the model.

It is very common to train the LeNet architecture over the MNIST dataset. The latter is a database of handwritten digits, containing a total of 70,000 digit images. This dataset splits the data into 60,000 images for training and 10,000 images for validation. Every digit image of this dataset has been size-normalized and centered in a 28×28 fixed-size image.

The simple architecture of LeNet combined with the compactness of the MNIST dataset are ideal to try new learning techniques and pattern-recognition methods on real-world data while spending minimal efforts on preprocessing and formatting. Excellent training results with accuracy exceeding 98% are possible to achieve in a very limited time, even without GPU acceleration.

The LeNet architecture is resilient to different types of transformations, thus it enables very robust character recognition. It provides (for reasonable transformations) many interesting properties, as illustrated in LeCun's website dedicated to LeNet [LeC]. Most notably, LeNet's robustness comes from the following properties:

- Translation invariance: this is mostly useful for vertical translations, as the positioning of the characters in a string is never perfect.
- Scale invariance: LeNet achieves this type of invariance over a wide range of sizes.
- Rotation invariance: the authors estimate that LeNet can recognize digits rotated by an angle of 40°.
- Squeezing invariance: LeNet shows robustness to variations of the aspect ratio.
- Stroke width invariance: this type of robustness is useful to limit the need of unreliable preprocessing methods such as line thinning.
- Noise robustness: LeNet is resilient against various types of noise added above the digits.

2.2.3.2 AlexNet Architecture

Among the most popular architectures, AlexNet was presented in 2012 by Krizhevsky et al. [KSH12] in the paper entitled "ImageNet Classification with Deep Convolutional Networks", which is widely regarded as one of the most influential publication in the field of Deep Learning. This architecture is designed to solve a difficult classification problem on the dataset ImageNet.

The dataset ImageNet is an image database organized according to the WordNet hierarchy, which is a lexical database for English words. Currently only the nouns from WordNet are considered in ImageNet. Each node of the hierarchy is depicted by hundreds or even thousands of images, with an average of over five hundred images per node. This dataset is well known as it is extremely rich. An annual contest centered around ImageNet was created: the ILSVRC (ImageNet Large-Scale Visual Recognition Challenge), in which people can evaluate and compare their architectures by training their CNNs on the ImageNet database to solve object detection and image classification problems at large scale. The ILSVRC contest has seen many CNN candidates become staples in the field of Computer Vision over the past few years.

The authors presented the AlexNet architecture to the 2012 ILSVRC and won the contest by far with a top 5 test error rate of 15.4%, when the second best entry achieved an error of 26.2%. This result was seen as an outstanding performance and amazed the Deep Learning and Computer Vision communities.

The AlexNet architecture consists of 60 millions parameters for 500,000 neurons, 5 convolutional layers, some of them followed by a max-pooling layer and two fully-connected layers followed by a softmax classifier of size 1000 to score the model.

The authors also introduced regularization to their architecture under the form of dropout. As discussed in previous section, this method prevents overfitting even for very deep neural networks.

2.2.3.3 GoogLeNet

In 2015, Google published a new architecture [SLJ+14]. GoogLeNet is a 22 layers CNN and was the winner of ILSVRC 2014 with a top 5 error rate of 6.7%. Along with its depth, what sets apart GoogleNet from most of the architectures back in 2014 is that it does not simply rely on an alternation of convolutional and pooling layers, which was the common usage since the introduction of the LeNet architecture. Instead of a sequential arrangement of the layers, the authors thought about the addition of parallelism in the architecture, with the introduction of 9 so-called "inception blocks" that are composed of different layers themselves, for a total of more than 100 layers. Figure 2.11 presents the architecture of an inception block.

GoogLeNet led the path to a new design philosophy in Deep Learning: the authors proved that a creative architecture could lead to improved performances and a computational efficiency. It opened the path to very creative CNN architecture designs.

2.2.3.4 Other Important Architectures

Many more CNNs architectures have proven to be staples in Deep Learning. Among them, we can cite:

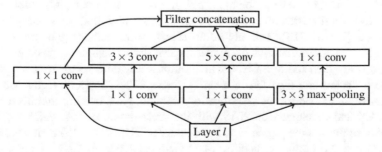

Fig. 2.11 GoogLeNet architecture: structure of an inception block

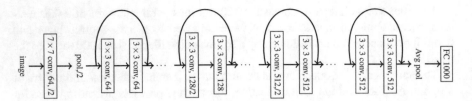

Fig. 2.12 Architecture of the ResNet network. It contains 34 layers and shortcuts each two convolutions

- ZF Net and DeConv Net (2013): Matthew Zeiler and Rob Fergus won the 2013 ILSVRC challenge with a CNN named ZF Net, which is a fine tuned version of AlexNet with ideas to improve overall performance. In their relative publication [ZF13], they also gave some rich insight regarding the intuition behind CNNs, and they also presented a special algorithm named DeConv Net which enables to visualize the response of the trained filters of a CNN when an image is fed to this CNN. This enables to examine what type of structures excite a given feature map and helps getting a deeper understanding of the way a specific architecture behaves.
- Microsoft ResNet (2015), see Fig. 2.12: Microsoft Research Asia proposed in 2015 a new architecture that broke many records [HZRS15]. ResNet was the deepest CNN when it was first presented, and it won the 2015 ILSVRC challenge, with an error rate of only 3.6%, which is usually lower than the error rate an average human would get. It introduced an architecture based on "residual blocks", which basically takes into account the input after a conv-relu-conv pattern, by adding the input to the output. According to the authors, "it is easier to optimize the residual mapping than to optimize the original, unreferenced mapping". Moreover, it helps dealing with the vanishing of the gradient. In many visual content recognition tasks ResNet with different amount of layers is used as the most efficient network.
- R-CNN (2013): In 2013, Ross Girshick and his group at UC Berkeley presented a new CNN architecture designed to solve the object recognition problem: R-CNN [GDDM13]. The problem is split into two parts: the region proposal step and the classification step. For the first step, an algorithm of selective search is used [UvdSGS13]. This algorithm selects using bounding boxes a certain number of regions that have the highest probability of containing an object. These regions are then fed to a CNN, which outputs a feature vector for each region. Those vectors are then fed to a set of linear SVM (Support Vector Machine) algorithms trained for the classification problem. Note that this architecture was revised and accelerated in 2015 with the presentation of Fast R-CNN [Gir15] which optimizes the pipeline, and of Faster R-CNN [RHGS15] a bit later which simplifies the complex pipeline of R-CNN and Fast R-CNN, thanks to the introduction of a region proposal network.

2.3   Optimization Methods

The machine learning models aim to construct a prediction function which minimizes the loss function. There are many algorithms which aim to minimize the loss function. Most of them are iterative and operate by decreasing the loss function following a descent direction. These methods solve the problem when the loss function is supposed to be convex. The main idea can be expressed simply as follows: starting from initial arbitrary (or randomly) chosen point in the parameter space, they allow the "descent" to the minimum of the loss function accordingly to the chosen set of directions. Here we discuss some of the most known and used optimization algorithms in this field.

The section presents the foundations of optimization methods. We use different notations from previous chapter but we will show how the methods can be applied for neural networks learning.

2.3.1   Gradient Descent

The use of gradient based algorithms has proven to be very effective in order to optimize the numerous parameters of neural networks, thus it is one of the most common approaches for network learning. As a result, many state-of-the-art Deep Learning libraries contain extensive implementation of the different forms of gradient algorithms. We briefly recall facts concerning the gradient descent algorithm.

Let us denote by θ the vector containing all the parameters of the neural network, and let $J(\theta, g(x), y)$ be the cost function which represents the error between the ground-truth labels $g(\mathbf{x})$ (associated with the training data $\mathbf{x}$) and the predicted data y estimated by the neural network under parameters θ. If one imagines the cost function representation as a valley, the idea behind gradient algorithm is to follow the slope of the mountain until we reach the bottom of the surface. Indeed, the direction given by the slope is the one which decreases the most locally, and it is actually the opposite to gradient. A thorough mathematical discussion about the gradient descent can be found in [WF12].

The iteration step for the gradient descent is given by:

$$\theta_{t+1} \leftarrow \theta_t - \eta \nabla_\theta J(\theta, g(x), y), \tag{2.6}$$

where η is a non negative constant called learning rate. This method works in spaces of any dimension (even infinite) and can be used for both linear and non-linear functions. To be well-posed, the gradient descent requires the function to be L-Lipschitz, that is

$$\forall u, v \in \mathbb{R}^N, \ \|\nabla J(u) - \nabla J(v)\|_2 \le L \|u - v\|_2, \tag{2.7}$$

where L is Lipschitz constant.

Moreover, the algorithm is only guaranteed to converge to the global solution in the case where the function J is strictly convex. If that is not the case, the algorithm will not even be guaranteed to find a local minimizer.

In the case of neural network learning, especially in deep learning, one wishes to optimize a huge number of parameters, that is the cost function arguments will be in a very high dimensional space, which leads to a proliferation of saddle points, known to be potentially harmful when looking for a minimizer, since they tend to be surrounded by high error plateaus [DPG+14].

Besides, choosing properly the learning rate is crucial for the method to be efficient: it has to be small enough to enable the convergence of the algorithm, though it should not be too small as it would dramatically slow down the process. Carefully and iteratively selecting this step size thus may help improving the results. A classical algorithm used for this purpose is the line search algorithm [Hau07].

2.3.2 Stochastic Gradient Descent

The problem encountered with gradient descent algorithms in the context of Deep Learning is that the mathematical properties required for the problem to be well-posed are not met in general. Indeed, in such very high dimensional context, the loss function can be non-convex and non-smooth, thus the convergence property is not satisfied, and convergence toward local minima can be extremely slow.

An excellent way to address those problems is to use the stochastic gradient descent algorithm. This algorithm is a stochastic approximation of the gradient descent algorithm. It aims at minimizing an objective function which can be written as a sum of differentiable functions (typically in the context of image processing, one function per image). Such process is done iteratively over batches of data (that is subsets of the whole dataset) randomly selected. Each objective function minimized this way approximates the "global" objective function. The following formula resumes this method:

$$\theta_{t+1} \leftarrow \theta_t - \eta \frac{1}{B_s} \sum_{i=1}^{B_s} \nabla_\theta J_t^i(\theta_t, x_t^i, y_t^i), \tag{2.8}$$

where B_s is the batch size, $B_t = (x_t^i, y_t^i)_{i \in [|1, B_s|]}$ is the batch of data associated with step t, where the x_t^i and the y_t^i are respectively the ground truth data and the estimated data in this batch, and $J_t = \frac{1}{B_s} \sum_{i=1}^{B_s} J_t^i$ is the stochastic approximation of the global cost function at step t over the batch B_t, decomposed into a sum of differentiable functions J_t^i associated to each pair (x_t^i, y_t^i).

Stochastic gradient descent (SGD) addresses most of the concerns encountered with gradient descent in the context of Deep Learning:

- The stochasticity of the method helps with the weak mathematical setting: although the problem is ill-posed with a high dimensional objective function that is non-convex and potentially non-smooth, stochasticity tends to improve convergence as it helps the objective function to pass through local minima and saddle points, which are known to be surrounded by very flat plateaus in high dimensional spaces.
- Convergence is much faster, as it is way less costly to perform many update steps over small batches of data than to perform a single update step over the whole dataset (in the context of Deep Learning, and specifically in visual data mining, datasets can contain several thousands of images or video frames).

This comes with its own drawbacks: at first, stochastic approximation of the gradient descent means that convergence toward the global minimizer cannot be guaranteed. Moreover, the smallest are the batches used for SGD, and the more variance can be observed in the results. As a consequence, many architectures are designed to take advantage of both sides, by selecting an average value of batch size.

2.3.3   Momentum Based SGD

The main purpose of momentum based approaches is to accelerate the gradient descent process. This is done by expanding the classical model with a velocity vector, which will build up iterations after iterations. A physical perspective is that such gradient acceleration mimics the increase in kinetic energy while a sphere is rolling down a valley.

In terms of optimization behaviour, it appears that (stochastic) gradient descent struggles to descend into a region where the surface of objective function curves more steeply in one direction than in another.

The iteration step for the momentum variant of the gradient descent is given below:

$$
\begin{aligned}
v_{t+1} &\leftarrow \mu v_t - \eta \nabla J(\theta_t), \\
\theta_{t+1} &\leftarrow v_{t+1} + \theta_t.
\end{aligned}
\tag{2.9}
$$

The vector v_{t+1} (initialized at zero) is computed first at each iteration and represents the update of the velocity of the "sphere rolling down the valley". The velocity stacks at each iteration, hence the need of the hyper-parameter μ, in order to damp the velocity when reaching a flat surface, otherwise the sphere would move too much near local extrema. A good strategy is to change the value of μ depending on learning stage.

In the velocity update, two terms are competing: the accumulated velocity μv_t and the negative gradient at the current point. The key idea is that in the scenario described in the previous paragraphs (when the surface curves more steeply in a direction than in another), then the two terms of the velocity will not have the same direction. This will prevent the gradient term from oscillating too much and thus will speed up the convergence. The importance of momentum and initialization in Deep Learning was discussed in a paper (2013) [SMDH13a] of Ilya Sutskever et al.

2.3.4 Nesterov Accelerated Gradient Descent

In 1983, Nesterov proposed in [Nes83] a slight modification of the "classical momentum" and showed that his algorithm had an improved theoretical convergence for convex function optimization. This approach has become very popular, as it usually performs better in practice than the classical momentum, and is still a good fit for gradient optimization even today.

The key difference between the Nesterov and the classical momentum algorithm is that the latter computes first the gradient at the current location θ_t and then performs a step in the direction of the accumulated velocity, whereas the Nesterov momentum first performs the step, which gives an approximation of the updated parameter which we call $\widetilde{\theta}_{t+1}$, and corrects this step by computing the gradient at this new location.

To understand the reasoning behind this difference, one can understand the gradient term $-\eta \nabla J(\theta_t)$ in the update v_{t+1} of the velocity as a correction term of the accumulated velocity μv_t. It makes more sense to correct an error (here the step performed by the accumulated velocity) after the error has been made, that is to compute the gradient at the location $\widetilde{\theta}_{t+1}$. The iterative step for the Nesterov momentum is:

$$
\begin{aligned}
\widetilde{\theta}_{t+1} &\leftarrow \theta_t + \mu v_t, \\
v_{t+1} &\leftarrow \mu v_t - \eta \nabla J(\widetilde{\theta}_{t+1}), \\
\theta_{t+1} &\leftarrow v_{t+1} + \theta_t.
\end{aligned}
\tag{2.10}
$$

2.3.5 Adaptative Learning Rate

We saw that in the stochastic gradient descent algorithm, the learning rate is defined as a non-negative constant. This can be a source of error: when the gradients are small but consistent (near local extrema for instance), a strong learning rate leads to oscillations in the valley, which prevents the method from converging properly. What we want is to move slowly in directions with strong but inconsistent gradients, and conversely to move quickly in directions with small but consistent gradients.

One can combine these two properties by adapting the learning rate adaptively. Some common annealing schedules used for this purpose are detailed below:

- Step decay: After each k epochs, multiply the learning rate by a constant $C < 1$.
- Polynomial decay: Set the learning rate as

$$\forall t \geq 0, \ \eta_t = \frac{a_0}{1 + b_0 t^n}, \ a_0, b_0 \in \mathbb{R}_+. \tag{2.11}$$

- Exponential decay: Set the learning rate as

$$\forall t \geq 0, \ \eta_t = a_0 e^{-b_0 t}, \ a_0, b_0 \in \mathbb{R}_+. \tag{2.12}$$

These strategies have the drawback to be arbitrary, and might not be suited for a specific learning problem. A thorough discussion regarding proper learning rate tuning for SGD methods was conducted by LeCun in [SZL13]. Another possibility is to propose an adaptive learning rate from the computation of the inverse of the Hessian matrix of the cost function (Newton and quasi-Newton approaches). The corresponding iterative step is

$$\widetilde{\theta}_{t+1} \leftarrow \theta_t - [HJ(\theta_t)]^{-1} \nabla J(\theta_t). \tag{2.13}$$

The latter method is an example of second order optimization procedure. More information regarding Newton's optimization method can be found in [Chu14].

In practice, it can be costly to compute the second order derivatives and to perform matrix inversion. Furthermore, stability problems in Hessian Matrix inversion can be encountered near local extrema. Therefore, this is not a well suited method in the context of Deep Learning.

2.3.6 Extensions of Gradient Descent

Here we recall briefly some of the existing expansions of the gradient descent:

Averaged Gradient Descent This particular version of the gradient descent, studied in the paper of Polyak [PBJ92] replaces the computation of the parameter values θ_t by the computation of the temporal mean values, from the updates obtained through gradient descent:

$$\overline{\theta}_T \leftarrow \frac{1}{T} \sum_{t=0}^{T} \theta_t. \tag{2.14}$$

Adagrad In this method first introduced in a 2011 paper [DHS11], the goal was to have the learning rate adjust itself, depending on the sparsity of the parameters.

In this context, the sparsity of a parameter means that this parameter has not been trained a lot during the iterative training process. Sparse parameters will learn faster while non-sparse ones will learn slower. This emphasizes the training of sparse features that would not have been trained properly in a classical gradient descent algorithm. The corresponding update step is different for each parameter $(\theta)_i$ in the parameter vector θ. It is given by

$$\forall i, \ (\theta_{t+1})_i \leftarrow (\theta_t)_i - \alpha \frac{\left(\nabla J(\theta_t)\right)_i}{\sqrt{\sum_{u=1}^{t} \left(\nabla J(\theta_u)\right)_i^2}}, \ \alpha > 0. \tag{2.15}$$

RMSProp Just like in the case of Adagrad, the RMSProp algorithm proposes to adjust automatically the learning rate of each parameter. It does so by running average of the magnitudes of recent gradients for that parameter. This algorithm was presented in the course [HSS12]. The corresponding update step is given by the formula below:

$$\begin{aligned}
\forall i, \ (\nabla_{t+1})_i &\leftarrow \delta(\nabla_t)_i + (1-\delta)\left(\nabla J(\theta_t)\right)_i^2, \\
\forall i, \ (\theta_{t+1})_i &\leftarrow (\theta_t)_i - \alpha \frac{\left(\nabla J(\theta_t)\right)_i}{\sqrt{(\nabla_{t+1})_i}}, \ \alpha > 0.
\end{aligned} \tag{2.16}$$

The parameter δ sets the confidence given either to the running average of the magnitudes of past gradients or to the magnitude of the last gradient computed.

Adam Adam algorithm is among the most recent and efficient first-order gradient descent based optimization algorithms. It was first presented in [KB14]. As in the case of Adagrad and RMSProp, it automatically adjusts the learning rate for each of the parameters. The particularity of this method is that it computes so called "adaptative moment estimations" (m_t, v_t). This method can be seen as a generalization of the Adagrad algorithm. Its corresponding update process is detailed below:

$$\begin{aligned}
\forall i, \ (m_{t+1})_i &\leftarrow \beta_1 \cdot (m_t)_i + (1-\beta_1) \cdot \left(\nabla J(\theta_t)\right)_i, \\
\forall i, \ (v_{t+1})_i &\leftarrow \beta_2 \cdot (v_t)_i + (1-\beta_2) \cdot \left(\nabla J(\theta_t)\right)_i^2, \\
\forall i, \ (\theta_{t+1})_i &\leftarrow (\theta_t)_i - \alpha \frac{\sqrt{1-\beta_2}}{1-\beta_1} \frac{(m_t)_i}{\sqrt{(v_t)_i} + \epsilon}, \ \alpha, \epsilon > 0 \text{ and } \beta_1, \beta_2 \in]0, 1[.
\end{aligned} \tag{2.17}$$

In the previous equation, ϵ is used as a precision parameter. The parameters β_1 and β_2 are used to perform running averages over the so-called moments m_t and v_t respectively.

2.4 Gradient Estimation in Neural Networks

The previous sections introduce general optimization methods. In this section, we consider this methods used for neural networks optimization. That is the cost function J is the loss function $\mathcal{L}$, the parameters θ correspond to the parameters of the neural network:

$$\theta = \left(w^{(l)}, b^{(l+1)} \right), \quad \text{for } l \in \{0, 1, \cdots, L - 1\}.$$

One of the main reasons for the resurgence of neural networks was the development of an algorithm which efficiently computes the gradient of the cost function, that is all the partial derivatives with respect to weights and biases in the neural network.

The key idea behind this algorithm, in the case of feed-forward neural networks, is that a slight modification of the weights and the bias in a layer l has a (slight) impact on the next layer, which cascades up to the output layer. Thus, in order to compute the partial derivatives with respect to weights and biases, we focus on the analysis of the errors (the slight modifications of the outputs), and we do it in a backward manner (hence the name backpropagation), since the loss function of our neural network depends directly on the activations in the output layer. In other words, we try to understand how the errors in the output layer propagate iteratively from a layer to the previous layer.

The backpropagation algorithm uses gradient descent to find the parameters that minimize the loss function. It is a recursive method based on chain rule. In the sequel, we consider a neural network with hidden layers and sigmoid as activation function for all the neurons (including the output layer). We also consider that the loss function is the cross-entropy function.

We use the same notation as in Fig. 2.3. Thus, the output of the network is a vector $\hat{\mathbf{y}} = (\hat{y}_1, \hat{y}_2, \cdots, \hat{y}_{n_L})$ whose elements are given by:

$$\hat{y}_i = y_i^{(L)} = \sigma \left(z_i^{(L)} \right)$$

$$= \sigma \left(\sum_{k=1}^{n_{L-1}} w_{k,i}^{(L-1)} y_k^{(L-1)} + b_i^{(L)} \right). \tag{2.18}$$

The cross-entropy for a single example is given by the sum:

$$\mathcal{L} = -\sum_{i=1}^{n_L} \left(y_i \ln(\hat{y}_i) + (1 - y_i) \ln(1 - \hat{y}_i) \right).$$

To use the gradient descent iteration (see Eq. (2.6)), we need to compute the derivatives of the loss function accordingly to parameters. We first consider the parameters in the output layer (layer L).

Let $w = w_{k,i}^{(L-1)}$ be a weight between a unit in layer $L-1$ and a unit i in the output layer. We need to compute the value of $\frac{\partial \mathcal{L}}{\partial w}$. For this, we use the chain rule:

$$\frac{\partial \mathcal{L}}{\partial w} = \frac{\partial \mathcal{L}}{\partial \hat{y}_i} \times \frac{\partial \hat{y}_i}{\partial z_i^{(L)}} \times \frac{\partial z_i^{(L)}}{\partial w}.$$

We also have:

$$\frac{\partial \mathcal{L}}{\partial \hat{y}_i} = -\frac{y_i}{\hat{y}_i} + \frac{1 - y_i}{1 - \hat{y}_i} = \frac{\hat{y}_i - y_i}{\hat{y}_i(1 - \hat{y}_i)}. \tag{2.19}$$

Since

$$\frac{\partial \hat{y}_i}{\partial z_i^{(L)}} = \hat{y}_i(1 - \hat{y}_i), \tag{2.20}$$

and $\frac{\partial z_i^{(L)}}{\partial w} = y_k^{(L-1)}$, we finally obtain:

$$\frac{\partial \mathcal{L}}{\partial w} = (\hat{y}_i - y_i)\, \hat{y}_k^{(L-1)}. \tag{2.21}$$

A similar calculation yields to:

$$\frac{\partial \mathcal{L}}{\partial b_i^{(L)}} = (\hat{y}_i - y_i). \tag{2.22}$$

The above gives the gradients with respect to the parameters in the last layer of the network. Computing the gradients with respect to the parameters in hidden layers requires another application of the chain rule.

Let $w = w_{k,i}^{(l)}$ be a weight between a unit k in a hidden layer l and a unit i in layer $l+1$ for $l \in \{1, 2, \cdots, L-2\}$. Then:

$$\frac{\partial \mathcal{L}}{\partial w} = \frac{\partial \mathcal{L}}{\partial y_i^{(l+1)}} \times \frac{\partial y_i^{(l+1)}}{\partial z_i^{(l+1)}} \times \frac{\partial z_i^{(l+1)}}{\partial w}. \tag{2.23}$$

Let $\{u_1, u_2, \cdots, u_m\}$ be the set of units in layer $l+2$ connected to unit i. We consider $\mathcal{L}$ as a function $\mathcal{L}\left(y_1^{(l+2)}, y_2^{(l+2)}, \cdots, y_m^{(l+2)}\right)$ of the outputs of units u_i in the layer $l+2$. Then:

$$\frac{\partial \mathcal{L}}{\partial y_i^{(l+1)}} = \sum_{j=1}^{m} \frac{\partial \mathcal{L}}{\partial y_j^{(l+2)}} \times \frac{\partial y_j^{(l+2)}}{\partial y_i^{(l+1)}}$$

$$= \sum_{j=1}^{m} \frac{\partial \mathcal{L}}{\partial y_j^{(l+2)}} \times \frac{\partial y_j^{(l+2)}}{\partial z_j^{(l+2)}} \times \frac{\partial z_j^{(l+2)}}{\partial y_i^{(l+1)}}$$

$$= \sum_{j=1}^{m} \frac{\partial \mathcal{L}}{\partial y_j^{(l+2)}} \times y_j^{(l+2)}(1 - y_j^{(l+2)}) \times w_{i,j}^{(l+1)}. \qquad (2.24)$$

Since $\frac{\partial y_i^{(l+1)}}{\partial z_i^{(l+1)}} = y_i^{(l+1)}(1 - y_i^{(l+1)})$ and $\frac{\partial z_i^{(l+1)}}{\partial w} = y_k^{(l)}$, we can derive the desired derivative:

$$\frac{\partial \mathcal{L}}{\partial w} = y_i^{(l+1)}(1 - y_i^{(l+1)})y_k^{(l)} \times \sum_{j=1}^{m} \frac{\partial \mathcal{L}}{\partial y_j^{(l+2)}} \times y_j^{(l+2)}(1 - y_j^{(l+2)}) \times w_{i,j}^{(l+1)}.$$

$$(2.25)$$

Equation (2.25) means that the derivative with respect to weights in the hidden layers can be calculated if all the derivatives with respect to the outputs of the next layer are known. This defines a recursive algorithm.

A similar computation can be done to get the derivatives with respect to the bias $b = b_i^{(l+1)}$:

$$\frac{\partial \mathcal{L}}{\partial b} = y_i^{(l+1)}(1 - y_i^{(l+1)})y_k^{(l)} \times \sum_{j=1}^{m} \frac{\partial \mathcal{L}}{\partial y_j^{(l+2)}} \times y_j^{(l+2)}(1 - y_j^{(l+2)}). \quad (2.26)$$

2.5 Recurrent Neural Networks

2.5.1 General Principles

Recurrent neural networks (RNN for short) take as input the current example they see and what they have received previously. The decision of an RNN at time t is affected by the decision made in step time $t - 1$. That is, recurrent neural networks have two sources of input: the present and the past. More formally, if we denote $y^{(t)}$ the response of the network at time t, then:

$$y^{(t)} = f\left(y^{(t-1)}, x^{(t)}\right).$$

Figure 2.13 explains intuitively the difference between RNNs and "classical" neural networks. In addition to the connections between nodes in a layer and the ones in its following layer, there are loops on the units of a layer.

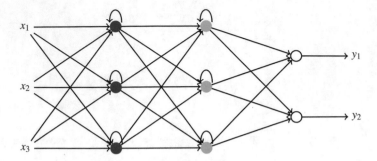

Fig. 2.13 A very simple representation of a recurrent neural network

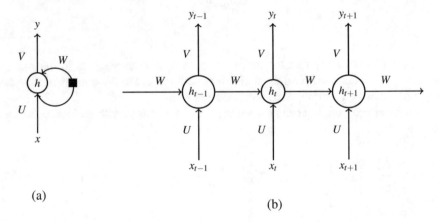

(a)

(b)

Fig. 2.14 A recurrent network and ints unfolding: (**a**) The network, (**b**) Unfolding in time

From the discussion above, one can explain the idea behind recurrent neural networks: we aim to make use of sequential information. In the previous architectures, we assume that inputs and outputs are independent, i.e., the output at time t have no impact on the output of time $t + 1$. But for many machine learning tasks, and in particular for video mining, this is not true. If one wants to annotate a video frame, it is necessary to consider which frames have been seen before.

Recurrent neural networks were introduced to overcome this problem: they can perform tasks on sequences and thus the output at time t is considered in the input at time $t + 1$. We say that they have a memory which can be used to store information about what has been calculated so far.

Figure 2.14a gives a simple representation of a neural network: the input x, combined with a matrix U (w.l.o.g., we consider the bias as a part of this matrix), is used to compute the value h of a hidden layer, then, this value is stored in the memory of the layer before it is passed to the output layer, using another matrix V.

Figure 2.14b gives another representation of the same concepts as in (a). It shows a recurrent neural network into a full network. This is called *unfolding*. By unfolding, we mean that we write out the network for the complete sequence. For example, if the sequence we care about is a video of 24 frames, the network would be unfolded into a 24-layer neural network, one layer for each frame. The following notation are used:

- x_t is the input at time t. This can be a one-hot vector encoding of a frame,
- h_t is the hidden state at time step t. It corresponds to the memory of the network. It is calculated using previous hidden value and the input at time step t:

$$h_t = f(Ux_t + Wh_{t-1}).$$

the function f is any activation function. It is usually a non-linear function such as tanh or *Relu*. The first hidden value h_0 is initialized to zeroes.
- y_t is the output at time step t. As for previous networks, it is expressed in terms of softmax:

$$y_t = \text{softmax}(Vh_t).$$

Training an RNN

As for any neural network training task, it is necessary to define a loss function which evaluates the performance of the network by comparing the predicted classes with the actual ones. For RNN, one can use any classical loss function (such as cross-entropy, etc.). Then, it is necessary to initialize the values of the parameters of the network (weights and biases). This step is dependent on the classification problem under consideration and on the input data. One can also use transfer learning to set the initial values. However, in practice, it is shown [SMDH13b, PMB13] that a Gaussian drawn with a standard deviation of 0.001 or 0.01 is a good choice.

Once the parameters initialized, it is necessary to use a method to train the network. Gradient descent based methods can be used. However, the time aspect in the structure of RNN makes classical methods (the ones we have seen for neural networks and for convolutional neural networks) not efficient. Thus, one needs to extend them to capture the time aspect of the RNN structure.

There are many methods that can be used to train an RNN. One can cite back-propagation through time [JH08], Kalman Filter-based learning methods [MJ98],... In this section, we will discuss the principle of the back-propagation through time (BPTT) method since it is an adaptation of the gradient descent method.

BPTT is a generalization of back-propagation for feed-forward networks. The main idea of the standard BPTT method for recurrent neural networks is to unfold the network in time (see Fig. 2.14) and propagate error signals backwards through time. As explained in Fig. 2.14, the set of parameters θ is defined by W, U, V and of course the set of biases (which, in our case, are encoded in the matrices). Computing the derivatives of the loss function with respect to V is roughly the same as for feed-forward networks. The extension is mainly done for the computations of the derivatives for U and W. Indeed, it is necessary to sum up the contributions of each time step to the gradient. In other words, because W is used in every step up to the output we care about, we need to backpropagate gradients through the network all the way.

Recurrent neural networks have difficulties learning long-time dependencies. This is due to the vanishing of the gradient. This happens when the derivatives of the activation function (sigmoid or tanh) approach 0. The corresponding neuron is then saturated and drive other gradients in previous layers towards 0. The small values in the matrices and the multiple matrix multiplications yield to a very fast gradient vanishing. Intuitively, this means that the gradient contributions from far away steps become 0. This problem is not limited to recurrent neural networks and is also present in (very) deep feed-forward networks.

There are solutions to avoid vanishing gradient problem. The most common solution is to use the *ReLu* as an activation function instead of the sigmoid or the tanh functions. A more suitable solution is to use the Long-Short Term Memory (LSTM) networks.

2.5.2 Long-Short Term Memory Networks

Recurrent networks are useful to capture the sequential dependencies in the inputs. However, as explained in the previous section, training such networks suffers from the exploding or the vanishing of the gradient. Thus the recurrent networks are not sufficiently efficient when used to learn long-term sequential dependencies in data. In 1997, Hochreiter and Schmidhuber [HS97] introduce the *Long Short-Term Memory* networks (LSTM for short). This architecture, and its variants become largely used in last years and proved its efficiency in many fields including speech recognition, text processing etc.

LSTM is a recurrent neural network where the units are more complex than in simple RNN. Figure 2.15 presents a typical unit of an LSTM.

Information comes to the unit from the current data x_t and from the previous value h_{t-1} of the hidden layer. The unit computes the value of the *forget gate*:

$$f_t = \sigma \left(W_f.[h_{t-1}, x_t] + b_f \right).$$

This corresponds to the information which is going to be thrown from the cell state C_t.

The same computation is done for the *input gate*:

$$i_t = \sigma \left(W_i.[h_{t-1}, x_t] + b_i \right),$$

and another value:

$$c_t = \tanh \left(W_C.[h_{t-1}, x_t] + b_C \right).$$

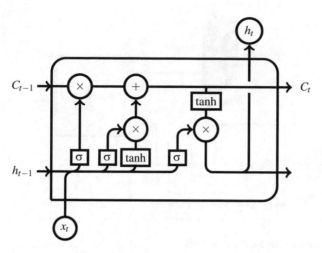

Fig. 2.15 A unit of a Short Long-Term Memory network

These values are then used to compute the new value C_t of the cell *state*:

$$C_t = f_t C_{t-1} + i_t c_t.$$

The last thing to decide is h_t, the output of the cell. This is done in two steps. First we compute the value of the (intermediate) output:

$$o_t = \sigma \left(W_o.[h_{t-1}, x_t] + b_o \right),$$

then:

$$h_t = o_t \tanh(C_t).$$

There are many other variants of the LSTM networks. The *Gated Recurrent Unit*, or GRU, introduced by Cho et al. [CVMBB14] aim to simplify the computation and combines the forget and the input gates into a single gate: the *update gate*. It also merges the cell state and the hidden state.

2.6 Generative Adversary Networks

Generative models are used to learn data distribution. The main goal is to learn true data distribution of a set so they can generate new data points from the learned distribution. There are mainly two efficient approaches both are based on neural networks: the Variational Autoencoders (see next Section) and Generative Adversarial Networks.

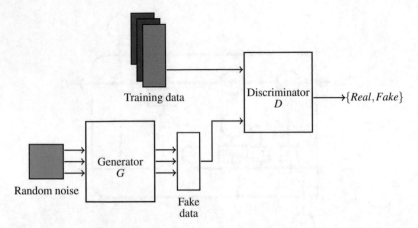

Fig. 2.16 A typical architecture of a GAN

Generative Neural Networks (GAN) [IJM+41], are based on game theory approach and are composed of two networks, see Fig. 2.16, the Generator and the Discriminator. The objective is to sample from a given distribution, uniform for example, and then learn to transform the noise to training data distribution using a universal function such as neural network.

The term adversarial comes from the way these two networks are trained. The generator G learns the data distribution and the discriminator D estimates the probability that a sample came from the training distribution rather that from the generator. Thus, the task of the last is to generate data similar to the training data (think to images which look like natural) and the task of the discriminator is to decide whether the generated data is real or fake. This can be seen as a min-max two player game where one, G, tries to fool the other, D, by generating real data as far as possible and the second tries to not get fooled by improving the discriminative capability.

2.7 Autoencoders

An Autoencoder (AEnc) [JMG95], is a neural network approach created to build identity for the training data while trying to compress the features space into smaller representation as shown in Fig. 2.17. Hence, an AEnc should be able to reproduce the inputs at the output layer. An AEnc is composed of two main components: an encoder that maps the input to a hidden representation and a decoder that reconstructs the original input space from the hidden representation by the same transformation as the encoder. The smallest hidden representation is called the latent space, the bottleneck. The training of the AEnc aims to minimize the difference between the input and the output i.e. minimize the reconstruction error (RE). After

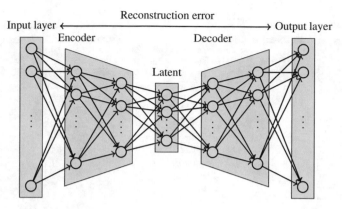

Fig. 2.17 Auto-encoder architecture

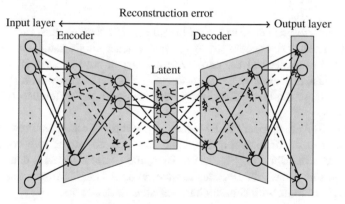

Fig. 2.18 Sparse auto-encoder architecture

being trained, the model must be able to rebuild previously unseen instances that are of the same data distribution as the training set. If the new input does not belong to the same distribution, the RE will be high. Many reconstruction error functions could be used such as the Mean Squared Error (MSE) which reflects the average squared difference between the estimated values (the output of the AEnc) and the actual value (the input).

Sparse Autoencoder (SAE) [KPC18], is a type of AEnc where we add a sparsity constraint on the hidden layers. In other words, the weights of the AEnc are penalized in the cost function. This technique is mainly used to avoid overfitting. In AEnc, the number of neurons in hidden layers is less than the neurons of the input/output layers. Meanwhile, in SAE, the number could be less or greater. However, thanks to the sparsity constraint, not all the neurons will be "active". Some of them will be disabled as in Fig. 2.18 (gray links). A neuron is considered "active" or as "firing" if its output value is close to 1. If it is close to 0, then it is being "inactive". Mathematically, an extra penalty term (penalizing activations of

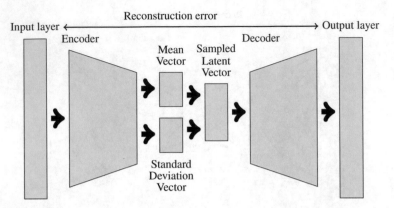

Fig. 2.19 Variational auto-encoder architecture

hidden layers) will be added to the optimization objective (the cost function) so that only a few nodes are encouraged to activate when a single sample is fed into the network. Many choices of the penalty term exist in the literature [JRP+15] such as L1 Regularizer (based on the L1 norm), L2 Regularizer (based on the L2 norm) and the Kullback-Leibler (KL) divergence (measures the difference between two distributions).

Variational Autoencoder (VAE) [LRJY18], inherits the architecture of traditional AEncs but instead of encoding the input into a fixed vector (the bottleneck), it maps it into a distribution (Fig. 2.19). Thus, the encoder learns a data generating distribution that allows the decoder to take random samples from the latent space and generate outputs with similar characteristics to the inputs.

2.8 Siamese Neural Networks

Siamese Convolutional Neural Networks (Siamese CNN) were first introduced by Chopra et al. in [CHL05]. Such a model is composed of two identical CNNs that share all their weights.

Let $X = (X_1, X_2)$ be a pair of inputs. The aim of the network, when given the two inputs X_1 and X_2, is to return a value $Y \in \{0, 1\}$ ($Y = 0$ if they match and 1 otherwise).

Each of the two components computes a result $G_W(X_i)$, than the euclidean distance between the two results is computed by:

$$E_W = \|G_W(X_1) - G_W(X_2)\|.$$

When training the network, instead of using cross-entropy loss function, the *contrastive* [HCL06], loss function is more suitable for training the siamese CNNs.

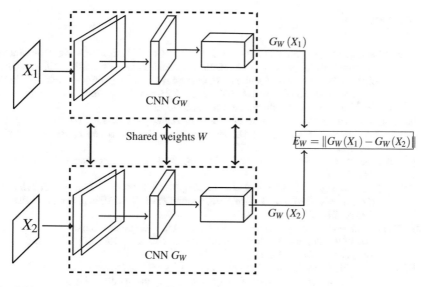

Fig. 2.20 Architecture of a Siamese CNN, as defined in [CHL05]

It is defined as follows:

$$\mathcal{L}(X, Y) = (1 - Y)\frac{1}{2}E_W^2 + Y\frac{1}{2}\{\max\{0, m - E_W\}\}^2, \qquad (2.27)$$

where m is a margin. This idea is to encourage the neural network to learn a embedding to place inputs with the same labels close to each other, and to distance the inputs with different labels. Figure 2.20 presents an example of such an architecture.

2.9 Conclusion

In this chapter, we introduced Artificial Neural Networks and Convolutional Neural Networks: two possibly feedforward architectures which proved their efficiency in many fields and, in particular, in image analysis and Computer Vision. Then we presented general gradient descent based methods used to train such networks. We also discussed architectures which capture temporality: recurrent neural networks and LSTM, we briefly discuss how gradient descent methods can be adapted to train such networks. The last section was dedicated to short presentations of recent specific and powerful architectures: Generative Adversarial Networks, Autoencoders and Siamese CNNs. The next chapters of the current book will illustrate the use of the presented architectures.

References

[CHL05] Sumit Chopra, Raia Hadsell, and Yann LeCun. Learning a similarity metric discriminatively, with application to face verification. In *2005 IEEE Computer Society Conference on Computer Vision and Pattern Recognition (CVPR 2005), 20–26 June 2005, San Diego, CA, USA*, pages 539–546. IEEE Computer Society, 2005.

[Chu14] Wei-Ta Chu. An introduction to optimization. 2014.

[CVMBB14] Kyunghyun Cho, Bart Van Merriënboer, Dzmitry Bahdanau, and Yoshua Bengio. On the properties of neural machine translation: Encoder-decoder approaches. *arXiv preprint arXiv:1409.1259*, 2014.

[Cyb89] George Cybenko. Approximation by superpositions of a sigmoidal function. *MCSS*, 2(4):303–314, 1989.

[DBL13] *Proceedings of the 30th International Conference on Machine Learning, ICML 2013, Atlanta, GA, USA, 16–21 June 2013*, volume 28 of *JMLR Workshop and Conference Proceedings*. JMLR.org, 2013.

[DHS11] John Duchi, Elad Hazan, and Yoram Singer. Adaptive subgradient methods for online learning and stochastic optimization. 12:2121–2159, July 2011. http://jmlr.org/papers/volume12/duchi11a/duchi11a.pdf.

[DPG+14] Yann Dauphin, Razvan Pascanu, Çaglar Gülçehre, Kyunghyun Cho, Surya Ganguli, and Yoshua Bengio. Identifying and attacking the saddle point problem in high-dimensional non-convex optimization. *CoRR*, abs/1406.2572, 2014. http://arxiv.org/abs/1406.2572.

[GBZN18] Pierre Gillot, Jenny Benois-Pineau, Akka Zemmari, and Yurii E. Nesterov. Increasing training stability for deep CNNS. In *2018 IEEE International Conference on Image Processing, ICIP 2018, Athens, Greece, October 7–10, 2018*, pages 3423–3427. IEEE, 2018.

[GDDM13] Ross B. Girshick, Jeff Donahue, Trevor Darrell, and Jitendra Malik. Rich feature hierarchies for accurate object detection and semantic segmentation. *CoRR*, abs/1311.2524, 2013. http://arxiv.org/abs/1311.2524.

[Gir15] Ross B. Girshick. Fast R-CNN. *CoRR*, abs/1504.08083, 2015. http://arxiv.org/abs/1504.08083.

[Hau07] Raphael Hauser. Line search methods for unconstrained optimisation. *Lecture 8, Numerical Linear Algebra and Optimisation Oxford University Computing Laboratory*, 2007.

[HCL06] Raia Hadsell, Sumit Chopra, and Yann LeCun. Dimensionality reduction by learning an invariant mapping. In *2006 IEEE Computer Society Conference on Computer Vision and Pattern Recognition (CVPR 2006), 17–22 June 2006, New York, NY, USA*, pages 1735–1742. IEEE Computer Society, 2006.

[HS97] Sepp Hochreiter and Jürgen Schmidhuber. Long short-term memory. *Neural Computation*, 9(8):1735–1780, 1997.

[HSS12] Geoffrey Hinton, Nitish Srivastava, and Kevin Swersky. Neural networks for machine learning - lecture 6a - overview of mini-batch gradient descent. 2012.

[HW68] D. H. HUBEL and T. N. WIESEL. Receptive fields and functional architecture of monkey striate cortex. 195:215–243, 1968. http://hubel.med.harvard.edu/papers/HubelWiesel1968Jphysiol.pdf.

[HZRS15] Kaiming He, Xiangyu Zhang, Shaoqing Ren, and Jian Sun. Deep residual learning for image recognition. *CoRR*, abs/1512.03385, 2015. http://arxiv.org/abs/1512.03385.

[IJM+41] Goodfellow I., Pouget-Abadie J., Mirza M., Xu B., Warde-Farley D., Ozair S., Courville A., and Bengio Y. Generative adversarial nets. *Advances in Neural Information Processing Systems*, pages 2672–2680, 2041.

[JH08] Orlando De Jesus and Martin T. Hagan. Backpropagation through time for general dynamic networks. In Hamid R. Arabnia and Youngsong Mun, editors, *Proceedings of the 2008 International Conference on Artificial Intelligence, ICAI 2008, July 14–17, 2008, Las Vegas, Nevada, USA, 2 Volumes (includes the 2008 International Conference on Machine Learning; Models, Technologies and Applications)*, pages 45–51. CSREA Press, 2008.

[JMG95] Nathalie Japkowicz, Catherine Myers, and Mark A. Gluck. A novelty detection approach to classification. In *Proceedings of the Fourteenth International Joint Conference on Artificial Intelligence, IJCAI 95, Montréal Québec, Canada, August 20–25 1995, 2 Volumes*, pages 518–523. Morgan Kaufmann, 1995.

[JRP+15] Nan Jiang, Wenge Rong, Baolin Peng, Yifan Nie, and Zhang Xiong. An empirical analysis of different sparse penalties for autoencoder in unsupervised feature learning. In *2015 International Joint Conference on Neural Networks, IJCNN 2015, Killarney, Ireland, July 12–17, 2015*, pages 1–8. IEEE, 2015.

[KB14] Diederik P. Kingma and Jimmy Ba. Adam: A method for stochastic optimization. *CoRR*, abs/1412.6980, 2014. https://arxiv.org/pdf/1412.6980.pdf.

[KPC18] Rafal Kozik, Marek Pawlicki, and Michal Choras. Sparse autoencoders for unsupervised netflow data classification. In Michal Choras and Ryszard S. Choras, editors, *Image Processing and Communications Challenges 10 - 10th International Conference, IP&C'2018, Bydgoszcz, Poland, 14–16 November 2018, Proceedings*, volume 892 of *Advances in Intelligent Systems and Computing*, pages 192–199. Springer, 2018.

[KSH12] Alex Krizhevsky, Ilya Sutskever, and Geoffrey E Hinton. Imagenet classification with deep convolutional neural networks. In *Advances in neural information processing systems*, pages 1097–1105, 2012.

[LBBH98] Yann Lecun, Léon Bottou, Yoshua Bengio, and Patrick Haffner. Gradient-based learning applied to document recognition. pages 2278–2324, 1998. http://yann. lecun.com/exdb/publis/pdf/lecun-01a.pdf.

[LeC] Yann LeCun. MNIST Demos. *Yann LeCun's website*. http://yann.lecun.com/exdb/ lenet/index.html.

[LRJY18] Romain Lopez, Jeffrey Regier, Michael I. Jordan, and Nir Yosef. Information constraints on auto-encoding variational bayes. In Samy Bengio, Hanna M. Wallach, Hugo Larochelle, Kristen Grauman, Nicolò Cesa-Bianchi, and Roman Garnett, editors, *Advances in Neural Information Processing Systems 31: Annual Conference on Neural Information Processing Systems 2018, NeurIPS 2018, 3–8 December 2018, Montréal, Canada*, pages 6117–6128, 2018.

[MJ98] Sheng Ma and Chuanyi Ji. A unified approach on fast training of feedforward and recurrent networks using EM algorithm. *IEEE Trans. Signal Processing*, 46(8):2270–2274, 1998.

[MP43] W. S. McCulloch and W. Pitts. A logical calculus of the ideas immanent in nervous activity. *Bulletin of Mathematical Biophysics*, 5:115–133, 1943.

[Nes83] Yurii Nesterov. A method of solving a convex programming problem with convergence rate $O(1/k^2)$. *Soviet Mathematics Doklady (Vol. 27)*, 1983.

[PBJ92] Boris Polyak and A B. Juditsky. Acceleration of stochastic approximation by averaging. 30:838–855, 07 1992.

[PBPZ+20] Miltiadis Poursanidis, Jenny Benois-Pineau, Akka Zemmari, Boris Mansencal, and Aymar de Rugy. Move-to-data: A new continual learning approach with deep CNNS, application for image-class recognition, 2020.

[PMB13] Razvan Pascanu, Tomas Mikolov, and Yoshua Bengio. On the difficulty of training recurrent neural networks. In *Proceedings of the 30th International Conference on Machine Learning, ICML 2013, Atlanta, GA, USA, 16–21 June 2013* [DBL13], pages 1310–1318.

[RHGS15] Shaoqing Ren, Kaiming He, Ross B. Girshick, and Jian Sun. Faster R-CNN: towards real-time object detection with region proposal networks. *CoRR*, abs/1506.01497, 2015. http://arxiv.org/abs/1506.01497.

[SHh+14] Nitish Srivastava, Geoffrey Hinton, Alex hevsky, Ilya Sutskever, and Ruslan Salakhutdinov. Dropout: A simple way to prevent neural networks from overfitting. *Journal of Machine Learning Research*, 15:1929–1958, 2014.

[SLJ+14] Christian Szegedy, Wei Liu, Yangqing Jia, Pierre Sermanet, Scott E. Reed, Dragomir Anguelov, Dumitru Erhan, Vincent Vanhoucke, and Andrew Rabinovich. Going deeper with convolutions. *CoRR*, abs/1409.4842, 2014. http://arxiv.org/abs/1409.4842.

[SMDH13a] Ilya Sutskever, James Martens, George Dahl, and Geoffrey Hinton. On the importance of initialization and momentum in deep learning. pages III–1139–III–1147, 2013. http://dl.acm.org/citation.cfm?id=3042817.3043064.

[SMDH13b] Ilya Sutskever, James Martens, George E. Dahl, and Geoffrey E. Hinton. On the importance of initialization and momentum in deep learning. In *Proceedings of the 30th International Conference on Machine Learning, ICML 2013, Atlanta, GA, USA, 16–21 June 2013* [DBL13], pages 1139–1147.

[SZL13] Tom Schaul, Sixin Zhang, and Yann LeCun. No more pesky learning rates. 28(3):343–351, 2013. https://arxiv.org/pdf/1206.1106.pdf.

[UvdSGS13] J. R. R. Uijlings, K. E. A. van de Sande, T. Gevers, and A. W. M. Smeulders. Selective search for object recognition. *International Journal of Computer Vision*, 104(2):154–171, 2013.

[WF12] Gezheng Wen and Li Fan. Large scale optimization - lecture 4. 2012.

[ZB20] Akka Zemmari and Jenny Benois-Pineau. *Deep Learning in Mining of Visual Content*. Springer Briefs in Computer Science. Springer, 2020.

[ZF13] Matthew D. Zeiler and Rob Fergus. Visualizing and understanding convolutional networks. *CoRR*, abs/1311.2901, 2013. http://arxiv.org/abs/1311.2901.

Chapter 3
Deep Learning for Semantic Segmentation

Alexandre Benoit, Badih Ghattas, Emna Amri, Joris Fournel, and Patrick Lambert

3.1 Introduction

Semantic segmentation refers to the task of assigning a class label to each of the low-level components of a media content. In the image analysis community, it consists in classifying each of the pixels on standard 2D images, or voxels when dealing with volumetric images or spatio-temporal data such as video. It then consists in dividing a media into sets that can be summarized as homogeneous areas that represent the same concept or share the same properties. This task is then close to the identification of a hierarchy of spatial taxons that would create a taxonomy of the observed visual scene as proposed by Barghout [Bar14].

Semantic segmentation is then a preliminary task required for many high level applications that require precise image understanding such as autonomous vehicle control, robotics including human assistive technologies such as medical surgery.

More specifically, the aim is to classify each pixel of the entire input sample but do not differentiate between object instances. Considering a set of target class labels $\Omega = \{c_0, c_1, \ldots, c_l\}$ and an input image composed of n pixels $I = p_0, p_1, \ldots, p_n$, semantic segmentation aim is to find the mapping function that satisfies $F_s : I, p_i \rightarrow c_j$ for each of the image pixels. This task has however strong connections with global image classification, object instance detection and segmentation [HGDG17] that restrict image labeling on a specific portion of the image and differentiate between object instances. It also connects with the recently proposed panoptic segmentation [KHG+19] that aims at conducting full image

A. Benoit (✉) · E. Amri · P. Lambert
LISTIC - Université Savoie Mont Blanc, Chambéry, France
e-mail: alexandre.benoit@univ-smb.fr; emna.amri@univ-smb.fr; patrick.lambert@univ-smb.fr

B. Ghattas · J. Fournel
I2M - Université Aix Marseille, CNRS, Centrale Marseille, Marseille, France
e-mail: badih.ghattas@univ-amu.fr; joris.fournel@univ-amu.fr

© Springer Nature Switzerland AG 2021
J. Benois-Pineau, A. Zemmari (eds.), *Multi-faceted Deep Learning*,
https://doi.org/10.1007/978-3-030-74478-6_3

Fig. 3.1 For a given image (**a**), one can distinguish: (**b**) semantic segmentation with the foreground fish and background sea and ground classes, (**c**) foreground instance detection and segmentation and (**d**) panoptic segmentation that embrace both full image semantic segmentation as well as instance detection. Objects masks are best viewed in color

segmentation while differentiating between object instances. Figure 3.1 illustrates these different tasks. Semantic segmentation is then a difficult problem that should address local features analysis as well as context analysis and spatial reasoning. Several challenges must then be faced to propose an effective model able to generalise well on unseen data.

Semantic segmentation can be achieved in very different ways, depending on the knowledge associated with the data and the target semantic level. Preliminary contributions mostly relied on unsupervised strategies. Initially considering few samples I_i with few or no annotations, it consist in pixels clustering by considering low level information (pixel color, luminance, contour detection, and so on). However recent advances focus more on high semantic level analysis in order to locally classify semantic concepts (objects, scenes, action and so on). This is achieved by relying on large annotated datasets that enable supervised learning considering sample tuples $(I_i, masks_i)$ with $masks_i$ a matrix of the same size as input image I_i but reporting the expected pixel level labels defined in Ω. This strategy is the main driver of current research that is supported by specific benchmarks such as the one associated with the CityScapes dataset [COR+15]. Nevertheless, as we will see along the chapter, unsupervised strategies are still an interesting option that currently benefit from the recent supervised model proposals with increased semantic level segmentation.

Several survey papers have been published recently and provide extensive comparison between recent semantic segmentation models for supervised approaches [YYT+18, LR19] as well as semi or weakly supervised methods [ZZZ+19]. One can observe that state-of-the-art methods are currently all based on deep neural networks. The reader is invited to refer to such survey papers to get a detailed view. In contrast, this chapter is expected to provide a wider overview and make the link between the semantic segmentation task challenges, traditional methods to help understand the interest of deep neural network based approaches for the semantic segmentation task.

This chapter first describes, in Sect. 3.2, the main challenges that face the segmentation models. Then an overview of traditional approaches is provided in Sect. 3.3 to help situate the interest of deep learning approaches. Next, more insights will be provided on deep learning based models in Sect. 3.4. The training strategies will guide the discussion and typical networks will be described. A more specific Sect. 3.5 details advanced techniques and networks structures that improve performance levels for a variety of networks and training context. The chapter concludes with a discussion on the variety of application domains and related benchmarks in order to point some current issues and research directions.

3.2 Semantic Segmentation Challenges

Several challenges have to be faced for image semantic segmentation that are related to general image processing problems and some more specific to the task. Here is provided a brief overview that helps understand the variety of proposed approaches.

Semantic Concept Variability and Imbalance From a general point of view, visual appearance of target concepts is wide because of variability in shape, texturing, interactions with the neighbourhood, occlusions, and so on. This variability in the data is then expected to be well represented by experimental data dedicated to the model optimisation and its validation. This requirement, however, generally brings data imbalances or scarcity since some rare situations can only be captured from large data acquisition campaigns that over represent the most common ones. This imbalance is particularly enforced for semantic segmentation since the semantic concept distribution at the pixel level is significantly different than class distribution at the object instance level. For instance, on the sample image shown in Fig. 3.1, the background classes (sea and ground) clearly dominates the foreground objects (fishes) in terms of number of pixels despite the fact that their number of instance is high. Data selection and concept category distribution must then be precisely controlled in both the optimisation and validation processes of the algorithm.

Data Acquisition and Conditioning Distortions For a given raw data x to be captured, a monotonous and potentially nonlinear transformation A is applied by a sensor and following preprocessing such that one obtains $\tilde{x}(t) = A(x)$. For a given task, several sensors and data preprocessing methods can be employed, each one presenting different designs that impact the statistical distribution of the data and noise. In the specific case of pixel level classification, this factor is actually critical since the pixel content is directly related to the sensor and preprocessing behaviours. Robustness against such transformation must be addressed for a variety of configurations.

Objects Scales Variability Depending on the sensor resolution and the object to camera distance, the size of the semantic objects to segment can vary dramatically, from a far silhouette to a finely textured appearance at a very close distance. However, a pixel level classifier should always associate each of the pixels of an object to the same category whatever its scale. This actually points that a pixel p_i should be classified with respect to a spatial neighbourhood $N(p_i)$, a portion of the input media, such that $F_s : N(p_i) \rightarrow c_j$ is robust against scale change in the data.

The neighbourhood extent can then be potentially very large in order to capture the whole object specific features and its own neighbourhood thus providing some contextual information.

Image Level Variability and Imbalance In the specific case of semantic segmentation, algorithms have to deal with the large variability and data imbalance *within* the image. Local image behaviours can change dramatically because of local linear and non-linear changes in the data that cannot be compensated by global image preprocessing. Then, the visual appearance of similar concept instances can significantly change within the same image. Changes can relate to image linear pixel dynamic modifications such as luminance and contrast in bright and dark areas. In addition, nonlinear and non-regular changes can be introduced by physical phenomena that impact the texture contrasts depending on the spatial position and the object to camera distance. Haze distortion is a natural example that can impact several applications including autonomous driving. Robustness against such variability is then a challenge since the physical models that drive this variability are not perfectly known as for haze [EKTM16].

Spatial Reasoning Spatial relations are of major importance and their modeling can significantly improve semantic segmentation [LYZ+20]. This is a specific task that completes the local textures and objects representation and tries to identify the co-occurrence of concepts as well as their spatial organization. This expects the algorithm to be able to consider a potentially large field of view in order to process both local and contextual information.

Annotation Quality As for many tasks, a typical issue relates to the ground truth quality. In the specific case of semantic segmentation, such information is actually costly since each of the image pixels should be precisely classified, generally by human experts. Methods that penalise more boundaries segmentation errors such as [RFB15a] expects even higher annotations quality. This expects a precision regularity in the expertise that cannot be guaranteed. Indeed, labeling a single image can require dozen minutes depending on its size and generates fatigue that degrades quality along time. Also multiple annotators may be involved for faster delivery which generates irregularities because of different levels of expertise. This induces both noisy or weak annotations that impact the training data as well as the reference data used to validate the models. This point should be taken into account with care especially when considering supervised optimisation of segmentation models.

Post Processing Refinements The discussed challenge yields complex models whose predictions are subject to errors that can be moderated by post-processing steps. Errors may be related to local pixel class confusion or errors that originate from a two narrow field of view that limits spatial reasoning capability. Several post processing strategies can then be employed however, their tuning is generally specifically adapted to the model errors and thus must be tuned for each new model proposal. This actually contradicts with the 'end to end learning' approach proposed by deep learning and increases computational cost. A good model should make limited or no use of such strategies.

Real Time Inference Processing and Light Weight Models Some application domains such as online medical assisted technologies as well as autonomous driving expect real-time inference with low power consumption. However, high semantic level object segmentation expects to rely on a hierarchical representation of the image that increase computation cost. Specific strategies must then be investigated, from light and memory efficient model design [CKR+19] to model simplification along training [EGM+20].

In the sequel, we provide a brief overview of traditional image processing approaches for semantic segmentation and situate them with respect to those challenges to better understand their advantages and limitations. This will allow to understand why deep learning approaches have enabled the breakthrough in the domain.

3.3 Traditional Approaches for Semantic Segmentation

In this section we briefly present some classical non-deep learning based approaches used for image segmentation.

Early algorithms used for this task are generally divided into two groups: edge-based or region-based methods. Edge-based methods look for strong intensity variations between neighboring pixels while region-based methods are mainly based on some local uniformity criterion. The duality of these approaches can be noted, as region boundaries more or less correspond to edge detection results. Those approaches are unsupervised and quite simple.

Edge-based methods are generally performed in two steps: (1) edge point detection and (2) edge points grouping to get closed contours. Many solutions have been proposed for these two steps [BCC+95]. However, contours remain low-level features that rarely match the object contours. An object is indeed generally composed of different parts generating as many contours and it is difficult to decide what the real contours of the objects are.

There is also a large number of region-based segmentation methods. A first set of solutions works in the spatial domain, trying to build an image partition composed of regions that are homogenous according to a certain criterion. The process can be a bottom-up process like region growing approaches [Pav72], or a top-down process as it consists in split and merge steps. Mathematical morphology also provides tools, like watershed [BM93], to get regions. A second set of solutions consists in using thresholding or clustering. These solutions are defined in the pixel intensity (or feature) domain and require a post-processing step (i.e. connected components labelling) to get the regions in the spatial domain. Thresholding is a very simple solution limited to situations where the image contains homogeneous objects that are clearly distinguishable from a homogeneous background. Then, there exists a threshold separating the pixel intensities between objects and background. Otsu [Ots79] proposes a solution to automatically determine the optimal threshold.

Clustering approaches are more interesting as they allow to use a richer description of each pixel (intensity, color, texture features...). The clustering is performed in the feature space. K-means [HW79]) is probably the most widely used clustering approach in statistics and machine learning. It is a non-hierarchical iterative approach which needs the number of clusters to be fixed in advance (some heuristics may help to choose or optimize this parameter). Variants of K-means are mainly based on the choice of features used in the algorithm. The default approach consists in using pixel intensities in the RGB or HSV coordinates. One may include in the features sets, either first or second derivatives of theses intensities to account for discontinuities between clusters (edge detection), or features like pixels' spatial positions to constrain neighbor pixels to belong to the same cluster (Superpixels, [CGN19]).

However, the performance of all these approaches remains limited because they do not seek a semantic understanding of the image specifically the long-range dependency between pixels. To this end, probabilistic graphical models, such as Conditional Random Fields (CRF) [LMP01] can be used to get a semantic segmentation as proposed in [SJC08] or [LVZ11]. They used a fully connected pairwise CRF characterized by a Gibbs energy function with a unary and a pairwise potential function. The unary potential is computed independently for each pixel by a classifier that produces a first label assignment which is generally inconsistent. Then, the pairwise potential encourages the pairwise label prediction of to be consistent with the corresponding hand-crafted features.

3.4 Semantic Segmentation Deep Learning Approaches

As for numerous domains, Deep Neural Networks (DNN) have introduced a strong breakthrough for semantic segmentation. The idea behind the interest of deep neural networks for this task is that one can intuit an analogy between the visual scene hierarchy and the local to global feature hierarchies learnt by these models. Also, since high semantic level features can be encoded locally, one can expect deep neural networks to enable high level concepts semantic segmentation. A striking point is that since such hierarchical local features description is required in the first processing stages of a variety of image understanding problems, cross task model transfers can be expected.

A large variety of recent semantic segmentation deep learning models have been proposed and detailed in survey papers such as [YYT+18, LR19, ZZZ+19]. Model architectures highly depend on the case study, the available data and the learning strategy. We propose in the following an overview of the typical approaches considered for supervised and weakly supervised learning. Each one will be explained and illustrated by a typical neural network structure. However, since the presented approaches share common network refinement strategies, this discussion is presented in the more specific Sect. 3.5 in order to clarify the presentation and show the connections between the approaches. The proposed overview presents the main strategies to build and train deep learning models related to semantic segmentation. Their adaptation to more specific case studies such as the Zero Shot Learning, described in Chap. 6 can be considered relying on similar model structures but with the adaptation of parameter optimization techniques.

3.4.1 Supervised Learning Approaches

Supervised approaches are a reference in the optimization of deep neural networks. For semantic segmentation, it expects the availability of images and ground truth segmentation masks. This is a critical point since pixel-level annotation is particularly costly in terms of time and expertise. Thus, this introduces two major limitations: annotated data quantity as well as annotation quality. Besides, the computational cost increases as the training data collection size and the model's complexity increase. However, learning high-level abstract features with good generalization capabilities from the data expects those limitations to be overcome. Two main directions can then be drawn for supervised learning strategies. Transfer learning can be considered when dealing with limited data or computational budget. Otherwise, whole model training can be considered when more data and computational power are available. In the following, we present both approaches and illustrate some of the most commonly considered models.

3.4.1.1 Transfer Learning Based Networks

Transfer Learning (TL), also known as 'knowledge transfer', is a technique designed to apply knowledge learned from one domain to enhance learning in another. It is then intended to overcome the paradigm of isolated learning and reuse general knowledge across related problems. TL has been defined in [PY09] as a method to improve the learning task $\mathcal{T}_T$ of the target predictive function $f_T(.)$ in a target domain $\mathcal{D}_T$.

The improvement is derived from a preliminary learning task $\mathcal{T}_S$ in a source domain $\mathcal{D}_S$. The prediction function $f_S(.)$ is considered in this preliminary learning step as the composition of a general knowledge descriptor $f_k(.)$ and the specialized task head $f_s(.)$. Afterwards, the target predictive function $f_T(.)$ exploits the knowledge descriptor $f_k(.)$ in combination with a new specialized head $f_t(.)$.

This process can then be formalized by Eqs. 3.1:

$$f_S(.) = f_s(f_k(.)), T_S \in D_S$$
$$f_T(.) = f_t(f_k(.)), T_T \in D_T, \tag{3.1}$$

where $\mathcal{D}_T \neq \mathcal{D}_S$, and/or $\mathcal{T}_T \neq \mathcal{T}_S$.

The motivation behind feature transfer with deep networks is that the first model layers ($f_k(.)$) are generic enough to be relevant for many tasks, while the later layers ($f_s(.)$ and $f_t(.)$) are more specific to the domain and task.

The successful use of TL has been highlighted by numerous cases as reported in [ZQD+20] and various application domains such as medicine and multimedia. Nevertheless, it is important to understand that it is not always successful. Indeed, it can be counterproductive if the source and target domains or tasks do not share common knowledge, i.e. if $\mathcal{D}_T \cup \mathcal{D}_S \neq 0$ and $\mathcal{T}_T \cup \mathcal{T}_S \neq 0$.

TL has been proposed in the typical case studies illustrated in Fig. 3.2 and described in the sequel.

Feature Extraction and Fine-Tuning Strategies

Feature extraction and fine-tuning are typical approaches for TL and are generally considered jointly. The basic feature extraction strategy consists in obtaining high-level semantic features from a fixed general knowledge descriptor $f_k(.)$ to be finally processed by a high-level operator $f_t(.)$ specifically optimized on the target domain $\mathcal{D}_T$. $f_k(.)$ is derived from a preliminary optimization on a source task and domain ($\mathcal{T}_S, \mathcal{D}_S$). The purpose is to provide relevant features and maintain them unchanged.

Typical knowledge descriptors to be used for image classification as well as semantic segmentation tasks are all but the last layers of classifier networks such as VGG, AlexNet, ResNet. These networks are generally pretrained on the Imagenet dataset and the related ILSVRC challenges [RDS+15].

A common application of this semantic segmentation strategy is the encoder-decoder architecture mentioned in Chap. 2. As illustrated in Fig. 3.3, $f_k(.)$ is used as an encoder providing low resolution but high semantic level features to the decoder

Fig. 3.2 Deep transfer learning strategies

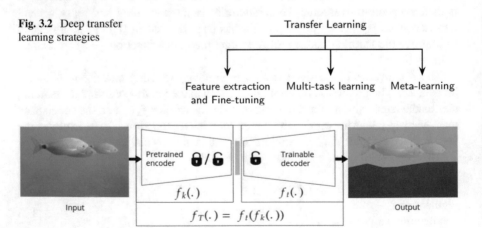

Fig. 3.3 Transfer learning model through an Encoder-Decoder architecture. The decoder is trainable, but the encoder parameters, relying on a transferred model can remain constant or be adapted to the target domain through fine-tuning

$f_t(.)$ that recovers the data size and maps the features to the target semantic labels. The degrees of freedom related to the complexity of $f_t(.)$ are constrained by the difficulty of the target task and related training data collection behaviors. However, this strategy expects the features provided by $f_k(.)$ to be sufficiently relevant and that $f_t(.)$ can exploit them effectively. These potential limits are overcome by the complementary fine-tuning strategy presented below.

Fine-tuning completes the feature extraction strategy by allowing for the optimization of the entire network on the target task and domain, including the transferred components. This strategy then enables domain adaptation of all or a subset of components of $f_k(.)$. Therefore, in the pre-trained encoder, it is possible to keep a variable number of first layers frozen and to retrain only the remaining high-level layers to help them adapt to the domain and to $f_t(.)$.

When applying transfer learning and fine-tuning together, the learning priority remains on the new head $f_t(.)$ while $f_k(.)$ generally adapts to the new domain at a lower learning rate. The objective is to preserve some of the representational structures learned in the original tasks. Using appropriate training hyper-parameters, the resulting model can outperform feature extraction approaches [GDDM14] [ARS+15] and models trained from scratch. The early observations made by Yosinski et al. detailed in [YCBL14] show the impact of fine-tuning and the depth of the transferred model for classification networks. Fine-tuning is indeed required to ensure the transferred model target domain adaptation and enhance compatibility with the new head.

However transfer and fine-tuning, when applied several times, can lead to a catastrophic forgetting effect in which the representation of the initial feature representation disappears along with the fine-tuning steps. This issue is currently an active area of research that is beyond the scope of this chapter, but some strategies such as [LH17] are already proposed.

When considering transfer and fine-tuning for semantic segmentation, the first breakthrough was proposed with the Fully Convolutional Networks(FCNs) [LSD15]. FCNs build on a standard image encoder $f_k(.)$ as the ones used for image classification. The decoding model section is a shallow function $f_t(.)$ that relies on a few deconvolution layers that upsample features and predict semantic labels at the initial resolution. Regarding TL, Long et al. work [LSD15] showed that transferring contemporary classification networks such as AlexNet [KSH12] and VGG[SZ14] yields accurate segmentation masks. This TL approach is actually of real interest for many semantic segmentation tasks since the optimisation of $f_k(.)$ on a classification task is much cheaper: more annotated data are available and their annotation cost is lower. Several works next made use of this strategy such as [SBT+19], [IS18], [GPHLG+16].

A more recent and successful application of transfer learning for semantic segmentation is the DeepLab model family first presented in [CPK+17]. For these architectures, a pre-trained VGG-16 or ResNet classification model is transferred to be used as $f_k(.)$. The decoding model section $f_t(.)$ is more advanced compared to FCNs. It makes use of Atrous Spatial Pyramid Pooling (ASSP) that apply atrous convolution filters with a variety of dilation rates to increase the model receptive

Table 3.1 The most widely used models for transfer learning to semantic segmentation tasks. Transferable classifiers are generally used as feature extractor referred to as $f_k(.)$ in text while transferable semantic segmentation models can either be fully transferred except their pixel classification head, or can be fully adapted to a new domain

Name	Release	Number of parameters	Related TL works
Transferable classifier			
AlexNet	2012	60M	[KSH12], [TLQJ18]
VGG	2014	138M	[IS18], [PKST20]
ResNet	2015	55M	[JTK+18], [PKST20]
Inception	2017	22M	[GdNV20]
Transferable semantic segmentation model			
U-net	2015	31M	[OdS18]
DeepLab	2016	34M	[LCS+19]

field and introduce robustness against feature scales. This model significantly improves overall performance and especially the object boundaries segmentation quality.

Several studies then prove the usefulness of TL for semantic segmentation. The work of Orsic et al. [OKBS19] further shows its interest for real-time operations and that the highest accuracy and speed are obtained relying on a transferred ResNet encoder. One can also report transfer of semantic segmentation models from one task to the other. For instance, [OdS18] transfer a U-net model with successful results and minimal retraining cost.

Table 3.1 summarizes the most widely used models in TL based works applied to semantic segmentation tasks. One details their associated number of parameters, first release date and some TL work examples. One can observe that the number of parameters tends to decrease over time. This is also related to the increase in the inference speed provided by the more recent models that benefit to the semantic segmentation models.

Connections with Multi-Task Learning (MTL)

In this section, a relationship between transfer and multitask learning is proposed. A more detailed presentation of multi-task learning is discussed in Sect. 3.5.3.

With TL, tasks are learned sequentially, so that the transferable knowledge descriptor $f_k(.)$ is first learned on the source domain and task $(\mathcal{T}_S, \mathcal{D}_S)$ and then reused for a new task and domain $(\mathcal{T}_T, \mathcal{D}_T)$. With MTL, multiple tasks $\{T_1, T_2, .., T_n\}$ are learned jointly on the same target domain $\mathcal{D}_T$. Then, a common knowledge descriptor, if any, is directly trained to satisfy multiple aims by learning the relationship between the tasks [Rud17]. In a way, MTL can be seen as a TL problem where the source and target task are learned jointly [WP15].

For semantic segmentation, multi-task models are regularly proposed. A well know one is Mask-RCNN [HGDG17], detailed in Sect. 3.5.3, a model that jointly predicts object bounding boxes, classes and segmentation masks. This model is however particular since it relies on an encoder referred to as 'backbone' that can be obtained through transfer learning from a preliminary classification task and then fine tuned. Some recent contributions propose similar models trained exclusively in a multitask way on the target domain as for [ZTC+18].

Meta-learning

In several computer vision tasks, pre-trained models on the ImageNet dataset have been shown to perform well when transferred to new tasks and domains with few training data. However, in many real situations, especially when the target data significantly differs from the data used for pre-training, transfer learning does not work effectively. This is obvious when considering medical imaging, astrophysics, and so on where data distribution differs from the multimedia images of ImageNet.

In this case Meta-Learning is designed as a technique to better take advantage of the preceding learning steps, even when pre-training is performed on tiny data collections. Meta-learning is defined in [TP98]. Considering a set of tasks and domains $\{(T_1, \mathcal{D}_1), \ldots, (T_n, \mathcal{D}_n)\}$ for which a model can be trained, the corresponding set of knowledge descriptors $\{f_{k_1}, f_{k_2}, .., f_{k_n}\}$ is learned. Meta learning then consists in guiding the learning of the new knowledge descriptor f_{k_T} of a target task and related domain $(T_T, \mathcal{D}_T)$ in order to maximize performances on T_T, taking advantage of previous experiences and their related knowledge descriptors. New tasks can then be learned from small amounts of data.

Rakelly et al. outline in [RSD+18] a variety of segmentation tasks and shows how, through meta-learning, a fast, accurate and efficient segmentation of data is enabled. Such approaches are then of interest when multiple (small) data collections are available for which a general representation of the data needs to be learned and transferred to new problems. It therefore extends Transfer Learning to other case studies.

3.4.1.2 Learning Without Prior Knowledge

In situations where no prior knowledge is available or when there is a requirement to learn new features without the potential bias of weight transfer, full model training (end-to-end learning) may be considered. Compared to TL, end-to-end learning provides the ability to directly learn representations of the appropriate characteristics of the target problem. End-to-end learning is achieved with a single training procedure, which can be more costly in terms of data and processing budgets since it requires optimizing a larger number of parameters from scratch.

In the context of semantic segmentation, the design and training of End-to-End models have a large degree of freedom. We present a classification of the typical state of the art approaches in Fig. 3.4 and detail them in the following paragraphs.

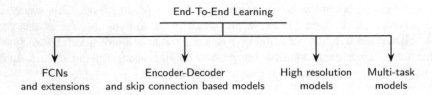

Fig. 3.4 Learning without prior knowledge

FCNs Models One of the first DL works for semantic segmentation is the use of FCNs. It can rely on transfer learning but can also be trained from scratch. The FCN is a network that does not contain Dense layers. Such layers are typically used in classification problems as final model layers in order to allow for all the information to be merged to make a global decision. As an alternative, and to comply with semantic segmentation that expects local classification, FCNs rely on 1×1 convolutions that locally merge all the features in order to perform pixel level predictions.

FCNs were proposed [LSD15] in the first time to realize an end-to-end semantic segmentation. FCNs rely on a classifier like backbone that gradually sub-sample features while increasing semantic level and introduce a last shallow up-sampling step. This latter makes use of deconvolutions as a way to up-sample the output feature maps to recover the spatial dimension of the target segmentation. This model introduces fusion of cross layer feature maps. It relies on summation with the output in order to take advantage of both the detailed spatial information (earlier layers) and the high discriminating semantic features of the output layers.

On year 2014, it achieved state of the art segmentation performance. This work is considered as a milestone in image segmentation, demonstrating that deep networks can be trained for semantic segmentation in an end-to-end manner. However, despite its popularity and effectiveness, the FCN model has some limitations. It is not fast enough for real-time inference, also, it does not take into account the global context information in an efficient way. However, upgrading the original model components with more recent ones presented in Sect. 3.5.1 can make the model more efficient.

Encoder-Decoder and Skip-Connection Based Models

Most of the deep models for image segmentation are actually based on the convolutional Encoder-Decoder architecture followed by a final pixel-wise classification layer. The encoder and decoder layers are mostly symmetrical to each other. The purpose of the encoder is to increase the semantic level of the representations and to extend the field of view, looking wider to help classify each pixel. However its internal sub-sampling or pooling operations results in a reduction in resolution, with a loss of detail, and can thus impact the quality of the segmentation on the class boundaries. Then, from the encoded data, the decoder purpose is to map the low resolution, high level semantic features to the original data resolution. Compared

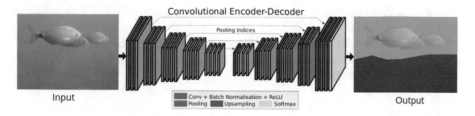

Fig. 3.5 Encoder-decoder architecture of SegNet. The pooling indices obtained at each maxpooling steps of the encoder are reported to the corresponding upsampling operators of the decoder

to the FCN networks, the features upscaling is accomplished in more steps thus enabling refined processing with higher model capacity.

One of the former proposal is the SegNet model [BKC17]. This model tries to resolve the resolution loss issue. Relying on a sub-sampling based on max pooling operators, it communicates the pooling indices from the encoder and uses them in the decoder upsampling operators as shown in Fig. 3.5.

Upsampling consists in copying the preceding neural activations only at the position of the pooled indices reported by the corresponding encoder step. Contrary to FCN models, upsampling thus reduces model complexity since no learnable parameter is required. However, features interpolation is next required. This is performed by stacks of convolutional layers that then allow for local feature enhancement. The final decoder output feature maps are fed to a softmax classifier for pixel-wise classification. The SegNet proposal was motivated by the fact that a much smaller and faster architecture than existing architectures was desired.

One of the most popular model in the encoder-decoder category is U-Net proposed by Ronneberger et al. [RFB15b]. It relies on the so called skip connexions that transfer sets of low level feature maps from the encoder to the decoder instead of using maximum grouping indices like SegNet. U-Net then concatenates, at the decoder level, encoder low-level abstract information with the decoder high-level semantic information resulting in a finer and more accurate prediction map. U-Net indeed makes full use of the details of the high resolution first model features. In addition, skip connections facilitate the model learning. Those connections indeed directly transmit error gradients to the first model layers and thus reduce gradient vanishing. However, because upsampling is carried through transposed convolutions, it has much more parameters to learn and is comparatively slower to train than SegNet.

The success of U-net has given rise to a large family of networks. The global U-Net structure can indeed benefit from several internal refinements to enhance task performance as well as processing cost. Section 3.5.1 highlights some of the most recognized block structures to be used at a given feature processing scale of the encoder and decoder. A well known extension is then FC-DenseNet that relies on densely connected layer blocks. In addition, the use of attention processes discussed

in Sect. 3.5.2 also yield good results as well as some explainability behaviors for the
SAUNet model [SDZW20].

Several other works adopt Encoder-Decoders for image segmentation, such as V-Net[FSG+18] that introduced a modified 3D version of U-Net with the customized
Dice coefficient loss function.

High Resolution Models

Instead of retrieving high-resolution from low-resolution representations as in Seg-Net and U-Net other model structures can process several feature scales in parallel.
The aim is then to increase robustness across scales and refine class boundaries. For
instance, HRNet [WSC+20] maintains high-resolution representations throughout
the whole network. The network is composed of sequential stages: each stage is
composed of parallel convolution streams with different resolutions; an exchange
unit is implemented at the end of each stage to fuse the information from all
resolutions through summation.

The main advantage is that there is no loss of spatial information (which is not
the case with encoder-decoders), especially useful for image segmentation. These
HRNet networks, combined with other techniques (such as OCRNet and SegFix
[YCW20]) have achieved the best results on many segmentation challenges such as
Cityscapes. In spite of its raw segmentation performances, the high resolution of the
feature maps kept throughout the network results in an inference time that can be
longer than with encoder-decoder segmentation networks.

Segmentation and Multitask Models

The models discussed above mainly focus on the semantic segmentation region
accuracy and do not explicitly focus on the segmentation quality of the region
boundaries. Boundaries detection, as well as other tasks connected to semantic
segmentation, can however be performed simultaneously with segmentation relying
on a multitask model. Multitask models can be trained in an end to end strategy in
order to reduce the overall computational cost and can also enhance the performance
of the segmentation. Section 3.5.3 presents an overview of the typical multitask
strategies and their interest for the specific case of semantic segmentation.

One can also report other works dedicated to semantic segmentation with very
specific multitasking strategies. An example is the Boundary Aware Salient Object
Detection (BASNet) model [QZH+20, QZH+19]. It consists of two components: a
semantic segmentation *prediction* network followed by a *residual* refinement model
head. The advantage is to capture large-scale structures with the first module and
then refine them with the second one. The prediction module provides multiscale
segmentation outputs with their own optimization criteria and the refinement
module has a single optimization criteria.

Table 3.2 Typical models trained end-to-end for semantic segmentation tasks

Name	Release date	Number of parameters	Publication
FCNs	2015	20M	[LSD15]
SegNet	2015	14.7M	[BKC17]
U-net	2015	31M	[OdS18]
FC-DenseNet	2017	15M	[JDV+17]
HRNet	2019	65.8M	[WSC+20]

End to End Learning Synthesis

As a final word on this section, one can observe the variety of model structures that can be trained end to end. Table 3.2 provides a brief summary of the presented models, showing their complexity in terms of number of parameters along with their temporal evolution.

End-to-end learning, like all approaches, has limitations which are studied by T. Glasmachers et al. [Gla17]. It shows how the end-to-end approach is framed in the deep learning context, presenting the reasons that make it an impractical option in some cases. One of these reasons is the need to have a huge amount of data when no prior knowledge is taken into account. A further difficulty lies in the model modification, where the hole model has to be replaced and re-trained.

3.4.1.3 Performance Metrics and Loss Functions

As far as semantic segmentation is concerned, improvements are based not only on the modification of the model structures but also the configuration such as the loss function, which plays an important role in the training of the model. Performance metrics are used to assess the quality of the segmentation. Some performance metrics may also be used as loss functions.

In the supervised framework, semantic segmentation stands for supervised classification. The ground truth, true pixels labels may be seen as a partition of the pixels within the image. The predicted classification may also be seen as another partition. Thus, metrics for comparing partitions may also be used to compare both segmentations, the ground truth and the predicted one.

Consider a ground truth image with N pixels, each of them having label $y_i \in \{1, .., J\}$, $i = 1..N$, where J is the number of labels considered in the image. When used in semantic segmentation, each pixel is presented as a one hot encoding vector, denoted $\mathbf{p}_i = (p_i^1, p_i^2, \ldots p_i^J)$, where all the components equal zero except the component corresponding to the true label. The output of the segmentation has a similar representation $\hat{\mathbf{p}}_i = (\hat{p}_i^1, \hat{p}_i^2, \ldots \hat{p}_i^J)$. We denote y_i the true label of a pixel i and $\hat{y}_i$ the predicted one, the label j which maximizes $\hat{p}_i^j$. Some metrics compare directly the labels y_i and $\hat{y}_i$ while others compare the probabilities vectors $\mathbf{p}_i$ and $\hat{\mathbf{p}}_i$. We give a brief description of such measures.

For simplification, we denote $\sum_i = \sum_{i=1}^N$, $\sum_j = \sum_{j=1}^J$ and $\sum_{i,j} = \sum_{i=1}^N \sum_{j=1}^J$.

Cross Entropy The most widely used loss function for performing the image segmentation task is the pixel-level cross entropy (CE) loss given by

$$CE = -\frac{1}{N} \sum_{i,j} p_i^j \cdot log(\hat{p}_i^j) \tag{3.2}$$

That is the average cross entropy over all the pixels of the image. It may be also averaged over a batch of images.

A problem can arise if the different classes are unbalanced; the most frequently occurring class may dominate the learning. In this case, a variant of CE called Weighted cross-entropy (WCE) is used. Each element in the sum in Eq. 3.2 is given a weight inversely proportional to the frequency of the corresponding label in the learning sample.

Semantic segmentation models usually consider the cross categorical entropy loss function during training. However, if there is an interest in obtaining the granular information of an image, more advanced loss functions are proposed. For instance, Focal loss (FL) [LGG+17] is an improvement to the standard cross-entropy criterion. The change realized in Eq. 3.3 aims to down-weighted the loss assigned to easy examples, enabling model to learn hard examples.

$$FL = -\frac{1}{N} \sum_{i,j} p_i^j \cdot (1 - \hat{p}_i^j)^\lambda log(\hat{p}_i^j) \tag{3.3}$$

Where $(1 - \hat{p}_i^j)$ is a modulating factor and λ is focusing parameter. Setting $\lambda > 0$ reduces the relatives loss for well classified examples ($\hat{p}_i^j > 0.5$), putting more focus on hard, miss-classified examples. The focusing parameter λ smoothly adjusts the rate at which easy examples are down-weighted.

Dice, Jaccard and IoU The Jaccard index often referred to as Intersection over Union (IoU) is the most used metric to evaluate models in natural image segmentation tasks.

Given two regions (sets of pixels corresponding to a specific label j) R_j and $\hat{R}_j$, their IoU evaluates the degree of overlapping between them. It is given by:

$$IoU(R_j, \hat{R}_j) = \frac{|R_j \cap \hat{R}_j|}{|R_j \cup \hat{R}_j|}$$

$|R_j|$ is number of pixels in R_j. $|R_j \cap \hat{R}_j|$ stands for the number of elements common to both regions. The IoU varies from 0 (when the regions are not overlapping at all, or disjoint) to 1 (when they are perfectly overlapping).

Another metric, the dice index, closely related to the Jaccard index is often preferred in medical image segmentation tasks. It is defined as:

$$DI(R_j, \hat{R}_j) = \frac{2|R_j \cap \hat{R}_j|}{|R_j| + |\hat{R}_j|} \qquad (3.4)$$

The Dice and Jaccard indexes may be averaged over the J labels. Their drawback is that they are biased toward object instances that cover a large image area. Hence, for tasks where large variations in scales can be observed for objects belonging to the same class, variants, such as instance-level IoU metrics, can be used. This is the case for example in the Cityscape dataset [COR+15], a street scene segmentation task, to evaluate how well the individual instances of traffic participants are segmented by the models.

Based on the dice index, the dice loss (DL, [SLV+17]) is often used when optimizing image segmentation:

$$DL = 1 - \frac{2\sum_{i,j} p_i^j \cdot \hat{p}_i^j + 1}{\sum_{i,j}(p_i^j)^2 + \sum_{i,j}(\hat{p}_i^j)^2 + 1} \qquad (3.5)$$

It may be seen as an approximation of the complementary to 1 of the dice index (1-DI). Here, 1 is added in numerator and denominator to ensure that the function is not undefined in edge case scenarios such as when $p_i^j = \hat{p}_i^j = 0$.

A generalized version of Dice Loss called Generalized Dice Loss (GDL) is a multi-class extension of Dice Loss where the weight of each class is inversely proportional to the square of the label frequencies [SLV+17].

Tversky loss [SEG17] presents a generalization of DL based on the regular Dice coefficient. It adds weight to FP (false positive) and FN (false negative) samples using the β coefficient, which is different from dice loss using equal weights for FN and FP.

$$TL = \frac{\sum_{i,j} p_i^j \cdot \hat{p}_i^j}{\sum_{i,j} p_i^j \cdot \hat{p}_i^j + \beta \sum_{i,j}(1 - p_i^j) \cdot \hat{p}_i^j + (1 - \beta) \sum_{i,j} p_i^j \cdot (1 - \hat{p}_i^j)} \qquad (3.6)$$

Hausdorff While the Dice index gives a global evaluation about the quality of the segmentation, the Hausdorff distance detects the outliers.

We first define a distance function between any point x of X and a region Y by:

$$d(x, Y) = \inf\{d(x, y) \mid y \in Y\}.$$

Then we define a distance between two regions X and Y:

$$d(X, Y) = \sup\{d(x, Y) \mid x \in X\}.$$

Finally the Hausdorff distance is given by:

$$d_{\mathrm{H}}(X, Y) = \max\{d(X, Y), d(Y, X)\}$$

The Hausdorff distance is often used in medical image segmentation together with the Dice index. However, for obvious reasons, this distance is not adapted to scattered regions.

Volume Similarity and Clinical Metrics Another metric worth to be mentioned is the Volume Similarity (VS):

$$VS(X, Y) = 1 - abs(|X| - |Y|)/(|X| + |Y|)$$

Volume related metrics are especially useful when the segmentations are used to quantify the volume and masses (given a density) for the considered region (e.g. in medical image segmentation volumes and masses of organs are often computed from a segmentation of the organ and pixel spacing properties).

In medical image segmentation tasks, clinical metrics are sometimes used jointly with the segmentation metrics we have just seen. In this paradigm, automated and manual segmentations are first used to quantify the volumes and masses for the considered organs, then errors are computed directly on the quantified informations using more classical measures of errors (mean absolute error, biases, . . .).

Classical Metrics Used to Compare Partitions Such metrics are often classified in two categories: measures based on counting pairs (*Rand index*, *Adjusted Rand* index, *Jaccard index*) and measures based on mutual information (*Mutual Information*, *Normalized Mutual Information*, *Variation of Information*, [Mei05]). Those metrics do not account for spatial distribution of pixels among images.

Compound Loss Combining multiple losses is also possible. A quite popular combination is the association of CE and DL. Another combination of FL and DL is realized in [ZHD+19]. Another loss combination is introduced in [AK19], which combines Focal and Tversky losses.

There have been several attempts to improve loss by integrating the relationship between pixels such as [ZWYC19]. It introduces in a region a mutual loss of information (MIL). Compared to the pixel-level loss which treats pixels as independent samples, MIL models each pixel while taking into account the pixel's dependencies concerning its neighbors. The proposed loss is based on CE which calculates the similarity between the pixel intensity, and the mutual information which measures the structural similarity between two images.

3.4.2 *Unsupervised and Weakly Supervised Learning*

Autoencoders and Restricted Boltzmann machines are among the most common approaches used in deep learning to tackle unsupervised problems. The first serves to encode images, and the second for clustering (of images or pixels). As the main common problem of semantic segmentation is the scarcity of pixel level labelling, a totally automatic unsupervised segmentation for images would be of great help.

Some recent approaches suggest such solutions. The W-net [CGN19] is a convolutional autoencoder network where both encoder and decoder are U-net structures. The output of the encoder is a mask which is processed (using condition random fields, and hierarchical clustering) to get the final output (the segmented image). The decoder takes the mask as input and outputs an image which should be the closest possible to the encoders input. The loss function is a combination of a reconstruction loss (commonly used in autoencoders) and a soft cut loss applied to the mask. This loss may be seen as a quite classical criterion similar to intra-variance of the clusters represented by the mask, thus connecting to the classical approaches presented in Sect. 3.3.

Other unsupervised methods rely on generative approaches such as ReDO [CAD19]. This model first identifies an image partition and then tries to generate the content of each partition separately before fusing them to create a generated image resembling the input. This approach is of interest since it allows the distribution of each partition to be identified.

Finally some semi supervised approaches have also been suggested for semantic segmentation, see for instance [OHT20]. In this context, unlike the classical semi supervised paradigm, strong labels learning approaches (using pixel level labeled images) is combined with weak labels learning (using image level labels) in various ways.

Unsupervised and semi supervised approaches are subject to increasing interest for semantic segmentation because of the limits of the available annotated data collections. Nevertheless, the associated internal model substructures generally follow the ones proposed with supervised methods while the global structure is adapted to comply with the chosen learning approach.

3.5 Model Refinements

The different models presented in the above sections provide an overview of state of the art model architectures. However several enhancements have been proposed and can be applied on most of them. This section describes some of the most advanced ones.

3.5.1 Block Level Enhancement

Features extraction at each scale is performed by neural layers substructures. As for other image analysis tasks such as classification, state of the art methods rely on layer structures with increased communication. As presented in Chap. 2, most of the recent models consider residual blocks and their variants. The basic idea is to rely on gradual data transformation such that $\mathbf{y} = \mathbf{x} + f(\mathbf{x}; \theta)$ where f is a non linear transformation with trainable parameters θ, performed by a few convolutional layers, typically three with hundreds of convolution kernels. By cascading such residual layers to build up very deep networks, highly transferable and relevant features can be learnt at the cost of model complexity. Extensions such as Xception [Cho17] and ResNeXt [XGD+17] introduce depthwise separable and grouped convolutions in f in order to find the best compromise between representation capability and computational cost. On the apposite, more shallow architectures with significantly fewer parameters but equivalent performance levels have been proposed. For instance, dense blocks proposed in [HLVDMW17] increase communication within block layers. In this approach, each layer generate few, typically less than 50, activation maps $\mathbf{y_l} = f(\mathbf{x_l}; \theta_l)$ where $\mathbf{x_l} = H([x0, x1, \ldots, xl - 1])$ is the concatenation of all the preceding layer activations of the sub structure. It then maximizes features reuse and limits the need of deep structures and wide layers with many neurons. FC-DenseNet [JDV+17] applies this concept to semantic segmentation and proved to be a top performing structure pattern. As a summary, a variety of recent operators can be considered to design efficient processing blocks within deep learning model. The choice impacts on model feature relevance, structure depth, connectivity and number of parameters. However, it also strongly impacts on the processing time for both training and inference modes. We propose in the following a discussion on this topic and identify some recent proposals that enable faster processing.

Deep learning model neural layers and their connections actually involve both high number of multiply-accumulate operations (MAC) and memory traffic that impacts processing latency. As show in Sect. 3.4, semantic segmentation models and their specific encoder-decoder structure present numerous long range cross layer (skip) connections and high resolution feature processing blocks. From the global structure to neural layers, metrics and strategies must be considered to reduce the computational budget and enable real-time inference capability. [CKR+19] suggest to consider the ratio between the number of MAC (or floating point operations) and memory traffic as a complementary performance metric. At the layer level, MAC is controlled by the number of neurons and the input size. Memory traffic is approximated by a simple metric, Convolution Input Output (CIO), that sums the input and output tensor sizes of each convolution layers of the network. Considering h, w, c, respectively, height, width and the number of channels of tensors exchanged across the network layer l, the overall network CIO can be estimated by Eq. 3.7.

$$CIO = \sum_{l} h_{in}^{(l)} \times w_{in}^{(l)} \times c_{in}^{(l)} + h_{out}^{(l)} \times w_{out}^{(l)} \times c_{out}^{(l)} \qquad (3.7)$$

Then, the computational density $MoC = MAC/CIO$ can be considered to design efficient network structures with respect to the inference platform. Low MoC should suggest high traffic cost with respect to the representation capacity of a given layer. Then, for instance 1×1 convolution layers generally used for dimension reduction should be avoided as well as excessive use of feature concatenations. Following this idea, recent work propose enhanced neural architectures such as the HardNet model proposed in [CKR+19]. This updated design of DenseNet involves 40% less traffic while delivering similar task performance levels.

One can conclude this discussion by considering that semantic segmentation makes use of general purpose network substructures described in Chap. 2. However, the high computational cost of semantic segmentation models being related to its deep encoder-decoder design, computationally and memory traffic efficient structures should be preferred for both inference and energy consumption rates.

3.5.2 Attention Processes

Attention mechanisms are processes inspired by biological models that guides perception. From the mid twentieth century, neuroscience researchers have proposed several models to explain human visual search, spatial attention and feature integration such as Triesman work [TG80]. Deep learning models started to integrate such ideas, starting from text translation [BCB15]. Currently, attention with deep learning is rather considered as a tool which aims are twofold. It is first intended to help the model to adapt its response to specific context to improve prediction accuracy. In addition, the attention maps can provide explanation on the predictions by highlighting the features that contributed the most for a given input stimuli.

A variety of attention models have been proposed to improve deep learning models for a variety of tasks as illustrated in this book. We focus in this chapter on the ones that are more adapted to the semantic segmentation task with respect to its challenges. The aim here is mainly to consider spatial attention and feature integration in order to increase the importance of areas within the image or a subset of features at a given processing level. This is introduced as specific trainable modules inserted within neural architectures. They are generally considered in the decoding or upsampling section of the network in order to improve features aggregation coming from different scales and semantic levels. Since those attention modules introduce an additional processing and memory cost in an already large model, lightweight strategies are preferred. Some recent paper propose simple yet efficient attention modules that rely on a features weighting principle summarized by Eq. 3.8 where a set of feature maps $\mathbf{f_c}$ is weighted by a scaling factor $\mathbf{a}$ whose dimensions and value depend on the attention model aim and design. The complementary ϵ scalar modulates the effect of the attention module. If $\epsilon = 0$, irrelevant features contributions can then be annihilated. On the contrary, if $\epsilon > 0$, and typically being 1, the attention module enforces the feature map of interest while keeping the initial contributions of the less relevant ones.

$$\tilde{\mathbf{f}}_{\mathbf{c}} = \mathbf{f}_{\mathbf{c}} \odot (\mathbf{a} + \epsilon) \tag{3.8}$$

In this vein, the Squeeze and Excitation attention module [HSS18] is a well adopted features integration attention model that highlights feature channels dependencies. Following Eq. 3.9 and considering a set of n feature channels to be processed, it considers the attention factor $\mathbf{a_c}$ (c for channel) as a vector of n scalars that globally scale each feature map. $\mathbf{a_c}$ builds on the global activity vector $\mathbf{z}$ whose components z_c are the reductions (squeeze) of each of the feature maps f_c of size H, W, relying on their global average pooling. Spatial information is thus not taken into account such that this method does not distinguish between short and long range spatial dependencies. Finally, a non linear operator g_f is considered to identify the channel dependencies. Its output activation relies on the Sigmoid operator that allows for multiple channels activation (excitation). In this approach, feature maps can be masked, then $\epsilon = 0$.

$$\mathbf{a_c} = \sigma(g_f(\mathbf{z}, \theta_f)) \ with \ \mathbf{z} = \{z_1, \ldots z_n\} \ and \ z_c = \frac{1}{W.H} \sum_{x=1}^{W} \sum_{y=1}^{H} f_c(x, y) \tag{3.9}$$

Spatial attention can be defined similarly. One now rely on, $\mathbf{a_s}$ (s for spatial), a 2D weight matrix of size H, W that is combined with each feature channel $f_{c,i}$ relying on the Hadamard (element-wise) product. [SDZW20] proposes a low complexity spatial attention module that computes $\mathbf{a_s}$ relying on g_{s,θ_s}, a 1×1 convolution based compression of the input feature maps $\mathbf{f_c}$ according to Eq. 3.10.

$$\mathbf{a_s} = \sigma(g_s(\mathbf{f_c}, \theta_s)) \tag{3.10}$$

Those two attention modules can be combined following Eq. 3.11 as proposed in [SDZW20] in order to improve performance levels while providing explanation on the prediction from the analysis of the $\mathbf{a_c}$ and $\mathbf{a_s}$ attention values.

$$\tilde{\mathbf{f}}_{\mathbf{c}} = (\mathbf{f_c} \odot \mathbf{a_c}) \odot (\mathbf{a_s} + 1) \tag{3.11}$$

Similar approaches with different combination strategies have been proposed such as CBAM [WPLSK18] and [LWHY19]. The latter processes features by parallel neural branches dealing with different scales that are finally modulated by a unified attention module. However, most of these proposals are not dedicated to semantic segmentation and are applicable to a variety of image analysis tasks. This task indeed expects connections between neighbor pixels, i.e. spatial grouping that is not taken into account with those attention mechanisms. This direction has however been recently initiated by the SANet [ZLB+20] model that propose to consider attention on subsampled features in order to explicitly introduce local pixel grouping.

3.5.3   Multi Task Learning

Semantic segmentation task is a fundamental problem but generally not the final objective of a process. It can be associated to other tasks to complete the main analysis aims such as object boundaries detection, object detection and so on. Conversely, other tasks can be associated to semantic segmentation in order to guide and refine the segmentation quality. Those ideas are related to the concept of multitask learning. It is actually expected to increase model generalization capability since it biases the model to learn a representation suitable for a set of tasks [Car97]. From a model structure point of view, it generally consists in a common features extraction path that generates features $a_m = f_m(x, \theta_m)$. Then, considering a set of tasks $T \in t1, \ldots tn$, each one has its own specialized operators that led to a specific network head hti. They can be fed by the sole shared features path thus being formalized as $y_{hti} = f_{hti}(a_m, \theta_{hti})$. This is the most used hard parameter sharing architecture. More sophisticated strategies are also proposed and are referred as partial or soft parameters sharing architectures. In this case, each task dedicated path sti is fed by both the shared features and the original input data thus allowing for more complex and dedicated data representation and are formalized as $y_{sti} = f_{sti}(x, a_m, \theta_{sti})$. From an optimization point of view, all the model paths can be optimized simultaneously. Relying on an optimisation criteria L_{ti} associated to each of the tasks, a global loss is minimized. This global loss is generally the weighted sum formalized in Eq. 3.12 where β_{ti} are task weights that can be manually adjusted and remain static along the training process or be dynamically adjusted to balance task importance [KGC18].

$$L = \sum_T \beta_{ti} L_{ti} \tag{3.12}$$

From a computational point of view, a striking point is that the shared features are extracted a single time before feeding each of the task specific model operators. Consequently, this significantly reduces the processing and memory cost for both train and inference time compared to the use of a variety of task specific models.

In the following are presented typical multitask architectures that relate to semantic segmentation.

3.5.3.1   Semantic Segmentation as an Auxiliary Task

A typical case study is the association of semantic segmentation with tasks related to object instance detection. This latter has been introduced by the popular Region-

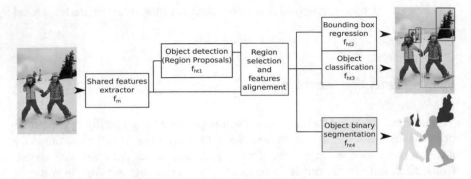

Fig. 3.6 The Mask RCNN proposed in [HGDG17], the semantic segmentation module completes the object instance detection, classification and bounding box regression

CNN model and more specifically its refined version Faster-RCNN [Gir15]. This is a multitask model able to detect, classify and regress object instance bounding boxes. Semantic segmentation capability has been added with the Mask RCNN architecture [HGDG17] illustrated in Fig. 3.6. In this approach, semantic segmentation is limited to the prediction of each object instance *binary* mask. This is processed in parallel to predicting the class and box. This decouples mask and class prediction thus reducing inter task dependencies. More into the details, given a K class object detection and segmentation problem, the semantic segmentation path output consists in K binary masks. The semantic segmentation head is composed of a few set of 3×3 convolution layers, typically between 3 and 6 with 256 neurons each except the last layer that is limited to K neurons, one for each target class. Along optimization, the loss relies on binary cross entropy computed between the true class mask and the related predicted mask. This loss is finally added to the other tasks loss with unit weight $\beta_{segmentation} = 1$, as for the other tasks. This popular model presents interesting performance rates on the standard benchmarks. There are however some limitations. First, inference speed is limited by the architecture complexity. In addition, segmentation mask boundaries lack of precision which particularly impacts small and thin objects detection. Recent extensions of the Mask-RCNN model include PointRend [KWHG20] that aims at refining segmentation quality on object boundaries.

3.5.3.2 Auxiliary Tasks for Improved Semantic Segmentation

One of the main difficulties with semantic segmentation is indeed related to the classification accuracy on the class pixel boundaries. Several reasons can be mentioned. From a data perspective, this can be explained by the limited precision of the boundaries reported by the annotations when relying on supervised training. Also, the combination of neighboring objects boundaries can present a higher variability than the objects themselves such that training is more difficult in those regions. From a model point of view, as discussed in Sect. 3.4, state of the art approaches generally rely on semantic features extraction obtained at a coarse resolution after several feature subsampling steps. This actually impact boundaries prediction accuracy and skip connections are then considered to compensate. However, no explicit constrains on the boundaries are generally proposed and one still rely on a single task system dedicated to pixel classification. Resulting models remain complex and limit the understanding of the predictions since object colors, textures and shapes are handled by the model in an obscure way.

An alternative is to guide semantic segmentation with an auxiliary task that allows to disentangle the factors related to the problem. The classical contours segmentation task is then an interesting option. Such an approach has been introduced by the Gated-SCNN model [TAJF19] and next improved for instance with the SAUNet model [SDZW20] illustrated in Fig. 3.7. The main contributions are the following. First the model relies on two processing paths. The former is dedicated to semantic segmentation and remains similar to the state of the art models presented above. The latter is more specific and is dedicated to shape description. Its very first layers are shared with the segmentation path and it then extracts features at the initial input image resolution in order to keep detailed shape information. This path has a specific optimisation criteria that constrains its output to generate shape masks, relying on the binary cross entropy loss between its output and reference shape masks. The issue is actually to extract only the relevant object boundary contours of the image. To this end, an attention mechanism called Gated convolution is proposed in order to let the shape path be modulated by the segmentation path in the aim to reduce the contribution of inner object contours. Such modulation with gated convolutions proposed in [TAJF19] and [SDZW20] can be summarized as

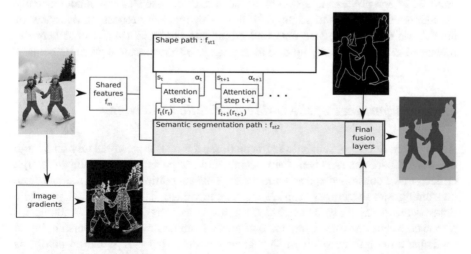

Fig. 3.7 The SAUNet model [SDZW20], a dual task deep neural network for semantic segmentation and boundaries detection. Semantic contours extraction is specifically optimised and help refine semantic segmentation

follows: for a given gating step t at a given processing scale, gated convolutions first aggregate the incoming shape s_t and semantic r_t feature streams by concatenation. Bilinear upscaling is applied to the semantic channel features when necessary to match shapes. Then an attention map is built, following Eq. 3.13.

$$\alpha_t = \sigma(C_{1x1}(s_t \| f_t(r_t)))$$ (3.13)

where C_{1x1} is a 1×1 convolution of the concatenated path features. The additional transformation f_t applied on r_t can be limited to bilinear interpolation as for [TAJF19] or introduce an additional 1×1 convolution in [SDZW20]. This attention map is applied on the shape stream using the Hadamard product. This approach permits to identify where image contours are of interest with respect to the semantic segmentation task. The output of the shape path is however a coarse map of the contours of interest. Then, in order to refine boundaries, it is combined with a traditional computer vision contours extractor such as the Canny filter [Can86] before being fused with the segmentation path to produce the final semantic segmentation predictions. This approach however induce high processing costs because of the processing of the shape path applied on high resolution features.

3.6 Data, Benchmarks and Model Evolution

This chapter provides an overview of the deep learning based semantic segmentation methods. The variety of model structures and training strategies highlights the difficulty of the problem. Even if deep learning approaches significantly outperform classical image processing approaches, the variety of case studies cannot currently be addressed by a unified strategy. In the following, we propose an overview of application domains, related data collections and challenges that would help the interested reader to identify up-to-date strategies with respect to a given domain.

3.6.1 Importance of the Data and Public Collections

As shown in Sect. 3.4, semantic segmentation deep neural networks rely on a large number of parameters. Then, their optimization expects the availability of large collections of controlled and annotated data. This annotation process is tedious, time consuming and sometimes requires expert knowledge. This is where large public datasets are particularly interesting. They are constructed and made available for public competitions for a wide range of areas, from natural scene understanding to medicine and earth observation. A regularly updated list is reported on platforms such as https://grand-challenge.org/challenges/ and https://evalai.cloudcv.org/.

An illustration of the current most considered application domains is shown in Fig. 3.8.

Fig. 3.8 Some typical application domains: multimedia, remote sensing, medicine

Beyond allowing for a comparison of new models to state-of-the-art under fair conditions, such datasets can be involved in other case studies. From a pedagogical point of view, they are tools to learn the mastery of deep learning. From a research and development perspective, they can be used to help models converge on specific domains. Transfer learning described in Sect. 3.4.1.1 is the classical approach but public datasets can also be fused with specific data for other learning strategies.

3.6.2   A Way to Follow the State of the Art

Other that the raw data, those competitions offer some useful features to the deep learning practitioner. Since deep learning is in constant evolution, it is not rare that a new model or method renders obsolete previous approaches: this evolution can be difficult to follow. In particular, the choice of the type of architecture to use on a specific task can be tricky (for example U-Net architecture is very often used in medical image segmentation tasks, where improvements are more to be obtained through augmenting the dataset size and quality). This is where the leaderboards of these competitions are precious: they allow to visualize easily what type of methods are working on a given task, what modifications historically lead to improvements and the link to the associated research paper is generally given on the website of the competition. Table 3.3 illustrates the main steps of the evolution of the IoU performance levels along time and shows which model structure and related innovation allowed for the significant performance increase.

3.6.3   Typical Benchmarks

The Cityscapes Dataset [COR+15] is a major public dataset in semantic understanding of urban street scenes. Most of the researchers that aim to develop a new state of the art model in natural image segmentation test their model on it and include a

Table 3.3 Evolution of the IoU performance metric on the CityScapes dataset along time. The main models and innovations that led to significant performance increase are reported, their refined and improved versions are not reported

Architecture	Release	IoU	Remarkable innovation
HRNetV2 [YCW20]	2019	0.818	Maintains high-resolution representations
DANet [FLT+18]	2019	0.815	Attention fuses local and global features
DeepLabv3 [CZP+18]	2018	0.813	ASPP and encoder-decoder structure
DeepLabv2+CRF [CPK+17]	2016	0.704	Atrous Spatial Pyramid Pooling (ASPP)
SegNet [BKC17]	2015	0.570	Non-linear upsampling in the decoder
FCN-8 [LSD15]	2015	0.653	Deep encoder and shallow upsampling

comparison of their results with the leaderboard results in their research paper. Such was the case for a number of pioneers architectures, some of which are displayed in Table 3.3. This challenge remains accessible for most researchers to evaluate new proposals in fair conditions. More recent ones such as LVIS [GDG19] that is more focused on object instance segmentation, significantly expand the quantity of data and semantic concepts to be detected. They allow for a refine evaluation of the generalization capabilities at larger scale with extended participation rules.

Other domains regularly propose challenges. One can cite the yearly data fusion context proposed by IEEE GRSS that focus on Earth Observation. Large scale tasks are regularly proposed with close connections to semantic segmentation, typically land cover mapping, but at the Earth Observation scale level. It allows new strategies to be developed, and take advantage of multi-modal and multi-temporal data, relying on a variety of ground based, aerial, and satellite sensors. Medical domain also regularly propose challenges that can be tracked on the https://grand-challenge.org/ platform for a variety of tasks often related to biological concepts segmentation such as organs, biological anomalies and so on. It highlights the difficulty related to the variety of acquisition devices, scarce annotations and variability in the biological data.

3.7 Conclusion

This chapter presents deep neural network application to semantic segmentation problems. A wide variety of model structures and optimization strategy has been discussed. Compared to traditional approaches, deep neural networks allow the semantic level of the segmented concepts to be increased, moving from low level pixel information to object level segmentation.

Semantic segmentation is a key component for advanced real world and higher level applications. Autonomous driving, robotics, medical imaging are typical examples as other ones as indexing and querying systems such as [WDYT19] that makes use of semantic segmentation for interactive actor and action segmentation in video sequences.

Current state of the art present impressive performance levels on now standardized large scale datasets. However still many challenges have to be faced, are more specifically real-time and high resolution segmentation in order to address large data processing, including video and volumetric data.

Another challenge is related to model explainability and prediction justification. The complexity of the model structures and the pixel level prediction task actually make this more complicated. The multitask models and attention strategies presented in the chapter draw the first step to explainability. Nevertheless more accurate justifications are required. The availability of large data collections with annotated objects parts would be of interest to provide refined expertise on the data that would help prediction understanding. Complementarily, explainability methods able to highlight the input data features that activate the most a given model component will complete the explainability toolkit. Such work have been recently initiated for semantic segmentation with [HMK+19] and [VDM20].

As a final word, it is important to highlight the complementarity of the semantic segmentation task with other tasks such as data sample global classification as well as object instance detection. The recently proposed Panoptic segmentation task defends this idea. Then interested reader should track model advances in those domains as well.

References

[AK19]   Nabila Abraham and Naimul Mefraz Khan. A novel focal tversky loss function with improved attention U-Net for lesion segmentation. In *2019 IEEE 16th International Symposium on Biomedical Imaging (ISBI 2019)*, pages 683–687. IEEE, 2019.

[ARS+15] Hossein Azizpour, Ali Sharif Razavian, Josephine Sullivan, Atsuto Maki, and Stefan Carlsson. Factors of transferability for a generic convnet representation. *IEEE transactions on pattern analysis and machine intelligence*, 38(9):1790–1802, 2015.

[Bar14]   Lauren Barghout. Visual taxometric approach to image segmentation using fuzzy-spatial taxon cut yields contextually relevant regions. In *International Conference on Information Processing and Management of Uncertainty in Knowledge-Based Systems*, pages 163–173. Springer, 2014.

[BCB15]   Dzmitry Bahdanau, Kyunghyun Cho, and Yoshua Bengio. Neural machine translation by jointly learning to align and translate. January 2015. 3rd International Conference on Learning Representations, ICLR 2015.

[BCC+95] Philippe Bolon, Jean-Marc Chassery, Jean-Pierre Cocquerez, Didier Demigny, Christine Graffigne, Annick Montanvert, Sylvie Philipp, Rachid Zéboudj, Josiane Zerubia, and Henri Maître. *Analyse d'images : Filtrage et segmentation*. Enseignement de la physique. MASSON, October 1995. Ouvrage publié avec l'aide du Ministère des affaires étrangères, direction de la coopération scientifique et technique. AVERTISSEMENT Le livre publié en 1995 chez MASSON (EAN13 : 9782225849237) est épuisé. Cette version pdf est une version élaborée à partie de la version préliminaire transmise à l'éditeur. La mise en page est légèrement différente de celle du livre. Malheureusement quelques figures de l'annexe C ont été perdues.

[BKC17] Vijay Badrinarayanan, Alex Kendall, and Roberto Cipolla. Segnet: A deep convolutional encoder-decoder architecture for image segmentation. *IEEE transactions on pattern analysis and machine intelligence*, 39(12):2481–2495, 2017.

[BM93] S. Beucher and F. Meyer. The morphological approach to segmentation: The watershed transformation. 1993.

[CAD19] Mickaël Chen, Thierry Artières, and Ludovic Denoyer. Unsupervised object segmentation by redrawing. In Hanna M. Wallach, Hugo Larochelle, Alina Beygelzimer, Florence d'Alché-Buc, Emily B. Fox, and Roman Garnett, editors, *Advances in Neural Information Processing Systems 32: Annual Conference on Neural Information Processing Systems 2019, NeurIPS 2019, December 8–14, 2019, Vancouver, BC, Canada*, pages 12705–12716, 2019.

[Can86] John Canny. A computational approach to edge detection. *IEEE Transactions on pattern analysis and machine intelligence*, (6):679–698, 1986.

[Car97] Rich Caruana. Multitask learning. *Machine learning*, 28(1):41–75, 1997.

[CGN19] Renju Chandran, Gopakumar, and Shyma S. Nair. A survey on different methods for superpixel segmentation. *International Journal of Science & Engineering Development Research*, 4:2:115–120, 2019.

[Cho17] François Chollet. Xception: Deep learning with depthwise separable convolutions. In *Proceedings of the IEEE conference on computer vision and pattern recognition*, pages 1251–1258, 2017.

[CKR+19] P. Chao, C. Kao, Y. Ruan, C. Huang, and Y. Lin. Hardnet: A low memory traffic network. In *2019 IEEE/CVF International Conference on Computer Vision (ICCV)*, pages 3551–3560, 2019.

[COR+15] Marius Cordts, Mohamed Omran, Sebastian Ramos, Timo Scharwächter, Markus Enzweiler, Rodrigo Benenson, Uwe Franke, Stefan Roth, and Bernt Schiele. The cityscapes dataset. In *CVPR Workshop on the Future of Datasets in Vision*, volume 2, 2015.

[CPK+17] Liang-Chieh Chen, George Papandreou, Iasonas Kokkinos, Kevin Murphy, and Alan L Yuille. Deeplab: Semantic image segmentation with deep convolutional nets, atrous convolution, and fully connected CRFS. *IEEE transactions on pattern analysis and machine intelligence*, 40(4):834–848, 2017.

[CZP+18] Liang-Chieh Chen, Yukun Zhu, George Papandreou, Florian Schroff, and Hartwig Adam. Encoder-decoder with atrous separable convolution for semantic image segmentation. *CoRR*, abs/1802.02611, 2018.

[EGM+20] Utku Evci, Trevor Gale, Jacob Menick, Pablo Samuel Castro, and Erich Elsen. Rigging the lottery: Making all tickets winners. Proceedings of Machine Learning Research. PMLR, 2020.

[EKTM16] Jessica El Khoury, Jean-Baptiste Thomas, and Alamin Mansouri. A color image database for haze model and dehazing methods evaluation. In Alamin Mansouri, Fathallah Nouboud, Alain Chalifour, Driss Mammass, Jean Meunier, and Abderrahim Elmoataz, editors, *Image and Signal Processing*, pages 109–117, Cham, 2016. Springer International Publishing.

[FLT+18] Jun Fu, Jing Liu, Haijie Tian, Zhiwei Fang, and Hanqing Lu. Dual attention network for scene segmentation. *CoRR*, abs/1809.02983, 2018.

[FSG+18] Filipe T Ferreira, Patrick Sousa, Adrian Galdran, Marta R Sousa, and Aurélio Campilho. End-to-end supervised lung lobe segmentation. In *2018 International Joint Conference on Neural Networks (IJCNN)*, pages 1–8. IEEE, 2018.

[GDDM14] Ross Girshick, Jeff Donahue, Trevor Darrell, and Jitendra Malik. Rich feature hierarchies for accurate object detection and semantic segmentation. In *Proceedings of the IEEE conference on computer vision and pattern recognition*, pages 580–587, 2014.

[GDG19] Agrim Gupta, Piotr Dollar, and Ross Girshick. LVIS: A dataset for large vocabulary instance segmentation. In *Proceedings of the IEEE Conference on Computer Vision and Pattern Recognition*, pages 5356–5364, 2019.

[GdNV20] Renato Giorgiani do Nascimento and Felipe Viana. Satellite image classification and segmentation with transfer learning. In *AIAA Scitech 2020 Forum*, page 1864, 2020.

[Gir15] Ross Girshick. Fast R-CNN. In *Proceedings of the IEEE international conference on computer vision*, pages 1440–1448, 2015.

[Gla17] Tobias Glasmachers. Limits of end-to-end learning. *arXiv preprint arXiv:1704.08305*, 2017.

[GPHLG+16] Luis C García-Peraza-Herrera, Wenqi Li, Caspar Gruijthuijsen, Alain Devreker, George Attilakos, Jan Deprest, Emmanuel Vander Poorten, Danail Stoyanov, Tom Vercauteren, and Sébastien Ourselin. Real-time segmentation of non-rigid surgical tools based on deep learning and tracking. In *International Workshop on Computer-Assisted and Robotic Endoscopy*, pages 84–95. Springer, 2016.

[HGDG17] Kaiming He, Georgia Gkioxari, Piotr Dollár, and Ross Girshick. Mask R-CNN. In *Proceedings of the IEEE international conference on computer vision*, pages 2961–2969, 2017.

[HLVDMW17] Gao Huang, Zhuang Liu, Laurens Van Der Maaten, and Kilian Q Weinberger. Densely connected convolutional networks. In *Proceedings of the IEEE conference on computer vision and pattern recognition*, pages 4700–4708, 2017.

[HMK+19] Lukas Hoyer, Mauricio Munoz, Prateek Katiyar, Anna Khoreva, and Volker Fischer. Grid saliency for context explanations of semantic segmentation. In *Advances in Neural Information Processing Systems*, pages 6462–6473, 2019.

[HSS18] Jie Hu, Li Shen, and Gang Sun. Squeeze-and-excitation networks. *2018 IEEE/CVF Conference on Computer Vision and Pattern Recognition*, pages 7132–7141, 2018.

[HW79] J. A. Hartigan and M. A. Wong. Algorithm AS 136: A K-Means clustering algorithm. *Applied Statistics*, 28(1):100–108, 1979.

[IS18] Vladimir Iglovikov and Alexey Shvets. Ternausnet: U-net with vgg11 encoder pre-trained on imagenet for image segmentation. *arXiv preprint arXiv:1801.05746*, 2018.

[JDV+17] Simon Jégou, Michal Drozdzal, David Vazquez, Adriana Romero, and Yoshua Bengio. The one hundred layers tiramisu: Fully convolutional densenets for semantic segmentation. In *Proceedings of the IEEE conference on computer vision and pattern recognition workshops*, pages 11–19, 2017.

[JTK+18] Mostafa Jahanifar, Neda Zamani Tajeddin, Navid Alemi Koohbanani, Ali Gooya, and Nasir Rajpoot. Segmentation of skin lesions and their attributes using multi-scale convolutional neural networks and domain specific augmentations. *arXiv preprint arXiv:1809.10243*, 2018.

[KGC18] Alex Kendall, Yarin Gal, and Roberto Cipolla. Multi-task learning using uncertainty to weigh losses for scene geometry and semantics. In *Proceedings of the IEEE Conference on Computer Vision and Pattern Recognition*, pages 7482–7491, 2018.

[KHG+19] Alexander Kirillov, Kaiming He, Ross Girshick, Carsten Rother, and Piotr Dollár. Panoptic segmentation. In *Proceedings of the IEEE conference on computer vision and pattern recognition*, pages 9404–9413, 2019.

[KSH12] Alex Krizhevsky, Ilya Sutskever, and Geoffrey E Hinton. Imagenet classification with deep convolutional neural networks. In *Advances in neural information processing systems*, pages 1097–1105, 2012.

[KWHG20] Alexander Kirillov, Yuxin Wu, Kaiming He, and Ross Girshick. Pointrend: Image segmentation as rendering. In *Proceedings of the IEEE/CVF Conference on Computer Vision and Pattern Recognition*, pages 9799–9808, 2020.

[LCS+19] Chenxi Liu, Liang-Chieh Chen, Florian Schroff, Hartwig Adam, Wei Hua, Alan L Yuille, and Li Fei-Fei. Auto-deeplab: Hierarchical neural architecture search for semantic image segmentation. In *Proceedings of the IEEE conference on computer vision and pattern recognition*, pages 82–92, 2019.

[LGG+17] Tsung-Yi Lin, Priya Goyal, Ross Girshick, Kaiming He, and Piotr Dollár. Focal loss for dense object detection. In *Proceedings of the IEEE international conference on computer vision*, pages 2980–2988, 2017.

[LH17] Zhizhong Li and Derek Hoiem. Learning without forgetting. *IEEE transactions on pattern analysis and machine intelligence*, 40(12):2935–2947, 2017.

[LMP01] John D. Lafferty, Andrew McCallum, and Fernando C. N. Pereira. Conditional random fields: Probabilistic models for segmenting and labeling sequence data. In *Proceedings of the Eighteenth International Conference on Machine Learning*, ICML '01, pages 282–289, San Francisco, CA, USA, 2001. Morgan Kaufmann Publishers Inc.

[LR19] Fahad Lateef and Yassine Ruichek. Survey on semantic segmentation using deep learning techniques. *Neurocomputing*, 338:321–348, 2019.

[LSD15] Jonathan Long, Evan Shelhamer, and Trevor Darrell. Fully convolutional networks for semantic segmentation. In *Proceedings of the IEEE conference on computer vision and pattern recognition*, pages 3431–3440, 2015.

[LVZ11] Victor S. Lempitsky, Andrea Vedaldi, and Andrew Zisserman. Pylon model for semantic segmentation. In John Shawe-Taylor, Richard S. Zemel, Peter L. Bartlett, Fernando C. N. Pereira, and Kilian Q. Weinberger, editors, *NIPS*, pages 1485–1493, 2011.

[LWHY19] Xiang Li, Wenhai Wang, Xiaolin Hu, and Jian Yang. Selective kernel networks. 2019.

[LYZ+20] Xia Li, Yibo Yang, Qijie Zhao, Tiancheng Shen, Zhouchen Lin, and Hong Liu. Spatial pyramid based graph reasoning for semantic segmentation. In *Proceedings of the IEEE/CVF Conference on Computer Vision and Pattern Recognition*, pages 8950–8959, 2020.

[Mei05] Marina Meilă. Comparing clusterings: an axiomatic view. In *Proceedings of the 22nd international conference on Machine learning*, pages 577–584. ACM, 2005.

[OdS18] Hugo Oliveira and Jefersson dos Santos. Deep transfer learning for segmentation of anatomical structures in chest radiographs. In *2018 31st SIBGRAPI Conference on Graphics, Patterns and Images (SIBGRAPI)*, pages 204–211. IEEE, 2018.

[OHT20] Y. Ouali, C. Hudelot, and M. Tami. Semi-supervised semantic segmentation with cross-consistency training. In *2020 IEEE/CVF Conference on Computer Vision and Pattern Recognition (CVPR)*, pages 12671–12681, Los Alamitos, CA, USA, Jun 2020. IEEE Computer Society.

[OKBS19] Marin Orsic, Ivan Kreso, Petra Bevandic, and Sinisa Segvic. In defense of pre-trained imagenet architectures for real-time semantic segmentation of road-driving images. In *Proceedings of the IEEE conference on computer vision and pattern recognition*, pages 12607–12616, 2019.

[Ots79] N. Otsu. A threshold selection method from gray-level histograms. *IEEE Transactions on Systems, Man, and Cybernetics*, 9(1):62–66, 1979.

[Pav72] T. Pavlidis. Segmentation of pictures and maps through functional approximation. *Comput. Graph. Image Process.*, 1:360–372, 1972.

[PKST20] Mohammad Pashaei, Hamid Kamangir, Michael J Starek, and Philippe Tissot. Review and evaluation of deep learning architectures for efficient land cover mapping with UAS hyper-spatial imagery: A case study over a wetland. *Remote Sensing*, 12(6):959, 2020.

[PY09] Sinno Jialin Pan and Qiang Yang. A survey on transfer learning. *IEEE Transactions on knowledge and data engineering*, 22(10):1345–1359, 2009.

[QZH+19] Xuebin Qin, Zichen Zhang, Chenyang Huang, Chao Gao, Masood Dehghan, and Martin Jagersand. Basnet: Boundary-aware salient object detection. In *The IEEE Conference on Computer Vision and Pattern Recognition (CVPR)*, June 2019.

[QZH+20] Xuebin Qin, Zichen Zhang, Chenyang Huang, Masood Dehghan, Osmar R. Zaiane, and Martin Jagersand. U2-net: Going deeper with nested u-structure for salient object detection. *Pattern Recognition*, 106:107404, 2020.

[RDS+15] Olga Russakovsky, Jia Deng, Hao Su, Jonathan Krause, Sanjeev Satheesh, Sean Ma, Zhiheng Huang, Andrej Karpathy, Aditya Khosla, Michael Bernstein, Alexander C. Berg, and Li Fei-Fei. ImageNet Large Scale Visual Recognition Challenge. *International Journal of Computer Vision (IJCV)*, 115(3):211–252, 2015.

[RFB15a] Olaf Ronneberger, Philipp Fischer, and Thomas Brox. U-net: Convolutional networks for biomedical image segmentation. In Nassir Navab, Joachim Hornegger, William M. Wells, and Alejandro F. Frangi, editors, *Medical Image Computing and Computer-Assisted Intervention – MICCAI 2015*, pages 234–241, Cham, 2015. Springer International Publishing.

[RFB15b] Olaf Ronneberger, Philipp Fischer, and Thomas Brox. U-net: Convolutional networks for biomedical image segmentation. In *International Conference on Medical image computing and computer-assisted intervention*, pages 234–241. Springer, 2015.

[RSD+18] Kate Rakelly, Evan Shelhamer, Trevor Darrell, Alexei A Efros, and Sergey Levine. Meta-learning to guide segmentation. 2018.

[Rud17] Sebastian Ruder. An overview of multi-task learning in deep neural networks. *arXiv preprint arXiv:1706.05098*, 2017.

[SBT+19] Suvash Sharma, John E Ball, Bo Tang, Daniel W Carruth, Matthew Doude, and Muhammad Aminul Islam. Semantic segmentation with transfer learning for off-road autonomous driving. *Sensors*, 19(11):2577, 2019.

[SDZW20] Jesse Sun, Fatemeh Darbehani, Mark Zaidi, and Bo Wang. Saunet: Shape attentive u-net for interpretable medical image segmentation, 2020.

[SEG17] Seyed Sadegh Mohseni Salehi, Deniz Erdogmus, and Ali Gholipour. Tversky loss function for image segmentation using 3d fully convolutional deep networks. In *International Workshop on Machine Learning in Medical Imaging*, pages 379–387. Springer, 2017.

[SJC08] J. Shotton, M. Johnson, and R. Cipolla. Semantic texton forests for image categorization and segmentation. In *2008 IEEE Conference on Computer Vision and Pattern Recognition*, pages 1–8, 2008.

[SLV+17] Carole H Sudre, Wenqi Li, Tom Vercauteren, Sebastien Ourselin, and M Jorge Cardoso. Generalised dice overlap as a deep learning loss function for highly unbalanced segmentations. In *Deep learning in medical image analysis and multimodal learning for clinical decision support*, pages 240–248. Springer, 2017.

[SZ14] Karen Simonyan and Andrew Zisserman. Very deep convolutional networks for large-scale image recognition. *arXiv preprint arXiv:1409.1556*, 2014.

[TAJF19] Towaki Takikawa, David Acuna, Varun Jampani, and Sanja Fidler. Gated-SCNN: Gated shape CNNS for semantic segmentation. In *Proceedings of the IEEE International Conference on Computer Vision*, pages 5229–5238, 2019.

[TG80] Anne M Treisman and Garry Gelade. A feature-integration theory of attention. *Cognitive psychology*, 12(1):97–136, 1980.

[TLQJ18] Hu Tao, Weihua Li, Xianxiang Qin, and Dan Jia. Image semantic segmentation based on convolutional neural network and conditional random field. In *2018 Tenth International Conference on Advanced Computational Intelligence (ICACI)*, pages 568–572. IEEE, 2018.

[TP98] Sebastian Thrun and Lorien Pratt. Learning to learn: Introduction and overview. In *Learning to learn*, pages 3–17. Springer, 1998.

[VDM20] Kira Vinogradova, Alexandr Dibrov, and Gene Myers. Towards interpretable semantic segmentation via gradient-weighted class activation mapping (student abstract). In *AAAI*, pages 13943–13944, 2020.

[WDYT19] Hao Wang, Cheng Deng, Junchi Yan, and Dacheng Tao. Asymmetric cross-guided attention network for actor and action video segmentation from natural language query. In *Proceedings of the IEEE International Conference on Computer Vision*, pages 3939–3948, 2019.

[WP15] Boyu Wang and Joelle Pineau. Online boosting algorithms for anytime transfer and multitask learning. In *Twenty-Ninth AAAI Conference on Artificial Intelligence*, 2015.

[WPLSK18] Sanghyun Woo, Jongchan Park, Joon-Young Lee, and In So Kweon. CBAM: Convolutional block attention module. In *Proceedings of the European conference on computer vision (ECCV)*, pages 3–19, 2018.

[WSC+20] Jingdong Wang, Ke Sun, Tianheng Cheng, Borui Jiang, Chaorui Deng, Yang Zhao, Dong Liu, Yadong Mu, Mingkui Tan, Xinggang Wang, et al. Deep high-resolution representation learning for visual recognition. *IEEE transactions on pattern analysis and machine intelligence*, 2020.

[XGD+17] Saining Xie, Ross Girshick, Piotr Dollár, Zhuowen Tu, and Kaiming He. Aggregated residual transformations for deep neural networks. In *Proceedings of the IEEE conference on computer vision and pattern recognition*, pages 1492–1500, 2017.

[YCBL14] Jason Yosinski, Jeff Clune, Yoshua Bengio, and Hod Lipson. How transferable are features in deep neural networks? In *Advances in neural information processing systems*, pages 3320–3328, 2014.

[YCW20] Yuhui Yuan, Xilin Chen, and Jingdong Wang. Object-contextual representations for semantic segmentation. In *16th European Conference Computer Vision (ECCV 2020)*, August 2020.

[YYT+18] Hongshan Yu, Zhengeng Yang, Lei Tan, Yaonan Wang, Wei Sun, Mingui Sun, and Yandong Tang. Methods and datasets on semantic segmentation: A review. *Neurocomputing*, 304:82–103, 2018.

[ZHD+19] Yongjin Zhou, Weijian Huang, Pei Dong, Yong Xia, and Shanshan Wang. D-UNET: a dimension-fusion u shape network for chronic stroke lesion segmentation. *IEEE/ACM transactions on computational biology and bioinformatics*, 2019.

[ZLB+20] Zilong Zhong, Zhong Qiu Lin, Rene Bidart, Xiaodan Hu, Ibrahim Ben Daya, Zhifeng Li, Wei-Shi Zheng, Jonathan Li, and Alexander Wong. Squeeze-and-attention networks for semantic segmentation. In *The IEEE/CVF Conference on Computer Vision and Pattern Recognition (CVPR)*, June 2020.

[ZQD+20] Fuzhen Zhuang, Zhiyuan Qi, Keyu Duan, Dongbo Xi, Yongchun Zhu, Hengshu Zhu, Hui Xiong, and Qing He. A comprehensive survey on transfer learning. *Proceedings of the IEEE*, 2020.

[ZTC+18] Zichen Zhang, Min Tang, Dana Cobzas, Dornoosh Zonoobi, Martin Jagersand, and Jacob L Jaremko. End-to-end detection-segmentation network with ROI convolution. In *2018 IEEE 15th International Symposium on Biomedical Imaging (ISBI 2018)*, pages 1509–1512. IEEE, 2018.

[ZWYC19] Shuai Zhao, Yang Wang, Zheng Yang, and Deng Cai. Region mutual information loss for semantic segmentation. In *Advances in Neural Information Processing Systems*, pages 11117–11127, 2019.

[ZZZ+19] Man Zhang, Yong Zhou, Jiaqi Zhao, Yiyun Man, Bing Liu, and Rui Yao. A survey of semi-and weakly supervised semantic segmentation of images. *Artificial Intelligence Review*, pages 1–30, 2019.

Chapter 4
Beyond Full Supervision in Deep Learning

Nicolas Thome

4.1 Context

Deep learning and convolutional neural networks (ConvNets) [LBD+89] have an old history in artificial intelligence and are among the oldest predictive models in machine learning. Although the community has been reluctant on using deep learning techniques in the 1990s and the early 2000s, we are witnessing a huge success of deep neural networks for about a decade now. Since the outstanding of ConvNets at the ImageNet ILSVRC'12 challenge [KSH12a], deep learning has become state-of-the art for any visual recognition tasks, and essentially any data science field, e.g. speech recognition, natural language processing, games and robotics, etc.

However, deep neural networks need a huge amount of annotated data to reach optimal performances. This is a main limitation of the widespread use of these models, especially when the labeling cost is high, e.g. because it requites a high level of expertise as in healthcare. Therefore, designing learning scheme for deep models able to deal with a limited number of annotations is currently a very extensive research field.

In computer vision, "Deep Features" consists in leveraging deep ConvNets trained on large scale datasets (e.g. ImageNet) to provide universal visual representations that are effective for various visual recognition tasks [ARS+16]. Note that Deep Features are robust to large domain shift, and are e.g. commonly used in remote sensing, healthcare, etc. We now present more general solutions to use deep learning models with few labeled samples.

N. Thome (✉)
CEDRIC, Conservatoire National des Arts et Métiers, Paris, France
e-mail: nicolas.thome@cnam.fr

© Springer Nature Switzerland AG 2021
J. Benois-Pineau, A. Zemmari (eds.), *Multi-faceted Deep Learning*,
https://doi.org/10.1007/978-3-030-74478-6_4

4.1.1 Weakly Supervised Learning (WSL)

Weakly supervised Learning (WSL) consists in training models to predict accurate labels (e.g. dense prediction in image segmentation), while trained on coarser ones (e.g. global image labels). Relaxing the requirement of expensive manual and accurate annotations of training data offers the possibility to build large scale databases at reasonable cost. For example, in the computer vision field, annotating images with a global label makes it possible to build databases containing several millions of examples, whereas annotations at the pixel level (i.e. segmentation) are much more expensive which explains that only moderate-size datasets (around thousands of images) are available. On the other hand, handling weakly labeled data generally requires to expand the representation space with latent variables to model hidden factors and compensate for the weak supervision.

The seminal works on WSL follow the Multiple Instance Learning (MIL) paradigm [DLLP97]: each example is represented as a bag of instances, and the weak supervision consists in providing a single label for each bag. This issue has been extensively studied during the last 15 years in several contexts: drug activity recognition in the seminal work of [DLLP97], text classification [ATH03], content-based image retrieval [CW04], etc. The main MIL assumption is related to the relationship between bag and instance labels: a bag is positive if it contains at least one positive instance, and negative if it contains only negative instances. A classical toy example consists in viewing a bag as set of keys: a bag is labeled positive if it contains a key able to open the door, and negative if none of the keys can. The MIL approaches can be classified into two categories: bag [MR98, CW04, GFKS02, KOB13] vs instance [ATH03, BM07, GC07, DF10, JB12] approaches. Bag approaches embed each bag into a feature space, where standard supervised learning techniques are used, whereas instance approaches learn a classification function in the instance space. More recent approaches have questioned the MIL assumption, e.g. in Learning with Label Proportion (LLP) framework [QSCL09, Rue10, YLK+13, LYCC14], which generalizes MIL, or in symmetric MIL variants [DTC18, DTC15, LV15a, DTC16, DMTC17].

In the deep learning era, several methods have been proposed for WSL of deep ConvNets with image-level labels [OBLS15, PKS15, PC15, PSLD15, SPC+16, ZKL+16]. The key issue is to determine how to pool the regions to have a score per class. The output of the ConvNet is a detection map for each category, so to train it with standard classification loss, it is necessary to aggregate the maps into a global prediction for each class. This pooling issue is also present in WSL structured models [YJ09, QWM+07, PLI14, SHPU12]. The most popular pooling is the max pooling [OBLS15, PKS15, YJ09], which selects the best region to perform prediction. In the case of binary classification, this pooling is an instantiation of the Multiple Instance Learning (MIL) paradigm [DLLP97]. In another way, some pooling strategies propose to use all regions to perform prediction, by marginalizing over the regions [ZKL+16, QWM+07, PLI14]. In Sect. 4.2, we detail the pooling function proposed in [DTC19], which is based on negative evidence and accounts for top scoring and least scoring regions.

4.1.2 Semi Supervised Learning (SSL)

In Semi supervised learning (SSL), one aims at leveraging (at lot of) unlabelled data in addition to (few) labeled data. For training deep models with relatively small annotated datasets, SSL can thus be used as a regularization technique to take advantage of unlabeled data for improving generalization performances of deep ConvNets [Zhu05].

One standard goal followed when training deep models with unlabeled data consists in designing models which fit input data well. Reconstruction error is the standard criterion used in (possibly denoising) Auto-Encoders [BLPL07, RS08, RHBL07, VLBM08], while maximum likelihood is used with generative models, e.g. Restricted Boltzmann Machines and Deep Belief Networks [HS06, RPCL07, LB08]. This unsupervised training framework was generally used as a pre-training before supervised learning with back-propagation [EBC+10], potentially with an intermediate step [GTCL13]. The currently very popular Generative Adversarial Networks [GPAM+14] also falls into this category. With modern ConvNets, regularization with unlabeled data is generally formulated as a multi-task learning problem, where reconstruction and classification objectives are combined during training [ZMGL16, ZLL16]. In these architectures, the encoder used for classification is regularized by a decoder dedicated to reconstruction.

This strategy of classification and reconstruction with an Auto-Encoder is however questionable, since classification and reconstruction may play contradictory roles in terms of feature extraction. Classification arguably aims at extracting class-invariant features, improving sample complexity of the learned model [HTF09], therefore inducing an information loss which prevents exact reconstruction. Some recent approaches have proposed SSL method, where the unsupervised term depart from the standard reconstruction criterion. In Ladder Networks [RBH+15], the previously mentioned conflict between reconstruction and classification is addressed by designing Auto-Encoders capable of discarding information. Reconstruction is made possible by means of skip connections between upper layers and a noisy version of lower layers. In HybridNet [RTC18], which is presented in Sect. 4.3, a two-branch network is introduced. The first branch is devoted to classification, and can thus discard information, whereas the second branch is designed to recover the information lost by the classification branch. In this way, the two networks can cooperate and leverage unlabelled data for performing reconstruction.

Another interesting regularization criterion relies on stability or smoothness of the prediction function, e.g. historically used in Slow Feature Analysis [TTC13]. Adding stability to the prediction function was studied in Adversarial Training [GSS15] and Virtual Adversarial Training [MMK+16]. Other recent semi-supervised models incorporate a stability-based regularizer on the prediction. The idea was first introduced by [SJT16] and proposes to make the prediction vector stable toward data augmentation (e.g. translation, rotation) and model stochasticity (dropout) for a given input. Following work [LA17, TV17] improves upon it by proposing variants on the way to compute stability targets to increase their consistency and better adapt to the model's evolution over training.

4.1.3   Self-training

In self-training, or self-supervised learning, the goal is to use a given model to automatically annotate data. In the context of semi-supervised-learning, self-training can be used to automatically annotate unlabeled data based on a model trained on labeled data. The re-labeling step can be iterative as in self-paced learning [KPK10], which consists in re-annotating examples gradually, from the easiest to the hardest. This is inspired by the general principle of curriculum learning [BLCW09], where the rationale is to make a student learn from easy examples first, so that building a coarse prediction model, before refining it with harder ones.

While performing iterative re-labeling, model confidence is generally used to select the pseudo-labeled examples. Confidence estimation with deep learning is a crucial yet complex problem. The most naive approach consists in using the Maximum Class Probability (MCP) [HG17]. Although this baseline is widely used in practice, it also suffers from fundamental drawbacks, e.g. probability calibration issues [GPSW17]. In Bayesian deep learning, Monte-Carlo Dropout [GG16] is commonly used. On the other hand, confidence estimation in misclassification is to properly separate correct predictions from errors, and some recent work introduce specific metrics and methods, e.g. trust score [JKGG18] or ConfidNet [CTBH+19].

Self-supervised learning have been increasingly used in last few years for dealing with datasets with few labels. For example, it has been explored for image classification [DMM19], semantic segmentation in medical images [BOS+17, ZWT+19, ZZSL19], as the method [PTC+18] detailed in Sect. 4.4, and in domain adaptation [LYV19, ZYVKW18, ZYL+19].

4.2   Negative Evidence Models for WSL

We present here the latent structured model based on negative evidence proposed in [DTC19]. We begin by introducing the notations, then our prediction function, the learning formulation and the intuitions. Finally, we compare our model with exiting latent models.

4.2.1 Notations

We first give some basic notations used in the (latent) structured output learning framework. We consider an input space $\mathcal{X}$, that can be arbitrary, and a structured output space $\mathcal{Y}$. For $(\mathbf{x}, \mathbf{y}) \in \mathcal{X} \times \mathcal{Y}$, we are interested in the problem of learning a discriminant function of the form: $f : \mathcal{X} \to \mathcal{Y}$. In order to incorporate hidden parameters that are not available at training time, we augment the description between an input/output pair with a latent variable $\mathbf{h} \in \mathcal{H}$. We define a scoring function $F_{\mathbf{w}}(\mathbf{x}, \mathbf{y}, \mathbf{h})$, with depends on the input data $\mathbf{x} \in \mathcal{X}$, the output $\mathbf{y} \in \mathcal{Y}$, the latent variable $\mathbf{h} \in \mathcal{H}$ and some parameters $\mathbf{w} \in \mathbb{R}^d$. Our goal is to learn a prediction function $f_{\mathbf{w}}$, parametrized by $\mathbf{w}$, so that the predicted output $\hat{\mathbf{y}}$ depends on $F_{\mathbf{w}}(\mathbf{x}, \mathbf{y}, \mathbf{h}) \in \mathbb{R}$. During training, we assume that we are given a set of N training pairs $\mathcal{D} = \{(\mathbf{x}_i, \mathbf{y}_i^\star) \in \mathcal{X} \times \mathcal{Y} : i \in \{1, \dots, N\}\}$, where $\mathbf{y}_i^\star$ is the ground-truth label of example i. Our goal is to optimize $\mathbf{w}$ in order to minimize a user-supplied loss function $\Delta(\mathbf{y}_i^\star, \mathbf{y})$ over the training set.

4.2.2 Negative Evidence Model

The main intuition of our negative evidence model is to equip each possible output $\mathbf{y} \in \mathcal{Y}$ with a pair of latent variables $(\mathbf{h}_{i,\mathbf{y}}^+, \mathbf{h}_{i,\mathbf{y}}^-)$. $\mathbf{h}_{i,\mathbf{y}}^+$ (resp. $\mathbf{h}_{i,\mathbf{y}}^-$) corresponding to the maximum (resp. minimum) scoring latent value, for input $\mathbf{x}_i$ and output $\mathbf{y}$:

$$\mathbf{h}_{i,\mathbf{y}}^+ = \arg\max_{\mathbf{h} \in \mathcal{H}} F_{\mathbf{w}}(\mathbf{x}_i, \mathbf{y}, \mathbf{h}) \tag{4.1}$$

$$\mathbf{h}_{i,\mathbf{y}}^- = \arg\min_{\mathbf{h} \in \mathcal{H}} F_{\mathbf{w}}(\mathbf{x}_i, \mathbf{y}, \mathbf{h}) \tag{4.2}$$

For an input/output pair $(\mathbf{x}_i, \mathbf{y})$, the scoring of the model, $s_{\mathbf{w}}(\mathbf{x}_i, \mathbf{y})$, sums $\mathbf{h}_{i,\mathbf{y}}^+$ and $\mathbf{h}_{i,\mathbf{y}}^-$ scores, as follows:

$$s_{\mathbf{w}}(\mathbf{x}_i, \mathbf{y}) = \frac{1}{2} \left(F_{\mathbf{w}}(\mathbf{x}_i, \mathbf{y}, \mathbf{h}_{i,\mathbf{y}}^+) + F_{\mathbf{w}}(\mathbf{x}_i, \mathbf{y}, \mathbf{h}_{i,\mathbf{y}}^-) \right) \tag{4.3}$$

Finally, our prediction is:

$$\hat{\mathbf{y}} = f_{\mathbf{w}}(\mathbf{x}_i) = \arg\max_{\mathbf{y} \in \mathcal{Y}} s_{\mathbf{w}}(\mathbf{x}_i, \mathbf{y}) \tag{4.4}$$

This maximization in Eq. (4.4) is known as the inference problem. We consider here deep ConvNets models for $F_{\mathbf{w}}$ and thus the scoring function in Eq. (4.3).

4.2.3   Learning Formulation

During training, we enforce the following constraints:

$$\forall \mathbf{y} \neq \mathbf{y}_i^\star, \quad s_{\mathbf{w}}(\mathbf{x}_i, \mathbf{y}_i^\star) \geq \Delta(\mathbf{y}_i^\star, \mathbf{y}) + s_{\mathbf{w}}(\mathbf{x}_i, \mathbf{y}) \qquad (4.5)$$

Each constraint in Eq. (4.5) requires the scoring value $s_{\mathbf{w}}(\mathbf{x}_i, \mathbf{y}_i^\star)$ for the correct output $\mathbf{y}_i^\star$ to be larger than the scoring value $s_{\mathbf{w}}(\mathbf{x}_i, \mathbf{y})$ for each incorrect output $\mathbf{y} \neq \mathbf{y}_i^\star$, plus a margin of $\Delta(\mathbf{y}_i^\star, \mathbf{y})$. $\Delta(\mathbf{y}_i^\star, \mathbf{y})$, a user-specified loss, makes it possible to incorporate domain knowledge into the penalization.

To give some insights of how the model parameters can be adjusted to fulfill constraints in Eq. (4.5), let us notice that:

- $s_{\mathbf{w}}(\mathbf{x}_i, \mathbf{y}_i^\star)$, i.e. the score for the correct output $\mathbf{y}_i^\star$, can be increased if we find high scoring variables $\mathbf{h}_{i,\mathbf{y}_i^\star}^+$ representing strong evidence for the presence of $\mathbf{y}_i^\star$, while $\mathbf{h}_{i,\mathbf{y}_i^\star}^-$ variables not having large negative scores.
- $s_{\mathbf{w}}(\mathbf{x}_i, \mathbf{y})$, i.e. the score for an incorrect output $\mathbf{y}$, can be decreased if we find low scoring variables $\mathbf{h}_{i,\mathbf{y}}^+$, limiting evidence of the presence of $\mathbf{y}$, while seeking $\mathbf{h}_{i,\mathbf{y}}^-$ variables with large negatives scores, supporting $\mathbf{y}$ absence.

To allow some constraints in Eq. (4.5) to be violated, we introduce the following loss function $\ell_{\mathbf{w}}(\mathbf{x}_i, \mathbf{y}_i^\star)$, which is shown in [DTC19] to be an upper bound of $\Delta(\hat{\mathbf{y}}, \mathbf{y}_i^\star)$:

$$\ell_{\mathbf{w}}(\mathbf{x}_i, \mathbf{y}_i) = \max_{\mathbf{y} \in \mathcal{Y}} \left[\Delta(\mathbf{y}_i^\star, \mathbf{y}) + s_{\mathbf{w}}(\mathbf{x}_i, \mathbf{y}) - s_{\mathbf{w}}(\mathbf{x}_i, \mathbf{y}_i^\star) \right] \qquad (4.6)$$

Using the standard max margin regularization term $\|\mathbf{w}\|^2$, our primal objective function is defined as follows, with the regularization parameter λ:

$$\mathcal{P}(\mathbf{w}) = \frac{\lambda}{2} \|\mathbf{w}\|^2 + \frac{1}{N} \sum_{i=1}^{N} \ell_{\mathbf{w}}(\mathbf{x}_i, \mathbf{y}_i^\star) \qquad (4.7)$$

4.2.4   Negative Evidence Intuition

To illustrate the rationale of the approach, let us consider a multi-class classification instantiation of our negative evidence model, where $\mathbf{x}$ is the image, $\mathbf{y}$ is the label and the latent variables $\mathbf{h}$ correspond to region locations.[1] $\mathbf{h}^+$ is the max scoring latent value for each class $\mathbf{y}$, i.e. the region which best represents class $\mathbf{y}$. $\mathbf{h}^-$ is the min

---

[1]This analysis can be extended to any problems where $\mathbf{h}$ is a part of $\mathbf{x}$, e.g. $\mathbf{h}$ is a paragraph of a text $\mathbf{x}$.

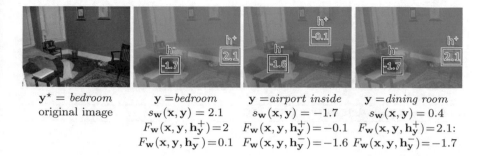

$$\begin{array}{llll}
\mathbf{y}^\star = bedroom & \mathbf{y} = bedroom & \mathbf{y} = airport\ inside & \mathbf{y} = dining\ room \\
\text{original image} & s_\mathbf{w}(\mathbf{x},\mathbf{y}) = 2.1 & s_\mathbf{w}(\mathbf{x},\mathbf{y}) = -1.7 & s_\mathbf{w}(\mathbf{x},\mathbf{y}) = 0.4 \\
& F_\mathbf{w}(\mathbf{x},\mathbf{y},\mathbf{h}_\mathbf{y}^+)=2 & F_\mathbf{w}(\mathbf{x},\mathbf{y},\mathbf{h}_\mathbf{y}^+)=-0.1 & F_\mathbf{w}(\mathbf{x},\mathbf{y},\mathbf{h}_\mathbf{y}^+)=2.1: \\
& F_\mathbf{w}(\mathbf{x},\mathbf{y},\mathbf{h}_\mathbf{y}^-)=0.1 & F_\mathbf{w}(\mathbf{x},\mathbf{y},\mathbf{h}_\mathbf{y}^-)=-1.6 & F_\mathbf{w}(\mathbf{x},\mathbf{y},\mathbf{h}_\mathbf{y}^-)=-1.7
\end{array}$$

Fig. 4.1 Negative evidence intuition. The heatmaps and the predicted regions ($\mathbf{h}_\mathbf{y}^+$ in green, $\mathbf{h}_\mathbf{y}^-$ in red and their scores $F_\mathbf{w}(\mathbf{x},\mathbf{y},\mathbf{h}_\mathbf{y}^+)$ and $F_\mathbf{w}(\mathbf{x},\mathbf{y},\mathbf{h}_\mathbf{y}^-)$) for different learned class models (*bedroom*, *airport inside* and *dining room*) are shown on a *bedroom* image $\mathbf{x}$. The *bedroom* and *dining room* models have high score for max regions because each model focus on objects discriminative for the class (bed for *bedroom* and chair for *dining room*). The min region brings complementary information: its score is low for *dining room* because the model has found a negative evidence (bed) for the absence of *dining room* class

scoring latent value, and can thus be regarded as an indicator of the absence of class $\mathbf{y}$ in the image.

To highlight the importance of the pair $(\mathbf{h}^+, \mathbf{h}^-)$, we show in Fig. 4.1, for an image of the class *bedroom* of MIT67 dataset [QT09], the heatmap representing the classification scores for each latent location using the *bedroom* classifier, the *airport inside* classifier and the *dining room* classifier. The $\mathbf{h}^+$ (resp. $\mathbf{h}^-$) regions are boxed in green (resp. red). As we can see, the prediction score for the correct class classifier (*bedroom*) is large, since the model finds strong local evidence $\mathbf{h}^+$ of it presence, and no clear evidence of its absence (medium score $F_\mathbf{w}(\mathbf{x},\mathbf{y} = bedroom, \mathbf{h}_\mathbf{y}^-) = 0.1$). For a wrong class very different like *airport inside*, the prediction score is very low, because there is not region similar to an airport. For a wrong class with similar objects like *dining room*, the maximum score for the *dining room* classifier is comparable with *bedroom* classifier: the model heavily fires on discriminative objects (bed for *bedroom* and chair for *dining room*). The prediction score $s_\mathbf{w}$ for the *dining room* classifier is significantly lower than for the *bedroom* classifier, because it also finds clear evidence of the absence of *dining room*, here bed ($F_\mathbf{w}(\mathbf{x},\mathbf{y} = dining\ room, \mathbf{h}_\mathbf{y}^-) = -1.7$). As a consequence, our negative evidence model correctly predicts the class *bedroom*.

Discussion To put into perspective connections between negative evidence and existing latent structured models, we introduce the following generalized scoring function, with "inverse temperature" β_h^+ and β_h^- parameters smoothing between

Table 4.1 Model comparison with corresponding parameters

Model	β_h^+	β_h^-
HCRF [QWM+07]/MSSVM [PLI14]	1	1
GAP [ZKL+16]	$\to 0$	$\to 0$
LSSVM [YJ09]/max [OBLS15]	$+\infty$	$+\infty$
Our model	$+\infty$	$-\infty$
ϵ-framework [SHPU12]/LSE [PC15, SPC+16]	$\beta_h^+ = \beta_h^- \in (1, +\infty[$	

max, softmax and average:

$$s_{\mathbf{w}}^{(\beta_h^+, \beta_h^-)}(\mathbf{x}, \mathbf{y}) = \frac{1}{2\beta_h^+} \log \frac{1}{|\mathcal{H}|} \sum_{\mathbf{h} \in \mathcal{H}} \exp[\beta_h^+ F_{\mathbf{w}}(\mathbf{x}, \mathbf{y}, \mathbf{h})] \qquad (4.8)$$

$$+ \frac{1}{2\beta_h^-} \log \frac{1}{|\mathcal{H}|} \sum_{\mathbf{h} \in \mathcal{H}} \exp[\beta_h^- F_{\mathbf{w}}(\mathbf{x}, \mathbf{y}, \mathbf{h})]$$

As shown in Table 4.1, the scoring function in Eq. (4.8) includes several state-of-the-art models as special cases.

4.2.5 ResNet-WELDON Network Architecture

Based on the model presented in previous section, we propose ResNet-WELDON, a new weakly supervised learning dedicated to learn localized visual features by using only image-level labels during training. The proposed network architecture is decomposed into two sub-networks: a deep feature extraction network based on Fully Convolutional Network (FCN) and a prediction network, as illustrated in Fig. 4.2.

Notation We note $F_{\mathbf{w}}^l(\mathbf{x}, \mathbf{y}, \mathbf{h})$ the output of the layer l at the location $\mathbf{h}$ of the feature map (or category) $\mathbf{y}$ for the input image $\mathbf{x}$. $\mathbf{w}$ are the parameters of the ConvNet.

4.2.5.1 Feature Extraction Network

The feature extraction network is dedicated to compute a fixed-size representation for any region of the input image. They convolutionalize standard classification networks (AlexNet, VGG16) by replacing fully connected by convolution layers.

To have a high spatial resolution on the top of the network, we use the recently introduced the ResNet-101 [HZRS16], which is by design fully convolutional. ResNet-101 has 100 convolutional layers followed by global average pooling and

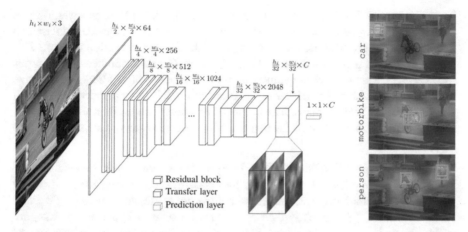

Fig. 4.2 ResNet-WELDON deep architecture is decomposed into two sub-networks: a feature extraction network (left) and a prediction network (right). The feature extraction network is based on ResNet-101 to extract local features from whole images with good spatial resolution. Then a transfer layer is used to learn class-specific heatmaps (*car*, *motorbike* and *person*), and a prediction layer aggregates the heatmaps to produce a single score for each class. We show for each class the 3 regions with the highest score on the right

fully-connected layer. To have feature map with spatial information, we remove the fully-connected layer (as usually done in the literature), and the global average pooling, which has not learnable parameter. The architecture with only the convolutional layers (and spatial pooling) allows to process images of arbitrary sizes, and the sharing of intermediate features over overlapping image regions. With this architecture, the spatial information is naturally preserved throughout the network: for an input image size of 224×224, the output size is 7×7. Spatial resolution impacts the localization and discriminability of the learned representations. We thus expect the resolution of the feature maps to be a key component for our model: finer maps keep more spatial resolution and lead to more specific regions. Moreover, ResNet is more effective at image classification while being parameter- and time-efficient than VGG16.

The input of the feature extraction network is an RGB image $h_i \times w_i$, where h_i (resp. w_i) is the height (resp. width). The output is a $h \times w \times 2048$ feature map, where $h = \frac{h_i}{32}$ and $w = \frac{w_i}{32}$ are number of sliding window positions in the horizontal and vertical direction in the image, respectively (see Fig. 4.2). The weights of the feature extraction network are initialized on ImageNet.

4.2.5.2   Prediction Network Design

This part aims at selecting relevant regions to properly predict the global (structured) image label.

Transfer Layer The first layer of the prediction network is a transfer layer. Its goal is to transfer weights of the feature extraction network from large scale datasets to new target datasets. It transforms the output of the feature extraction network F^{fe} into a feature map F^t of size $h \times w \times C$, where C is the number of categories (see Fig. 4.2). This layer is convolutional layer, composed of C filters, each of size $1 \times 1 \times 2048$. Due to the kernel size of the convolution, this layer preserves the spatial resolution of the feature maps. The output of this layer can be seen as localization heatmaps. In Fig. 4.2, we show the heatmaps for different categories: *car*, *motorbike* and *person*.

Weakly-Supervised Prediction (WSP) Layer The second layer is a spatial pooling layer s aggregates the score maps into classification scores: for each output $\mathbf{y} \in \{1, \ldots, C\}$, the score over the $h \times w$ regions are aggregated into a single scalar value. We note $F_{\mathbf{w}}^t(\mathbf{x}_i, \mathbf{y}, \mathbf{h})$ is the score of region $\mathbf{h}$ from image $\mathbf{x}_i$ for category $\mathbf{y}$, and $\mathcal{H} = \{1, \ldots, r_i\}$ the region index set, and r_i is the number of regions for image $\mathbf{x}_i$. The output s of the prediction layer is a vector $1 \times 1 \times C$. As mentioned in Sect. 4.1.1, the standard approach for WSL inherited from MIL is to select the max scoring region. We propose to improve this strategy in two complementary directions: use negative evidence and several instances.

WELDON Pooling This pooling improves max pooling by incorporate negative evidence. The prediction consists in summing the max and min scoring regions. Based on recent MIL insights on learning with top instances [LV15b], we also propose to extend the selection of a single region to multiple regions. Formally, let $h_z \in \{0, 1\}$ be the binary variable denoting the selection of the zth region from layer F^t. We propose the scoring function s^{top}, which selects the k^+ highest scoring regions as follows:

$$s_{\mathbf{w}, k^+}^{top}(F^t(\mathbf{x}_i, \mathbf{y})) = \frac{1}{k^+} \sum_{z=1}^{r_i} h_z^+ F_{\mathbf{w}}^t(\mathbf{x}_i, \mathbf{y}, z) \tag{4.9}$$

$$\text{where } \mathbf{h}^+ = \arg\max_{\mathbf{h} \in \{0,1\}^{r_i}} \sum_{z=1}^{r_i} h_z F_{\mathbf{w}}^t(\mathbf{x}_i, \mathbf{y}, z) \text{ s.t. } \sum_{z=1}^{r_i} h_z = k^+$$

where $F_{\mathbf{w}}^t(\mathbf{x}_i, \mathbf{y}, z)$ is the value of the zth region score for class y.

The intuition behind F^{top} is to provide a robust region selection strategy, since using a single area necessarily increases the risk of selecting outliers. To incorporate negative evidence in our prediction function, we propose the scoring function s^{low}, which selects the k^- lowest scoring regions as follows:

$$s_{\mathbf{w}, k^-}^{low}(F^t(\mathbf{x}_i, \mathbf{y})) = \frac{1}{k^-} \sum_{z=1}^{r_i} h_z^- F_{\mathbf{w}}^t(\mathbf{x}_i, \mathbf{y}, z) \tag{4.10}$$

$$\text{where } \mathbf{h}^- = \arg\min_{\mathbf{h} \in \{0,1\}^{r_i}} \sum_{z=1}^{r_i} h_z F_{\mathbf{w}}^t(\mathbf{x}_i, \mathbf{y}, z) \text{ s.t. } \sum_{z=1}^{r_i} h_z = k^-$$

The final prediction simply consists in summing F^{top} and F^{low}:

$$s_{\mathbf{w}}(\mathbf{x}_i, \mathbf{y}) = \frac{1}{2}\left(s^{top}_{\mathbf{w},k^+}(F^t(\mathbf{x}_i, \mathbf{y})) + s^{low}_{\mathbf{w},k^-}(F^t(\mathbf{x}_i, \mathbf{y}))\right) \tag{4.11}$$

This prediction function is equivalent to MANTRA prediction function whenever $k^+ = k^- = 1$.

4.2.6   Learning and Instantiations

As shown in Fig. 4.2, the WELDON model outputs $s \in \mathbb{R}^C$. This vector represents a structured output, which can be used in a classification framework (details in [DTC19]), but we also address the problem of optimizing ranking metrics, and especially Average Precision (AP).

We use a latent structured output ranking formulation, following [YFRJ07]: our input is a set of N training images $\mathbf{x} = \{x_i : i \in 1, \ldots, N\}$, with their binary labels y_i, and our goal is to predict a ranking matrix $\mathbf{y} \in \mathcal{Y}$ of size $N \times N$ providing an ordering of the training examples. Our ranking feature map for category c is expressed as:

$$F_{\mathbf{w}_c}(\mathbf{x},\mathbf{y},\mathbf{h}) = \sum_{p\in\mathcal{P}}\sum_{n\in\mathcal{N}} \mathbf{y}_{pn}(F^{fe}_{\mathbf{w}}(x_p,c,h^{pn}) - F^{fe}_{\mathbf{w}}(x_n,c,h^{np})) \tag{4.12}$$

where $\mathcal{P}$ (resp. $\mathcal{N}$) is the set of positive (resp. negative) examples. h^{pn} (resp. h^{np}) is a vector which represent the selected region for image x_p (resp x_n) when we consider the couple of image (p, n), and $\mathbf{h}$ is the set of selected regions for all pair of examples $(p, n) \subset \mathcal{P} \times \mathcal{N}$

$$\mathbf{h} = \{(h^{pn}, h^{np}) \in \{0, 1\}^{r_p} \times \{0, 1\}^{r_n}, \tag{4.13}$$

$$\sum_{z=1}^{r_p} h^{pn}_z = k, \ \sum_{z=1}^{r_n} h^{np}_z = k, \ (p,n) \in \mathcal{P} \times \mathcal{N}\}$$

where r_p is the number of regions for image x_p (resp. x_n). $F^{fe}_{\mathbf{w}}(x_p,c,h^{pn})$ is the score for category c of region h^{pn} of image x_p

During training, we aim at minimizing the following loss: $\Delta_{ap}(\mathbf{y}^\star, \mathbf{y}) = 1 - AP(\mathbf{y}^\star, \mathbf{y})$, where $\mathbf{y}^\star$ is the ground-truth ranking. Since AP is non-smooth, we use the following surrogate (upper-bound) loss:

$$\ell_{\mathbf{w}}(\mathbf{x}, \mathbf{y}^\star) = \max_{\mathbf{y}\in\mathcal{Y}}\left[\Delta_{ap}(\mathbf{y}^\star, \mathbf{y}) + s_{\mathbf{w}}(\mathbf{x}, \mathbf{y})\right] - s_{\mathbf{w}}(\mathbf{x}, \mathbf{y}^\star) \tag{4.14}$$

The maximization in Eq. (4.14) is generally referred to as Loss-Augmented Inference (LAI). Exhaustive maximization is intractable due to the huge size of the structured output space. The problem is even exacerbated in the WSL setting, see [BMJK15]. We exhibit here the following result for WELDON:

Proposition 1 *For each training example, let us denote* $s(i) = s_{w,k}^{top}(F_w^t(x_i, c)) +$ $s_{w,k}^{low}(F_w^t(x_i, c))$ *in Eq. (4.11). Inference and LAI for the WELDON ranking model can be solved exactly by sorting examples in descending order of* $s(i)$.

The proof is detailed in [DTC19] and comes from an elegant symmetrization of the problem due to the max + min operation. This reduces inference and LAI optimization to fully supervised problems. Inference solution directly corresponds to $s(i)$ sorting. It also allows to use our model with different loss functions, as soon as there is an algorithm to solve the loss-augmented inference in the fully supervised setting. To solve it with Δ_{ap}, we use the greedy algorithm proposed by [YFRJ07], but it is possible to use faster methods [MJK14] to address large-scale problem if required.

4.2.7 Experiments

The comparison of the proposed approach with respect to state-of-the-art methods is shown in Table 4.2. It shows the very good performances obtained by the ResNet-WELDON model for classification on various standard datasets. Detailed results analyzing the approach for detection of segmentation, or analyzing the impact of the proposed pooling function can be found in [DTC19].

4.3 Beyond Reconstruction in Semi-supervised Learning

Here, we present the semi-supervised learning (SSL) HybridNet model proposed in [RTC18]. The rationale of the approach is shown in Fig. 4.3: the goal is to leverage unlabeled data to improve the classification performances using a reconstruction cost. The majority of state-of-the-art approaches combine a classification cost and a reconstruction cost with a single model, e.g. using Auto-Encoders (AE)—see Sect. 4.1.2. The main limitation of these methods illustrated in Fig. 4.3 is that classification and reconstruction may arguably play contradictory roles, since classification aims at learning invariant intra-class features. In HybridNet, Instead of having an AE that will try to represent all the information in a single latent space $\mathbf{h}$, we propose to split this information into two complementary latent spaces $\mathbf{h}_c$ and $\mathbf{h}_u$.

Table 4.2 mAP results on object recognition datasets. ResNet-WELDON and state-of-the-art methods results are reported. Half at the top shows the performances using global image representation, whereas the half at the bottom shows performances for models based on regions selection

Method	VOC07	VOC12	MSCOCO
Return Devil [CSVZ14]	82.4	–	–
VGG16 [SZ15]	89.3	89.0	–
SPP net [HZRS14]	82.4	–	–
NUS-HCP [WXH+14]	85.2	84.2	–
Nonlinear Embeddings [SS15]	86.1	–	–
ResNet-101 [HZRS16][a]	89.8	89.2	72.5
DeepMIL [OBLS15]	–	86.3	62.8
MANTRA [DTC15]	85.8	–	–
WELDON [DTC16]	90.2	–	68.8
ProNet [SPC+16]	–	89.3	70.9
RRSVM [WH16]	92.9	–	–
SPLeaP [KJZ+16]	88.0	–	–
ResNet-max	92.0	90.9	78.9
ResNet-WELDON	**95.0**	**93.4**	**80.7**

[a] The results are obtained by fine-tuning the network on the dataset with the online code https://github.com/facebook/fb.resnet.torch

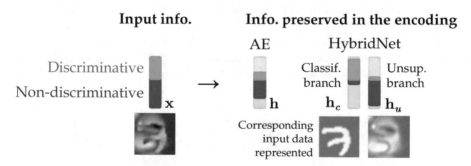

Input info.   **Info. preserved in the encoding**

Discriminative / Non-discriminative $\mathbf{x}$ AE $\mathbf{h}$ HybridNet — Classif. branch $\mathbf{h}_c$ / Unsup. branch $\mathbf{h}_u$

Corresponding input data represented

Fig. 4.3 Illustration of how an AE and HybridNet encode the information differently Considering an image contains discriminative and non-discriminative information represented on the left. On the right, we show the information preserved in the latent space of an AE and of HybridNet. The auto-encoder will retain as much information as possible for reconstruction, possibly keeping lots of non-discriminative information

4.3.1 Designing the HybridNet Architecture

General Architecture As we have seen, classification requires intra-class invariant features while reconstruction needs to retain all the information. To circumvent this issue, HybridNet is composed of two auto-encoding paths, the *discriminative path* (E_c and D_c) and the *unsupervised path* (E_u and D_u). Both encoders E_c and E_u take an input image $\mathbf{x}$ and produce representations $\mathbf{h}_c$ and $\mathbf{h}_u$, while decoders D_c

and D_u take respectively $\mathbf{h}_c$ and $\mathbf{h}_u$ as input to produce two partial reconstructions $\hat{\mathbf{x}}_c$ and $\hat{\mathbf{x}}_u$. Finally, a classifier C produces a class prediction using discriminative features only: $\hat{\mathbf{y}} = C(\mathbf{h}_c)$. The two paths should play different and complementary roles. The discriminative path must extract discriminative features $\mathbf{h}_c$ that should eventually be well crafted to perform a classification task effectively, and produce a purposely partial reconstruction $\hat{\mathbf{x}}_c$ that should not be perfect since preserving all the information is not a behavior we want to encourage. Consequently, the role of the unsupervised path is to be complementary to the discriminative branch by retaining in $\mathbf{h}_u$ the information lost in $\mathbf{h}_c$. This way, it can produce a complementary reconstruction $\hat{\mathbf{x}}_u$ so that, when merging $\hat{\mathbf{x}}_u$ and $\hat{\mathbf{x}}_c$, the final reconstruction $\hat{\mathbf{x}}$ is close to $\mathbf{x}$. The HybridNet architecture, visible on Fig. 4.4, can be described by the following equations:

$$
\begin{aligned}
\mathbf{h}_c &= E_c(\mathbf{x}) & \hat{\mathbf{x}}_c &= D_c(\mathbf{h}_c) & \hat{\mathbf{y}} &= C(\mathbf{h}_c) \\
\mathbf{h}_u &= E_u(\mathbf{x}) & \hat{\mathbf{x}}_u &= D_u(\mathbf{h}_u) & \hat{\mathbf{x}} &= \hat{\mathbf{x}}_c + \hat{\mathbf{x}}_u
\end{aligned}
\tag{4.15}
$$

Note that here, the end-role of reconstruction is to act as a regularizer for the discriminative encoder. For example, thanks to the additional data, it can make convolutional filters more robust by better grasping the variety of the input data.

The main challenge and contribution of this paper is to find a way to ensure that the two paths will in fact behave in this desired way. The two main issues that we tackle are the fact that we want the discriminative branch to focus on discriminative features, and that we want both branches to cooperate and contribute to the reconstruction. Indeed, with such an architecture, we could end up with two paths that work independently: a classification path $\hat{\mathbf{y}} = C(E_c(\mathbf{x}))$ and a reconstruction path $\hat{\mathbf{x}} = \hat{\mathbf{x}}_u = D_u(E_u(\mathbf{x}))$ and $\hat{\mathbf{x}}_c = 0$. We address both those

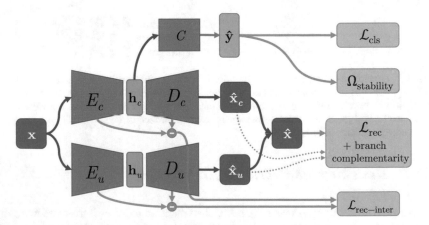

Fig. 4.4 General description of the HybridNet framework. E_c and C correspond to a classifier, E_c and D_c form an autoencoder that we call *discriminative path*, and E_u and D_u form a second autoencoder called *unsupervised path*. The various loss functions used to train HybridNet are also represented in yellow

issues through the design of the architecture of the encoders and decoders as well as an appropriate loss and training procedure.

Branches Design To design the HybridNet architecture, we start with a convolutional architecture adapted to the targeted dataset, for example a state-of-the-art ResNet architecture for CIFAR-10. This architecture is split into two modules: the discriminative encoder E_c and the classifier C. On top of this model, we add the discriminative decoder D_c. The location of the splitting point in the original network is free, but C will not be directly affected by the reconstruction loss. In our experiments, we choose $\mathbf{h}_c$ (E_c's output) to be the last intermediate representation before the final pooling that aggregates all the spatial information, leaving in C a global average pooling followed by one or more fully-connected layers. The decoder D_c is designed to be a "mirror" of the encoder's architecture, as commonly done in the literature, e.g. [ZMGL16, RBH+15, ZF14].

After constructing the discriminative branch, we add an unsupervised complementary branch. To ensure that both branches are "balanced" and behave in a similar way, the internal architecture of E_u and D_u is mostly the same as for E_c and D_c. The only difference remains in the mirroring of pooling layers, that can be reversed either by upsampling or unpooling. An upsampling will increase the spatial size of a feature map without any additional information while an unpooling, used in [ZMGL16, ZLL16], will use spatial information (*pooling switches*) from the corresponding max-pooling layer to do the upsampling. In our architecture, we propose to use upsampling in the discriminative branch because we want to encourage spatial invariance, and use unpooling in the unsupervised branch to compensate this information loss and favor the learning of spatial-dependent low-level information. An example of HybridNet architecture is presented in Fig. 4.5.

As mentioned previously, one key problem to tackle is to ensure that this model will behave as expected, i.e. by learning discriminative features in the discriminative encoder and non-discriminative features in the unsupervised one. This is encouraged in different ways by the design of the architecture. First, the fact that only $\mathbf{h}_c$ is used for classification means that E_c will be pushed by the classification loss to produce discriminative features. Thus, the unsupervised branch will naturally focus on information lost by E_c. Using upsampling in D_c and unpooling in D_u also encourages the unsupervised branch to focus on low-level information. In addition to this, the design of an adapted loss and training protocol is a major contribution to the efficient training of HybridNet.

4.3.2 Training HybridNet

The HybridNet architecture has two information paths with only one producing a class prediction and both producing partial reconstructions that should be combined. In this section, we will address the question of training this architecture efficiently. The complete loss is composed of various terms as illustrated on Fig. 4.4. It com-

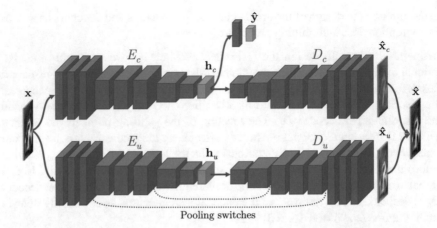

Fig. 4.5 Example of HybridNet architecture where an original classifier (ConvLarge) constitutes E_c and has been mirrored to create D_c and duplicated for E_u and D_u, with the addition of unpooling in the discriminative branch

prises terms for classification with $\mathcal{L}_{\text{cls}}$; final reconstruction with $\mathcal{L}_{\text{rec}}$; intermediate reconstructions with $\mathcal{L}_{\text{rec-inter}b,l}$ (for layer l and branch b); and stability with $\Omega_{\text{stability}}$. It is also accompanied by a branch complementarity training method. Each term is weighted by a corresponding parameter λ:

$$\mathcal{L} = \lambda_c \mathcal{L}_{\text{cls}} + \lambda_r \mathcal{L}_{\text{rec}} + \sum_{b \in \{c,u\},l} \lambda_{rb,l} \mathcal{L}_{\text{rec-inter}b,l} + \lambda_s \Omega_{\text{stability}} . \tag{4.16}$$

HybridNet can be trained on a partially labeled dataset, i.e. that is composed of labeled pairs $\mathcal{D}_{\text{sup}} = \{(x^{(k)}, y^{(k)})\}_{k=1..N_s}$ and unlabeled images $\mathcal{D}_{\text{unsup}} = \{x^{(k)}\}_{k=1..N_u}$. Each batch is composed of n samples, divided into n_s image-label pairs from $\mathcal{D}_{\text{sup}}$ and n_u unlabeled images from $\mathcal{D}_{\text{unsup}}$.

4.3.2.1 Classification

The classification term is a regular cross-entropy term, that is applied only on the n_s labeled samples of the batch and averaged over them:

$$\ell_{\text{cls}} = \ell_{\text{CE}}(\hat{\mathbf{y}}, \mathbf{y}) = - \sum_i \mathbf{y}_i \log \hat{\mathbf{y}}_i , \qquad \mathcal{L}_{\text{cls}} = \frac{1}{n_s} \sum_k \ell_{\text{cls}}(\hat{\mathbf{y}}^{(k)}, \mathbf{y}^{(k)}) . \tag{4.17}$$

4.3.2.2 Reconstruction Losses

In HybridNet, we chose to keep discriminative and unsupervised paths separate so that they produce two complementary reconstructions $(\hat{\mathbf{x}}_u, \hat{\mathbf{x}}_c)$ that we combine with an addition into $\hat{\mathbf{x}} = \hat{\mathbf{x}}_u + \hat{\mathbf{x}}_c$. Keeping the two paths independent until the reconstruction in pixel space, as well as the merge-by-addition strategy allows us to apply different treatments to them and influence their behavior efficiently. The merge by addition in pixel space is also analogous to wavelet decomposition where the signal is decomposed into low- and high-pass branches that are then decoded and summed in pixel space. The reconstruction loss that we use is a simple mean-squared error between the input and the sum of the partial reconstructions:

$$\ell_{\text{rec}} = ||\hat{\mathbf{x}} - \mathbf{x}||_2^2 = ||\hat{\mathbf{x}}_u + \hat{\mathbf{x}}_c - \mathbf{x}||_2^2, \qquad \mathcal{L}_{\text{rec}} = \frac{1}{n} \sum_k \ell_{\text{rec}}(\hat{\mathbf{x}}^{(k)}, \mathbf{x}^{(k)}) . \qquad (4.18)$$

In addition to the final reconstruction loss, we also add reconstruction costs between intermediate representations in the encoders and the decoders which is possible since encoders and decoders have mirrored structure. We apply these costs to the representations $\mathbf{h}_{b,l}$ (for branch b and layer l) produced just after pooling layers in the encoders and reconstructions $\hat{\mathbf{h}}_{b,l}$ produced just before the corresponding upsampling or unpooling layers in the decoders. This is common in the literature [ZMGL16, ZLL16, RBH+15] but is particularly important in our case: in addition to guiding the model to produce the right final reconstruction, it pushes the discriminative branch to produce a reconstruction and avoid the undesired situation where only the unsupervised branch would contribute to the final reconstruction. This is applied in both branches ($b \in \{c, u\}$):

$$\mathcal{L}_{\text{rec}-\text{inter}b,l} = \frac{1}{n} \sum_k ||\hat{\mathbf{h}}_{b,l}^{(k)} - \mathbf{h}_{b,l}^{(k)}||_2^2 . \qquad (4.19)$$

4.3.2.3   Branch Cooperation

As described previously, we want to ensure that both branches contribute to the final reconstruction, otherwise this would mean that the reconstruction is not helping to regularize E_c, which is our end-goal. Having both branches produce a partial reconstruction and using intermediate reconstructions already help with this goal. In addition, to balance their training even more, we propose a training technique such that the reconstruction loss is only backpropagated to the branch that contributes less to the final reconstruction of each sample. This is done by comparing $||\hat{\mathbf{x}}_c - \mathbf{x}||_2^2$ and $||\hat{\mathbf{x}}_u - \mathbf{x}||_2^2$ and only applying the final reconstruction loss to the branch with the higher error.

This can be implemented either in the gradient descent or simply by preventing gradient propagation in one branch or the other using features like `tf.stop_gradient` in Tensorflow or `.detach()` in PyTorch:

$$\ell_{\text{rec-balanced}} = \begin{cases} ||\hat{\mathbf{x}}_u + \text{stopgrad}(\hat{\mathbf{x}}_c) - \mathbf{x}||_2^2 & \text{if } ||\hat{\mathbf{x}}_u - \mathbf{x}||_2^2 \geq ||\hat{\mathbf{x}}_c - \mathbf{x}||_2^2 \\ ||\text{stopgrad}(\hat{\mathbf{x}}_u) + \hat{\mathbf{x}}_c - \mathbf{x}||_2^2 & \text{otherwise} \end{cases} .$$

$$(4.20)$$

4.3.2.4 Encouraging Invariance in the Discriminative Branch

We have seen that an important issue that needs to be addressed when training this model is to ensure that the discriminative branch will filter out information and learn invariant features. For now, the only signal that pushes the model to do so is the classification loss. However, in a semi-supervised context, when only a small portion of our dataset is labeled, this signal can be fairly weak and might not be sufficient to make the discriminative encoder focus on invariant features.

In order to further encourage this behavior, we propose to use a *stability regularizer*. Such a regularizer is currently at the core of the models that give state-of-the-art results in semi-supervised setting on the most common datasets [SJT16, LA17, TV17]. The principle is to encourage the classifier's output prediction $\hat{\mathbf{y}}^{(k)}$ for sample k to be invariant to different sources of randomness applied on the input (translation, horizontal flip, random noise, etc.) and in the network (e.g. dropout). This is done by minimizing the squared euclidean distance between the output $\hat{\mathbf{y}}^{(k)}$ and a "stability" target $\mathbf{z}^{(k)}$. Multiple methods have been proposed to compute such a target [SJT16, LA17, TV17], for example by using a second pass of the sample in the network with a different draw of random factors that will therefore produce a different output. We have:

$$\Omega_{\text{stability}} = \frac{1}{n} \sum_k ||\hat{\mathbf{y}}^{(k)} - \mathbf{z}^{(k)}||_2^2 . \qquad (4.21)$$

By applying this loss on $\hat{\mathbf{y}}$, we encourage E_c to find invariant patterns in the data, patterns that have more chances of being discriminative and useful for classification. Furthermore, this loss has the advantage of being applicable to both labeled and unlabeled images.

4.3.3 Experiments

Experiments have been carried out on several image classification datasets. In Table 4.3, we show an ablation study on CIFAR-10, comparing the performance of the classification model without unlabeled data, the performances obtained with the stability loss in [LA17], and the HybridNet performances. Additional results

Table 4.3 Ablation study performed on CIFAR-10 with ConvLarge architecture

Model	Labeled samples		
	1000	2000	4000
Classification	63.4	71.5	79.0
Classification and stability	65.6	74.6	81.3
HybridNet architecture	63.2	74.0	80.3
HybridNet architecture and full training loss	**74.1**	**81.6**	**86.6**

including a finer analysis of HybridNet components, visualizations and state-of-the-art comparisons can be found in [RTC18].

Note that the two-branch architecture has also recently been used for segmentation of medical images [CJP+18] and video prediction [GT20b, GT20a].

4.4   Medical Image Segmentation with Partial Labels

We present here the approach proposed in [PTC+18] for medical image segmentation with partial labels

Abdominal organ segmentation is a major challenge in medical imaging and computed-aided diagnosis. A good localization and delineation of the organs help physicians to focus on the region of interest by giving them information about the anatomy of the internal structures.

Currently, state-of-the-art methods for visual recognition rely on deep learning. Convolutional Neural Networks (ConvNets) [KSH12b] and more precisely Fully Convolutional Neural Networks (FCNs) [LSD15] have imposed themselves as standard solutions for semantic segmentation of generalist images. In the context of medical image segmentation, specific architectures such as U-Net and variants [RFB15, ÇAL+16, MNA16, LCQ+18] are standard choices showing optimal performances.

However, an important issue when training deep ConvNets relates to the need of having a large amount of annotated data. The problem is particularly pronounced for medical image segmentation, where the annotation process is extremely time-consuming and requires highly qualified professionals. As a consequence, large-scale and clean medical image datasets are seldom yet valuable. In abdominal organ segmentation, the manual annotation process often focuses on specific anatomical structures, e.g. the liver and its pathologies. Thus, large datasets containing partially labeled images are easier to obtain than a complete dataset containing all the abdominal organs.

In this paper, we address the problem of training deep convnets in such partially annotated datasets. Our training context is illustrated in Fig. 4.6: in this example, the input slice is partially annotated with 3 organ classes out of 7 for the unknown complete labeling.

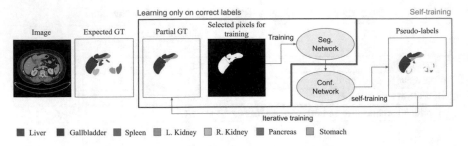

Fig. 4.6 The 3D CT-scan is partially annotated: in this slice, only 3 out of 7 organs are labeled. Naively using such partial ground-truth (GT) annotation is inappropriate since it includes wrong background labels for missing organs. Our SMILE approach is based on identifying pixels for which labels are correct, and ignore others.

To specifically handle the partial labeling problem, we introduce a method which encompasses two main contributions. Firstly, we propose a specific loss to train the segmentation network dedicated to include only correct labels, i.e. which selects pixels that could be learned and those that should be ignored during training (white vs black pixels in Fig. 4.6). The general motivation is to eliminate all pixels that are wrongly labeled as background for missing organs. Secondly, we propose a self-supervised scheme to iteratively relabel the missing organs by introducing pseudo-labels into the train set, in order to recover the unknown complete ground-truth annotations. Our overall approach is denoted as SMILE for Self-training of deep convnets for Medical Image segmentation with partial LabEls.

4.4.1 Training from Partial Annotations with SMILE

In this section we detail the proposed SMILE method for training deep ConvNets on partially annotated data.

We address the issue of learning on partially labeled data by a simple yet effective method. The first step consists in extracting the maximum of information from the partially labeled data. This is done by deducing from the labeled organs where there are ambiguities that should be handled. An illustration of the method is shown in Fig. 4.7.

Training Exclusively with Correct Labels We know by construction that if an organ is unlabeled, then it is the case for the entire volume, i.e. no intermediate slice contains this label. Thus, assuming that every patient has all the organs (e.g. no missing spleen or kidney), we can deduce beforehand the missing classes for every patient. However, we do not know where they are located and thus where the wrong labels are.

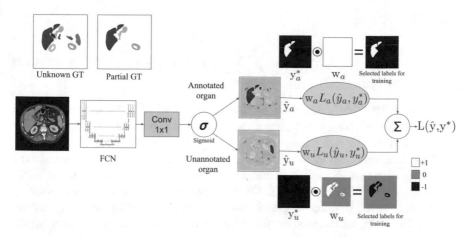

Fig. 4.7 Training SMILE on a partially annotated dataset. Each organ is predicted by a common FCN. Depending on the missing organs deduced by the available annotation, an ambiguity map w_k is created to ignore potential wrong labels in the loss. It acts as a weighting in the final loss function

However, if we want to exclusively use correct labels, we cannot use a classic softmax activation function and a multiclass loss. Indeed, in that configuration when at least one organ is missing no background label can be used. To address this problem, we transform the $(K + 1)$ multiclass classification problem into K binary classification problems where each organ is learned independently. The rationale behind this is to control the classes that are labeled and can be learned and those that are unlabeled and need to be ignored. By doing that we can learn features from the labeled classes for both the positives (the organs) and the negatives (the background) whereas for the unlabeled classes, both positives and negatives are ignored.

In practice, we replace the final softmax by a sigmoid activation function in the last 1×1 convolution layer. However, we still want to keep the exclusive aspect of the softmax: our class prediction is computed by taking, for each voxel, the class with the highest probability among all K classes - and the background label is assigned if it is lower than 0.5.

Training K binary classifiers requires adjustments, especially on the loss function. Actually, we have K losses, one for each class. We choose the binary cross entropy to train our model defined in Eq. 4.22 for each voxel i and class k:

$$l_{i,k}(\hat{y_{i,k}}, y_{i,k}) = -(y_{i,k} \, log(\hat{y_{i,k}}) + (1 - y_{i,k}) \, log(1 - \hat{y_{i,k}})) \qquad (4.22)$$

Let us denote as $\hat{\mathbf{Y}} \in \mathbb{R}^{H,W,K}$ the dense prediction of our model and $\mathbf{Y} \in \mathbb{R}^{H,W,K}$ as the ground truth. Then the K losses are aggregated to obtain one final loss in

Eq. 4.23:

$$\mathcal{L}_{ce}(\hat{\mathbf{Y}}, \mathbf{Y}) = \sum_{k=1}^{K} \sum_{i=1}^{N} w_{i,k}\, l_{i,k}(\hat{y_{i}}_{,k}, y_{i,k}) \tag{4.23}$$

where $\mathbf{W} \in \mathbb{R}^{H,W,K}$ is composed of K maps $\mathbf{w_k} \in \mathbb{R}^{H,W}$, a binary matrix which selects or discards for a given class k the voxels that should be learned, specifying on which voxels the back-propagation is applied during training.

This ambiguity map $\mathbf{W}$ is built beforehand based on the missing organs of each patient. As shown in Fig. 4.7, if an organ is labeled we can learn the entire associated classifier, so we fill w_k with ones. On the other hand, when an organ is missing w_k is set to zeros in order to ignore this organ during training for this particular patient. However, we can also use the other organs to add extra information, which are assigned as negative labels.

In the example of Fig. 4.7, three organs are labeled. However, when learning a missing organ like the spleen at the bottom branch of the example, we use an ambiguity map containing zeros everywhere except where the other organs for which the label is available. In that case the label of the organ is used to fill the ambiguity map of the spleen with ones.

4.4.2 Self-supervision and Pseudo-Labeling

The number of TP linearly decreases with respect to the ratio of missing organs α. We can improve the results by recovering new labels in unannotated training images. The idea is to iteratively add new positive target labels $y_{i,t}^{*} = 1$ in an image with missing annotations $\mathbf{x}_i$ for each class k,[2] using a curriculum strategy [BLCW09].

Iterative Relabeling Initially, the model is trained on all correct labels that can be regarded as "easy positive samples". Let us denote as $\hat{y}_i^{+}$, the pixels predicted as positive for a given unannotated image $\mathbf{x}_i$. The idea of SMILE is to recover $\oplus$ labels $y_{i,t}^{*,+}$ by selecting the top scoring pixels among $\hat{y}_i^{+}$. Then, the model is retrained with the new labels added to the train set.

This procedure is iteratively performed T times, by selecting a ratio $\gamma_t = \frac{t}{T}\gamma_{max}$ of top scoring pixels among the positives. The new labels incorporated at each step are the "hard examples" since they come from a self-supervised scheme that could introduces errors.

---

[2] We drop the dependence of class k in $y_{i,t}^{*}$ for clarity.

Algorithm 1 Training SMILE for class k

Input: $y^*_{i,0} = y^*_i$ // labels on fully annotated training set
Output: model m_T // trained after T self-training iterations
$m_0 \leftarrow$ FCN model trained on partially annotated data $y^*_{i,0}$
$N_u \leftarrow$ number of unannotated images for class k
for $t \leftarrow 0$ to $T - 1$ **do**
   $\gamma_t = \frac{t}{T}\gamma_{max}$
   for $i = 1$ to N_u **do**
      $\hat{y_i}^+ \leftarrow (m_t, x_i)$ // Find predicted positive pixels by m_t in image x_i
      $y^{*,+}_{i,t} \leftarrow (m_t, x_i, \gamma_t, \hat{y_k}^+)$ // Assign new $\oplus$ target labels
      $y^*_{i,t} = y^*_{i,t-1} \cup y^{*,+}_{i,t}$ // Augment training set
   end for
   $m_t = train(\{(x_i, y^*_{i,t})\})$ // Re-train model with augmented training set
end for
Return m_T

4.4.3   Experiments

We evaluate the proposed approach on a dataset with complete ground truth annotations for three organs: liver, pancreas and stomach, which gathers 72 3D volume CT-scans. We generate a partially annotated dataset by randomly removing $\alpha\%$ of organs in the volumes independently. The results in Fig. 4.8 show that SMILE can reach performances close to a model trained with the complete ground truth even when removing 70% of labels. More details can be found in [PTC+18].

4.5   Conclusion

In this chapter, we present solutions for limiting the need for huge sets of annotated datasets for training deep models. We present methods leveraging coarse annotations for performing accurate predictions, i.e. Weakly supervised learning (WSL) and methods dedicated to exploit unlabelled data, i.e. Semi-supervised learning (SSL). These approaches enables to improve deep learning performances in the low data regime. A main perspective for future work relates to the proper definition a unsupervised training criterion, which essentially remains an open question. Current approaches based on predictive coding and contrastive learning, e.g. transformers [VSP+17], are certainly an interesting path to follow.

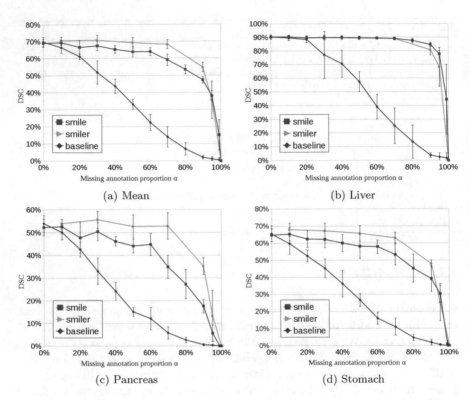

(a) Mean   (b) Liver

(c) Pancreas   (d) Stomach

Fig. 4.8 Dice score versus the proportion of missing annotations α. The baseline is represented in blue, SMILE in red and SMILEr in green

References

[ARS+16] Hossein Azizpour, Ali Sharif Razavian, Josephine Sullivan, Atsuto Maki, and Stefan Carlsson. Factors of transferability for a generic convnet representation. *TPAMI*, 2016.

[ATH03] Stuart Andrews, Ioannis Tsochantaridis, and Thomas Hofmann. Support vector machines for multiple-instance learning. In *NIPS*, 2003.

[BLCW09] Yoshua Bengio, Jérôme Louradour, Ronan Collobert, and Jason Weston. Curriculum learning. In *Proceedings of the 26th Annual International Conference on Machine Learning*, ICML '09, pages 41–48, New York, NY, USA, 2009. ACM.

[BLPL07] Yoshua Bengio, Pascal Lamblin, Dan Popovici, and Hugo Larochelle. Greedy layer-wise training of deep networks. In *Advances in Neural Information Processing Systems (NIPS)*, 2007.

[BM07] Razvan C. Bunescu and Raymond J. Mooney. Multiple instance learning for sparse positive bags. In *International Conference on Machine Learning*, 2007.

[BMJK15] Aseem Behl, Pritish Mohapatra, C. V. Jawahar, and M. Pawan Kumar. Optimizing average precision using weakly supervised data. *TPAMI*, 2015.

[BOS+17] Wenjia Bai, Ozan Oktay, Matthew Sinclair, Hideaki Suzuki, Martin Rajchl, Giacomo Tarroni, Ben Glocker, Andrew King, Paul M. Matthews, and Daniel Rueckert. Semi-supervised learning for network-based cardiac mr image segmentation. In Maxime

Descoteaux, Lena Maier-Hein, Alfred Franz, Pierre Jannin, D. Louis Collins, and Simon Duchesne, editors, *Medical Image Computing and Computer-Assisted Intervention - MICCAI 2017*, pages 253–260, Cham, 2017. Springer International Publishing.

[ÇAL+16] Özgün Çiçek, Ahmed Abdulkadir, Soeren S. Lienkamp, Thomas Brox, and Olaf Ronneberger. 3d U-Net: Learning dense volumetric segmentation from sparse annotation. In Sebastien Ourselin, Leo Joskowicz, Mert R. Sabuncu, Gozde Unal, and William Wells, editors, *Medical Image Computing and Computer-Assisted Intervention – MICCAI 2016*, pages 424–432, Cham, 2016. Springer International Publishing.

[CJP+18] Agisilaos Chartsias, Thomas Joyce, Giorgos Papanastasiou, Scott Semple, Michelle Williams, David Newby, Rohan Dharmakumar, and Sotirios A. Tsaftaris. Factorised spatial representation learning: Application in semi-supervised myocardial segmentation. In Alejandro F. Frangi, Julia A. Schnabel, Christos Davatzikos, Carlos Alberola-López, and Gabor Fichtinger, editors, *Medical Image Computing and Computer Assisted Intervention – MICCAI 2018*, pages 490–498, Cham, 2018. Springer International Publishing.

[CSVZ14] K. Chatfield, K. Simonyan, A. Vedaldi, and A. Zisserman. Return of the Devil in the Details: Delving Deep into Convolutional Nets. In *BMVC*, 2014.

[CTBH+19] Charles Corbière, Nicolas Thome, Avner Bar-Hen, Matthieu Cord, and Patrick Pérez. Addressing failure prediction by learning model confidence. In *Advances in Neural Information Processing Systems*, pages 2902–2913, 2019.

[CW04] Yixin Chen and James Z. Wang. Image categorization by learning and reasoning with regions. *J. Mach. Learn. Res.*, 5:913–939, December 2004.

[DF10] Thomas Deselaers and Vittorio Ferrari. A conditional random field for multiple-instance learning. In *International Conference on Machine Learning*, 2010.

[DLLP97] Thomas G. Dietterich, Richard H. Lathrop, and Tomás Lozano-Pérez. Solving the multiple instance problem with axis-parallel rectangles. *Artif. Intell.*, 1997.

[DMM19] Thibaut Durand, Nazanin Mehrasa, and Greg Mori. Learning a deep convnet for multi-label classification with partial labels. In *IEEE Conference on Computer Vision and Pattern Recognition, CVPR 2019, Long Beach, CA, USA, June 16–20, 2019*, pages 647–657. Computer Vision Foundation/IEEE, 2019.

[DMTC17] Thibaut Durand, Taylor Mordan, Nicolas Thome, and Matthieu Cord. Wildcat: Weakly supervised learning of deep convnets for image classification, pointwise localization and segmentation. In *Proceedings of the IEEE Conference on Computer Vision and Pattern Recognition (CVPR)*, July 2017.

[DTC15] Thibaut Durand, Nicolas Thome, and Matthieu Cord. MANTRA: Minimum Maximum Latent Structural SVM for Image Classification and Ranking. In *ICCV*, 2015.

[DTC16] Thibaut Durand, Nicolas Thome, and Matthieu Cord. WELDON: Weakly Supervised Learning of Deep Convolutional Neural Networks. In *CVPR*, 2016.

[DTC18] Thibaut Durand, Nicolas Thome, and Matthieu Cord. Symil: Minmax latent SVM for weakly labeled data. *IEEE Trans. Neural Networks Learn. Syst.*, 29(12):6099–6112, 2018.

[DTC19] T. Durand, N. Thome, and M. Cord. Exploiting negative evidence for deep latent structured models. *IEEE Transactions on Pattern Analysis and Machine Intelligence*, 41(2):337–351, 2019.

[EBC+10] Dumitru Erhan, Yoshua Bengio, Aaron Courville, Pierre-Antoine Manzagol, Pascal Vincent, and Samy Bengio. Why does unsupervised pre-training help deep learning? *Journal of Machine Learning Research (JMLR)*, 2010.

[GC07] Peter Gehler and Olivier Chapelle. Deterministic annealing for multiple-instance learning. In *Artificial Intelligence and Statistics*, 2007.

[GFKS02] Thomas Gärtner, Peter A. Flach, Adam Kowalczyk, and Alex J. Smola. Multi-instance kernels. In *International Conf. on Machine Learning*. Morgan Kaufmann, 2002.

[GG16] Yarin Gal and Zoubin Ghahramani. Dropout as a bayesian approximation: Representing model uncertainty in deep learning. In *international conference on machine learning*, pages 1050–1059, 2016.

[GPAM+14] Ian Goodfellow, Jean Pouget-Abadie, Mehdi Mirza, Bing Xu, David Warde-Farley, Sherjil Ozair, Aaron Courville, and Yoshua Bengio. Generative adversarial nets. In *Advances in Neural Information Processing Systems (NIPS)*, 2014.

[GPSW17] Chuan Guo, Geoff Pleiss, Yu Sun, and Kilian Q. Weinberger. On calibration of modern neural networks. volume 70 of *Proceedings of Machine Learning Research*, pages 1321–1330, International Convention Centre, Sydney, Australia, 06–11 Aug 2017. PMLR.

[GSS15] Ian J Goodfellow, Jonathon Shlens, and Christian Szegedy. Explaining and harnessing adversarial examples. In *International Conference on Learning Representations (ICLR)*, 2015.

[GT20a] Vincent Le Guen and Nicolas Thome. A deep physical model for solar irradiance forecasting with fisheye images. In *2020 IEEE/CVF Conference on Computer Vision and Pattern Recognition, CVPR Workshops 2020, Seattle, WA, USA, June 14–19, 2020*, pages 2685–2688. IEEE, 2020.

[GT20b] Vincent Le Guen and Nicolas Thome. Disentangling physical dynamics from unknown factors for unsupervised video prediction. In *2020 IEEE/CVF Conference on Computer Vision and Pattern Recognition, CVPR 2020, Seattle, WA, USA, June 13–19, 2020*, pages 11471–11481. IEEE, 2020.

[GTCL13] Hanlin Goh, Nicolas Thome, Matthieu Cord, and Joo-Hwee Lim. Top-down regularization of deep belief networks. In *Advances in Neural Information Processing Systems (NIPS)*, 2013.

[HG17] Dan Hendrycks and Kevin Gimpel. A baseline for detecting misclassified and out-of-distribution examples in neural networks. *Proceedings of International Conference on Learning Representations*, 2017.

[HS06] Geoffrey E Hinton and Ruslan R Salakhutdinov. Reducing the dimensionality of data with neural networks. *Science*, 2006.

[HTF09] Trevor Hastie, Robert Tibshirani, and Jerome Friedman. *The elements of statistical learning: data mining, inference and prediction*. Springer, 2009.

[HZRS14] Kaiming He, Xiangyu Zhang, Shaoqing Ren, and Jian Sun. Spatial pyramid pooling in deep convolutional networks for visual recognition. In *ECCV*, 2014.

[HZRS16] Kaiming He, Xiangyu Zhang, Shaoqing Ren, and Jian Sun. Deep residual learning for image recognition. In *CVPR*, 2016.

[JB12] Armand Joulin and Francis Bach. A convex relaxation for weakly supervised classifiers. In *International Conference on Machine Learning*, 2012.

[JKGG18] Heinrich Jiang, Been Kim, Melody Guan, and Maya Gupta. To trust or not to trust a classifier. In S. Bengio, H. Wallach, H. Larochelle, K. Grauman, N. Cesa-Bianchi, and R. Garnett, editors, *Advances in Neural Information Processing Systems 31*, pages 5541–5552. Curran Associates, Inc., 2018.

[KJZ+16] Praveen Kulkarni, Frédéric Jurie, Joaquin Zepeda, Patrick Pérez, and Louis Chevallier. Spleap: Soft pooling of learned parts for image classification. In *ECCV*, 2016.

[KOB13] Gabriel Krummenacher, Cheng S. Ong, and Joachim Buhmann. Ellipsoidal multiple instance learning. In *International Conference on Machine Learning*, 2013.

[KPK10] M. P. Kumar, Benjamin Packer, and Daphne Koller. Self-paced learning for latent variable models. In J. D. Lafferty, C. K. I. Williams, J. Shawe-Taylor, R. S. Zemel, and A. Culotta, editors, *Advances in Neural Information Processing Systems 23*, pages 1189–1197. Curran Associates, Inc., 2010.

[KSH12a] Alex Krizhevsky, Ilya Sutskever, and Geoff Hinton. Imagenet classification with deep convolutional neural networks. In *NIPS*. 2012.

[KSH12b] Alex Krizhevsky, Ilya Sutskever, and Geoffrey E Hinton. Imagenet classification with deep convolutional neural networks. In F. Pereira, C. J. C. Burges, L. Bottou, and K. Q. Weinberger, editors, *Advances in Neural Information Processing Systems 25*, pages 1097–1105. Curran Associates, Inc., 2012.

[LA17] Samuli Laine and Timo Aila. Temporal ensembling for semi-supervised learning. In *International Conference on Learning Representations (ICLR)*, 2017.

[LB08] Hugo Larochelle and Yoshua Bengio. Classification using discriminative restricted Boltzmann machines. In *International Conference on Machine Learning (ICML)*. ACM, 2008.

[LBD+89] Yann LeCun, Bernhard Boser, John S Denker, Donnie Henderson, Richard E Howard, Wayne Hubbard, and Lawrence D Jackel. Backpropagation applied to handwritten zip code recognition. *Neural computation*, 1(4):541–551, 1989.

[LCQ+18] Xiaomeng Li, Hao Chen, Xiaojuan Qi, Qi Dou, Chi-Wing Fu, and Pheng-Ann Heng. H-DenseUNet: Hybrid densely connected UNet for liver and tumor segmentation from CT volumes. *IEEE transactions on medical imaging*, 37(12):2663–2674, 2018.

[LSD15] Jonathan Long, Evan Shelhamer, and Trevor Darrell. Fully convolutional networks for semantic segmentation. In *Proceedings of the IEEE conference on computer vision and pattern recognition*, pages 3431–3440, 2015.

[LV15a] Wei-Xin Li and Nuno Vasconcelos. Multiple instance learning for soft bags via top instances. In *Conference on Computer Vision and Pattern Recognition*, 06 2015.

[LV15b] Weixin Li and Nuno Vasconcelos. Multiple Instance Learning for Soft Bags via Top Instances. In *CVPR*, 2015.

[LYCC14] Kuan-Ting Lai, Felix X. Yu, Ming-Syan Chen, and Shih-Fu Chang. Video event detection by inferring temporal instance labels. In *CVPR*, 2014.

[LYV19] Yunsheng Li, Lu Yuan, and Nuno Vasconcelos. Bidirectional learning for domain adaptation of semantic segmentation. In *Proceedings of the IEEE Conference on Computer Vision and Pattern Recognition*, pages 6936–6945, 2019.

[MJK14] Pritish Mohapatra, C.V. Jawahar, and M. Pawan Kumar. Efficient optimization for average precision SVM. In *NIPS*. 2014.

[MMK+16] Takeru Miyato, Shin-ichi Maeda, Masanori Koyama, Ken Nakae, and Shin Ishii. Distributional smoothing with virtual adversarial training. In *International Conference on Learning Representations (ICLR)*, 2016.

[MNA16] F. Milletari, N. Navab, and S. Ahmadi. V-net: Fully convolutional neural networks for volumetric medical image segmentation. In *2016 Fourth International Conference on 3D Vision (3DV)*, pages 565–571, 2016.

[MR98] Oded Maron and Aparna Lakshmi Ratan. Multiple-instance learning for natural scene classification. In *International Conference on Machine Learning*, 1998.

[OBLS15] M. Oquab, L. Bottou, I. Laptev, and J. Sivic. Is object localization for free? – Weakly-supervised learning with convolutional neural networks. In *CVPR*, 2015.

[PC15] Pedro O. Pinheiro and Ronan Collobert. From image-level to pixel-level labeling with convolutional networks. In *CVPR*, 2015.

[PKS15] George Papandreou, Iasonas Kokkinos, and Pierre-Andre Savalle. Modeling Local and Global Deformations in Deep Learning: Epitomic Convolution, Multiple Instance Learning, and Sliding Window Detection. In *CVPR*, 2015.

[PLI14] Wei Ping, Qiang Liu, and Alex Ihler. Marginal structured SVM with hidden variables. In *ICML*, 2014.

[PSLD15] Deepak Pathak, Evan Shelhamer, Jonathan Long, and Trevor Darrell. Fully Convolutional Multi-Class Multiple Instance Learning. In *ICLR (Workshop)*, 2015.

[PTC+18] Olivier Petit, Nicolas Thome, Arnaud Charnoz, Alexandre Hostettler, and Luc Soler. Handling missing annotations for semantic segmentation with deep convnets. In *Deep Learning in Medical Image Analysis and Multimodal Learning for Clinical Decision Support (DLMIA workshop MICCAI)*, pages 20–28. Springer, 2018.

[QSCL09] N. Quadrianto, A. Smola, T. Caetano, and Q. Le. Estimating labels from label proportions. *Journal of Machine Learning Research*, 10:2349–2374, 2009.

[QT09] Ariadna Quattoni and Antonio Torralba. Recognizing indoor scenes. In *CVPR*, 2009.

[QWM+07] Ariadna Quattoni, Sybor Wang, Louis-Philippe Morency, Michael Collins, and Trevor Darrell. Hidden conditional random fields. *TPAMI*, 2007.

[RBH+15] Antti Rasmus, Mathias Berglund, Mikko Honkala, Harri Valpola, and Tapani Raiko. Semi-supervised learning with ladder networks. In *Advances in Neural Information Processing Systems (NIPS)*, 2015.

[RFB15] Olaf Ronneberger, Philipp Fischer, and Thomas Brox. U-net: Convolutional networks for biomedical image segmentation. In Nassir Navab, Joachim Hornegger, William M. Wells, and Alejandro F. Frangi, editors, *Medical Image Computing and Computer-Assisted Intervention – MICCAI 2015*, pages 234–241, Cham, 2015. Springer International Publishing.

[RHBL07] M. Ranzato, F. J. Huang, Y. L. Boureau, and Y. LeCun. Unsupervised learning of invariant feature hierarchies with applications to object recognition. In *IEEE Conference on Computer Vision and Pattern Recognition (CVPR)*, June 2007.

[RPCL07] Marc'Aurelio Ranzato, Christopher Poultney, Sumit Chopra, and Yann Lecun. Efficient learning of sparse representations with an energy-based model. In *Advances in Neural Information Processing Systems (NIPS)*, 2007.

[RS08] Marc'Aurelio Ranzato and Martin Szummer. Semi-supervised learning of compact document representations with deep networks. In *International Conference on Machine Learning (ICML)*. ACM, 2008.

[RTC18] Thomas Robert, Nicolas Thome, and Matthieu Cord. Hybridnet: Classification and reconstruction cooperation for semi-supervised learning. In Vittorio Ferrari, Martial Hebert, Cristian Sminchisescu, and Yair Weiss, editors, *Computer Vision - ECCV 2018 - 15th European Conference, Munich, Germany, September 8–14, 2018, Proceedings, Part VII*, volume 11211 of *Lecture Notes in Computer Science*, pages 158–175. Springer, 2018.

[Rue10] Stefan Rueping. SVM classifier estimation from group probabilities. In *International Conference on Machine Learning*, 2010.

[SHPU12] A. G. Schwing, T. Hazan, M. Pollefeys, and R. Urtasun. Efficient Structured Prediction with Latent Variables for General Graphical Models. In *ICML*, 2012.

[SJT16] Mehdi Sajjadi, Mehran Javanmardi, and Tolga Tasdizen. Regularization with stochastic transformations and perturbations for deep semi-supervised learning. In *Advances in Neural Information Processing Systems (NIPS)*, 2016.

[SPC+16] Chen Sun, Manohar Paluri, Ronan Collobert, Ram Nevatia, and Lubomir Bourdev. ProNet: Learning to Propose Object-Specific Boxes for Cascaded Neural Networks. In *CVPR*, 2016.

[SS15] Gaurav Sharma and Bernt Schiele. Scalable nonlinear embeddings for semantic category-based image retrieval. In *ICCV*, 2015.

[SZ15] Karen Simonyan and Andrew Zisserman. Very Deep Convolutional Networks for Large-Scale Image Recognition. In *ICLR*, 2015.

[TTC13] Christian Thériault, Nicolas Thome, and Matthieu Cord. Dynamic Scene Classification: Learning Motion Descriptors with Slow Features Analysis. In *IEEE Conference on Computer Vision and Pattern Recognition (CVPR)*, 2013.

[TV17] Antti Tarvainen and Harri Valpola. Mean teachers are better role models: Weight-averaged consistency targets improve semi-supervised deep learning results. In *Advances in Neural Information Processing Systems (NIPS)*, 2017.

[VLBM08] Pascal Vincent, Hugo Larochelle, Yoshua Bengio, and Pierre-Antoine Manzagol. Extracting and composing robust features with denoising autoencoders. In *International Conference on Machine Learning (ICML)*. ACM, 2008.

[VSP+17] Ashish Vaswani, Noam Shazeer, Niki Parmar, Jakob Uszkoreit, Llion Jones, Aidan N Gomez, Ł ukasz Kaiser, and Illia Polosukhin. Attention is all you need. In I. Guyon, U. V. Luxburg, S. Bengio, H. Wallach, R. Fergus, S. Vishwanathan, and R. Garnett, editors, *Advances in Neural Information Processing Systems*, volume 30, pages 5998–6008. Curran Associates, Inc., 2017.

[WH16]   Zijun Wei and Minh Hoai. Region Ranking SVM for Image Classification. In *CVPR*, June 2016.

[WXH+14] Yunchao Wei, Wei Xia, Junshi Huang, Bingbing Ni, Jian Dong, Yao Zhao, and Shuicheng Yan. CNN: single-label to multi-label. *CoRR*, abs/1406.5726, 2014.

[YFRJ07] Yisong Yue, T. Finley, F. Radlinski, and T. Joachims. A support vector method for optimizing average precision. In *SIGIR*, 2007.

[YJ09]   Chun-Nam Yu and T. Joachims. Learning structural SVMs with latent variables. In *ICML*, 2009.

[YLK+13] Felix X. Yu, Dong Liu, Sanjiv Kumar, Tony Jebara, and Shih-Fu Chang. $\propto$SVM for learning with label proportions. In *ICML*, 2013.

[ZF14]   Matthew D Zeiler and Rob Fergus. Visualizing and understanding convolutional networks. In *European Conference on Computer Vision (ECCV)*, 2014.

[Zhu05]   Xiaojin Zhu. Semi-supervised learning literature survey. Technical Report 1530, Computer Sciences, University of Wisconsin-Madison, 2005.

[ZKL+16] Bolei Zhou, Aditya Khosla, Agata Lapedriza, Aude Oliva, and Antonio Torralba. Learning Deep Features for Discriminative Localization. In *CVPR*, 2016.

[ZLL16]   Yuting Zhang, Kibok Lee, and Honglak Lee. Augmenting supervised neural networks with unsupervised objectives for large-scale image classification. In *International Conference on Machine Learning (ICML)*, 2016.

[ZMGL16] Junbo Zhao, Michael Mathieu, Ross Goroshin, and Yann LeCun. Stacked What-Where Auto-encoders. In *International Conference on Learning Representations (ICLR) Workshop*, 2016.

[ZWT+19] Yuyin Zhou, Yan Wang, Peng Tang, Song Bai, Wei Shen, Elliot Fishman, and Alan Yuille. Semi-supervised 3d abdominal multi-organ segmentation via deep multi-planar co-training. In *2019 IEEE Winter Conference on Applications of Computer Vision (WACV)*, pages 121–140. IEEE, 2019.

[ZYL+19] Yang Zou, Zhiding Yu, Xiaofeng Liu, BVK Kumar, and Jinsong Wang. Confidence regularized self-training. In *Proceedings of the IEEE International Conference on Computer Vision*, pages 5982–5991, 2019.

[ZYVKW18] Yang Zou, Zhiding Yu, BVK Vijaya Kumar, and Jinsong Wang. Unsupervised domain adaptation for semantic segmentation via class-balanced self-training. In *Proceedings of the European conference on computer vision (ECCV)*, pages 289–305, 2018.

[ZZSL19] Yuan-Xing Zhao, Yan-Ming Zhang, Ming Song, and Cheng-Lin Liu. Multi-view semi-supervised 3d whole brain segmentation with a self-ensemble network. In *International Conference on Medical Image Computing and Computer-Assisted Intervention*, pages 256–265. Springer, 2019.

Chapter 5
Similarity Metric Learning

Stefan Duffner, Christophe Garcia, Khalid Idrissi, and Atilla Baskurt

5.1 Introduction

Traditionally, neural networks, and in particular deep neural networks, have been used in settings involving supervised learning, and they showed state-of-the-art performance in many applications where abundant annotated data are available. In these applications, it is usually required to know in advance the number of classes to predict, and their clear meaning, i.e. which instance belongs to them or not, or, in case of a regression problem: the values corresponding to the instances. Moreover, the relationship of different classes (or sub-classes) is not explicitly modelled, as this is mostly not useful in the given predictive scenarios. In this part, we will consider other applications that do not allow or are not suited for such supervised learning approaches. This is the case, for example, when:

- instance labels (e.g. positive/negative, foreground/background or a person identifier) are not available or too difficult to obtain for training or
- the number of classes is not fixed a priori or
- the classes to predict at test time are not the same as the ones available for training or
- the relationship or similarity between instances and categories of instances should be modelled and represented explicitly or
- a robust rejection strategy needs to be implemented for a classification task.

If no information on instance classes or their relationship is given at all, *unsupervised approaches*, such as clustering methods or auto-encoders, are most suitable to automatically learn and infer a general model of the data. However, in

S. Duffner (✉) · C. Garcia · K. Idrissi · A. Baskurt
Université de Lyon, INSALyon, LIRIS, Villeurbanne, France
e-mail: stefan.duffner@liris.cnrs.fr; christophe.garcia@liris.cnrs.fr; khalid.idrissi@liris.cnrs.fr;
atilla.baskurt@liris.cnrs.fr

© Springer Nature Switzerland AG 2021
J. Benois-Pineau, A. Zemmari (eds.), *Multi-faceted Deep Learning*,
https://doi.org/10.1007/978-3-030-74478-6_5

many settings, class labels for at least some of the training data is available, and one is interested in a generic model that is applicable for all data of the same type. In this case, *weakly supervised* or *semi-supervised learning* algorithms are commonly employed.

One such weakly supervised approach is to automatically learn a similarity metric between instances of a given category (e.g. faces). That is, the instance labels of a training dataset are not explicitly learnt but rather used to model the distance between similar and dissimilar instances—for example, by using pairs of instances. Note that sometimes the term *distance learning* is used. But many existing models, e.g. those based on neural networks described in this chapter, do not fulfil the mathematical requirements of a distance, especially the triangle inequality, and thus do not represent a metric neither.

The notion of similarity or dissimilarity differs from one application to another. Thus its definition depends on the learning problem and on the training dataset. For example, let us consider a set of face images: (1) for the task of face verification, two images of the same person are considered similar; (2) for the task of gender verification, two face images showing two males or two females are defined as similar; (3) for the task of kinship verification, a similar pair of face images must indicate a biological relationship such as father-son, mother-daughter etc.

In principle, most metric learning algorithms receive positive and negative pairs of instances, where a positive pair is formed of instances that are considered similar (e.g. belonging to the same category), and a negative pair is composed of instances that are considered different or dissimilar, although this may not be simple to determine in practice. By presenting all possible pairs, a good metric learning method must then be able to capture the intrinsic relationship between the concerned semantic contents of two objects. Figure 5.1 shows pairs of images of the same persons from the LFW dataset. These pairs can be used to define a similarity relationship, and a generic distance metric can be learnt and applied to unknown

Fig. 5.1 Pairs of images (LFW dataset) from the same person that can be used to learn a similarity relationship. Similarity metric learning aims at learning a metric space where similar examples (e.g. faces from the same person) are close and examples not considered similar have a large distance. This metric can then be used to verify if two unknown examples, possibly from unknown classes, belong to the same class (e.g. for face verification)

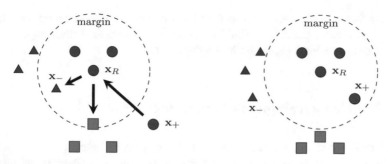

Fig. 5.2 Principle approach for similarity metric learning. Minimising an appropriate loss function tries to bring similar instances closer ($\mathbf{x}_R$, $\mathbf{x}_+$) and to separate dissimilar ones ($\mathbf{x}_R$, $\mathbf{x}_-$) (if they are closer than a certain margin). A *pair-wise* loss function iteratively optimises these objectives on all positive and negative pairs, whereas a *triplet*-based loss is defined on triplets ($\mathbf{x}_R$, $\mathbf{x}_+$, $\mathbf{x}_-$) (c.f. Sect. 5.2.3)

persons, for example for face verification. An alternative approach is to give triplets of instances, with one reference (also called "anchor"), one similar instance and one dissimilar instance w.r.t. to the reference. Then, the similar instance is considered more similar to the reference than the third one. Or, sometimes quadruplets are used, where the first pair is considered more similar than the second pair. Figure 5.2 illustrates the principle approach for similarity metric learning with pairs or triplets of instances.

In this chapter, we will present similarity metric learning methods and models based on Siamese Neural Networks (SNN). The original term "siamese" relates to the use of pairs of instances as introduced by Bromley et al. [BGL+94]. However, in the literature, this has been extended to triplet or tuple-based architectures. As we will outline in the next section, there are other models, for example, based on statistical projections or Support Vector Machines. However, feed-forward neural networks have several properties that make them interesting and particularly suitable for this problem:

- They can model a wide variety of linear and non-linear functions, c.f. the universal approximation theorem [HSW89], and well-established optimisation approaches can be used.
- By carefully specifying the architectures, the complexity of the models can be controlled (in terms of the number of parameters and the dimensions at different abstraction levels, considering it as a data processing pipeline).
- Through multi-layered architectures and the error back-propagation algorithm, one can jointly learn optimal features and projections into vector spaces that best represent semantic similarities.
- Using deep Convolutional Neural Network architectures, the resulting models are suitable and very powerful for (natural) image data.
- Finally, for a large variety of applications and data, neural networks showed a high generalisation capacity and robustness to different types of noise.

In the following, we will outline the different algorithms and neural models that exist for similarity metric learning and describe the principal SNN approach and variants that have been presented in the literature.

5.2 Metric Learning with Neural Networks

Most linear metric learning methods employ two types of metrics: the Mahalanobis distance or a more general similarity metric. In both cases, a linear transformation matrix W is learnt projecting input features into a target vector space. Typically, distance metric learning relates to a Mahalanobis-like distance function [XNJR03, WBS06]: $d_W(\mathbf{x}_1, \mathbf{x}_2) = \sqrt{(\mathbf{x}_1 - \mathbf{x}_2)^T W (\mathbf{x}_1 - \mathbf{x}_2)}$, where $\mathbf{x}_1$ and $\mathbf{x}_2$ are two example vectors, and W is not the (fixed) covariance matrix, as for the Mahalanobis distance, but is to be learnt by the algorithm. Note that when W is the identity matrix, $d_W(\mathbf{x}_1, \mathbf{x}_2)$ corresponds to the Euclidean distance. In contrast, similarity metric learning methods learn a function of a more general form: $s_W(\mathbf{x}_1, \mathbf{x}_2) = \mathbf{x}_1^T W \mathbf{x}_2 / N(\mathbf{x}_1, \mathbf{x}_2)$, where $N(\mathbf{x}_1, \mathbf{x}_2)$ is a normalisation term [QGCL08]. Specifically, when $N(\mathbf{x}_1, \mathbf{x}_2) = 1$, $s_W(\mathbf{x}_1, \mathbf{x}_2)$ is the bilinear similarity function [CSSB10]; and when $N(\mathbf{x}_1, \mathbf{x}_2) = \sqrt{\mathbf{x}_1^T W \mathbf{x}_1} \sqrt{\mathbf{x}_2^T W \mathbf{x}_2}$, $s_W(\mathbf{x}_1, \mathbf{x}_2)$ corresponds to the generalised cosine similarity function [HL11].

Non-linear metric learning methods are constructed by simply substituting the above linear projection with a non-linear transformation [HLT14, CHL05, KTS+12, YYGT16]. For example, Hu et al. [HLT14] and Chopra et al. [CHL05] employed neural networks to accomplish this. With these approaches, the learning algorithm optimises the parameters of a non-linear projection $\mathbf{o} = f(\mathbf{x})$ into a metric space such that a simple distance measure $d_W(f(\mathbf{x}_1), f(\mathbf{x}_2))$, mostly the Euclidean or cosine distance, reflects the semantic similarities between the instances. Such non-linear methods are subject to local optima and more inclined to overfit the training data but have the potential to outperform linear methods on many problems [BHS13, KTS+12]. Compared with linear models, non-linear models are usually preferred on large training sets to more accurately capture the underlying distribution of the data [LBOM12]. A detailed survey and review of metric learning approaches has been published recently by Bellet et al. [BHS13], and an experimental analysis and comparison by Moutafis et al. [MLK16]. We will concentrate here on Siamese Neural Networks (SNN) that can represent linear or non-linear projections depending on the used activation function and the number of layers. With larger and more complex models, the term *deep similarity metric learning* has often been employed in the literature.

As mentioned before, a SNN essentially distinguishes itself from classical feed-forward neural networks by its specific training strategy involving sets of examples labelled as similar or dissimilar. The capabilities of different SNN-based methods depend on four main points: the network architecture, the training set selection strategy, the loss function, and the training algorithm [BGL+94]. In the following, we will explain these points in more detail and give some examples.

5.2.1 Architectures

A SNN can be seen as two identical, parallel neural networks *NN* sharing the same set of weights W (see Fig. 5.3). Although the weights of the input branches are shared in most cases, this may not be the case for some specific applications, notably when the two input examples are of different type or from different sources (e.g. images taken from two different camera views). Then, each input branch of the neural network is dedicated to a specific input type, and the two cannot be interchanged. In any case, the sub-networks receive input examples x_1, x_2, and produce output feature vectors o_1, o_2 that are supposed to be *close* for examples from the same class and *far apart* for examples from different classes, according to some distance measure, such as the cosine similarity metric (c.f. Sect. 5.2.3). During the training phase, a loss function $\mathcal{L}_W$, defined using the chosen distance measure over the output of all input examples, is iteratively minimised (c.f. Sect. 5.2.3). The architecture of the neural network determines the type and complexity of the projection function. One crucial parameter is the size of the output layer as it defines the dimension of the vector space of the embedding. This is usually an empirical choice. It should be large enough to capture the similarity relationships of the training data (according to different aspects/axes) but not too large as the distance measures in too large vector spaces may become extremely small and thus meaningless.

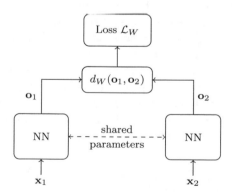

Fig. 5.3 The principal SNN architecture. Two identical neural networks NN receive two different input vectors x_1, x_2 producing two embeddings o_1 and o_2, respectively. The loss function $\mathcal{L}_W$ to minimise is based on a distance measure d_W between these embeddings and is used to update the shared weights of the neural network computing the gradient w.r.t. to both of them and back-propagating the error

Bromley et al. [BGL+94] introduced the Siamese architecture in 1994, using a Siamese CNN with two sub-networks for a signature verification system handling time-series of hand-crafted low-level features. In 2005, Chopra, Hadsell and LeCun [CHL05] formalised the Siamese architecture applying a CNN on raw images for face verification, before adapting it to a dimensionality reduction technique [HCL06] (see Fig. 5.4b for an illustration of a siamese CNN architecture). More recently, Siamese CNNs have been used successfully for various tasks, such as person re-identification [YLLL14], speaker verification [CS11], and face identification [SCWT14].

CNN-based architectures are more specific to image inputs, and several research works propose to use feed-forward Multi-Layer Perceptrons (MLP) to handle more general vector inputs (c.f. Fig. 5.4a). For example, Yih et al. [YTPM11] apply SNNs to learn similarities on text data, Bordes et al. [BWCB11] on entities in Knowledge Bases, and Masci et al. [MBBS14] on multi-modal data. Recently, siamese or triplet architectures have become popular and have been used for many different applications and types of data and combined with other types of neural models. For example, for sequence or time series modelling with recurrent neural networks as illustrated in Fig. 5.4c. Mueller and Thyagarajan [MT16] proposed a Siamese neural network for learning similarities between sentences using Long Short-Term Memory (LSTM) neural networks, where the distances are computed on the last hidden layer output. Similarly, the approach from Neculoiu et al. [NVR16] uses a Bidirectional LSTM to compute similarities between words. Here, the hidden layer outputs are averaged over the sequence in order to obtain two embedding vectors of the same dimension. SNNs have also been used with auto-encoder architectures [KBR16, YCS19, SL20, ASS20] (c.f. Fig. 5.4d). In that case, the two (or more) auto-encoders share their weights and the siamese or triplet loss is minimsed on the latent embedding (code) produced by the encoder, whereas the reconstruction loss is optimised at the same time. This approach has been extended by Compagnon et al. [CLDG19] for sequence metric learning with a LSTM-based sequence-to-sequence auto-encoder. Such multi-task learning structures and algorithms have very popular in the past years and successfully applied to various domains, notably Computer Vision. Recently, Graph Neural Networks (GNN) have been combined with pair-wise or triplet-based similarity metric learning for image retrieval [CBB19] or graph matching in general [LGD+19]. Here, the parameters are shared over several GNN and the similarity metric is learnt on the fixed-size graph embedding as illustrated in Fig. 5.4e.

5.2.2 Training Set Selection

5.2.2.1 Pairs

The selection strategy for training examples depends mostly on the application and the kind of knowledge about similarities that one wants to incorporate in the

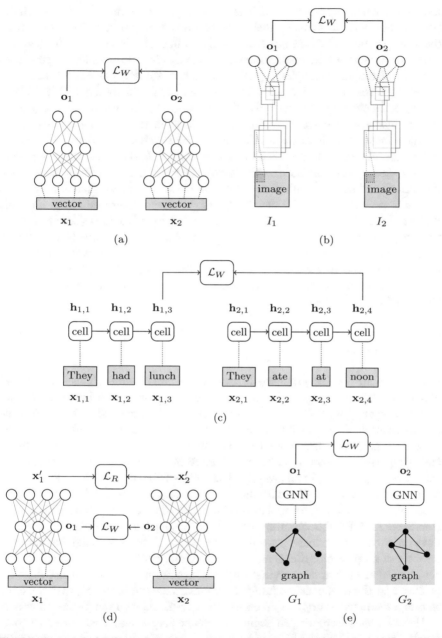

Fig. 5.4 Examples of different types of neural network models for deep similarity metric learning where two inputs $\mathbf{x}_1$ and $\mathbf{x}_2$ ($I_1, I_2, G_1, G2$, respectively) produce fixed-size output embeddings $\mathbf{o}_1$ and $\mathbf{o}_2$ that are used for minimising the loss function $\mathcal{L}_W$: (**a**) Multi-Layer Perceptrons (MLP) for vector data, (**b**) Convolutional Neural Network (CNN) for images, (**c**) Recurrent Neural Networks for variable-length sequential data (here: vectors of word embeddings), (**d**) auto-encoders minimising simultaneously the reconstruction loss $\mathcal{L}_R$ and (**e**) Graph Neural Networks operating on graph structures. For each of the architectures, only two branches are shown, but these models can be easily adapted to several inputs (triplets, quadruplets etc.)

model. For many applications, such as face or signature verification, the similarity between examples depends on their "real-world" origin, i.e. faces/signatures from the same person, and the neural network allows to determine the genuineness of a test example w.r.t. a reference by means of a binary classification. Most approaches use pairs of training examples (x_1, x_2) and a binary similarity relation which takes different values for similar and dissimilar pairs (c.f. illustration in Fig. 5.3). In principle, the overall loss is defined as the sum of loss terms over all possible pairs of training instances (see Sect. 5.2.3 for common loss functions). However, this creates a large imbalance as there are usually much more negative pairs than positive ones if the examples are uniformly distributed over classes. And this imbalance may lead to convergence problems because the learning algorithm tries to focus more on the dissimilarities than the similarities, which are not well defined. Thus it may fail to learn a meaningful similarity metric. A straightforward approach to alleviate this problem is to balance the number of positive and negative pairs (i.e. using over-sampling/under-sampling) in each training iteration, or alternatively introduce a weight term in the loss function [ZIG+15]. In some cases, it is even beneficial to learn only with positive examples [ZDI+18]. This ensures that during learning the dissimilar pairs do not impair the learnt projection that brings positive examples closer in the metric space.

5.2.2.2   Triplets

Lefebvre et al. [LG13] proposed to expand the information about the expected neighbourhood, and suggested a more symmetric representation based on the idea of Weinberger et al. [WBS06]: by considering a reference example x_R for each known relation, it is possible to define triplets (x_R, x_+, x_-), with x_+ forming a genuine pair with the reference x_R, while x_- is an example from another class (impostor)—sometimes also called the *anchor*, the *positive* and the *negative* examples, respectively. In this way, the learnt embedding may correspond to a more balanced representation of similarities and dissimilarities of the data. As for pair-wise SNNs, in principle, the global loss function is defined over all possible triplets (see Sect. 5.2.3 for the definition of different triplet loss functions). In practice, however, there are two many possible combinations. Thus, at each training iteration, a subset of all possible triplets are usually randomly and uniformly sampled. One can also define each training example as reference example x_R in turn, and draw the positive and negative ones randomly. Or if classes are not balanced, first randomly sample a reference example from each class in turn, and then sample the other two.

However, these sampling strategies may not be very efficient because, after some training iterations, most of the triplets are not contributing much (or not at all) to learning the similarity metric because they have already been integrated in the model, and the loss (and gradient) is thus very small or even zero. Therefore, different heuristics have been proposed to improve convergence and also the semantic expressiveness of the learnt similarity metric.

For example, Hermans et al. [HBL17] proposed the *hard-batch* triplet loss which is a modification of the common triplet loss (c.f. Eq. 5.8), where, for each reference example, only the hardest positive examples (i.e. the ones that are furthest from the reference) and the hardest negative examples (i.e. the closest ones) are selected for training. The hard-batch approach is sometimes referred to as the "top-ranking constraint". Wang et al. [WZL17] extended this idea with an efficient method for mining hard examples by learning a sub-space and clustering the identities, and they applied this approach for large-scale face recognition with 100,000 identities. Schroff et al. [SKP15] argued that learning with the hardest negatives may impair the convergence and the performance of the model and proposed a *semi-hard* batch mining strategy which consists in not using those negatives that have a smaller distance to the reference than the positive example in a mini-batch. The deep metric learning formulation from Yu and Tao [YT19] introduce a scaling scheme in their loss where hard triplets are up-weighted and easier triplets are down-weighted. An additional slack variable introduces a margin that prevents too hard negatives from gaining too much importance. Ge et al. [GHDS18] introduce an hierarchical triplet loss, which consists in dynamically adapting the margin m in Eq. 5.8 and thus incorporates a hierarchical structure in the embedding that is in turn constructed using the learnt distance metric. This allows to select meaningful hard negatives and to improve the convergence and performance. Another approach called RankTriplet has been proposed by Chen et al. [CDS+18] for ranking in a retrieval application. Here hard triplets are selected in subsets (training batches) according to their ranks w.r.t. to a query, i.e. only mis-ranked positive and negative examples are used. This allows to define a list-wise measure based on the similarity ranking. Also the effective mining of appropriate positive examples has been studied in the literature. Shi et al. [SYZ+16], for instance, propose a *moderate positive mining* strategy that excludes very hard positives examples as their Euclidean distance can differ considerably w.r.t. the geodesic distance on a complex manifold that the learnt metric should correspond to.

5.2.2.3 Tuples

Several SNN approaches that go beyond triplet architectures have been proposed in the past. For instance, the model introduced by Yih et al. [YTPM11] uses a loss based on two pairs of training examples (see Eq. 5.12), where the first pair is considered more similar than the second one. For some applications, this type of *relative* similarity may be easier to define than the binary relation between similar and dissimilar pairs. Chen et al. [CCZH17a] propose a quadruplet loss (c.f. Eq. 5.14) introducing an additional constraint to triplets that also pushes away negatives pairs from positive pairs w.r.t. to different reference images. They experimentally showed that this approach reduces the intra-class variance while increasing the inter-class variance, thus improving its discrimination capacity.

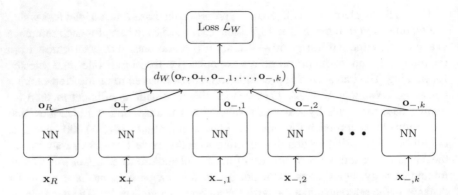

Fig. 5.5 A SNN architecture trained with tuples composed of a reference example $\mathbf{x}_R$, a positive example $\mathbf{x}_+$, and K negative examples $\mathbf{x}_{-,1}, \ldots, \mathbf{x}_{-,K}$. For $K = 1$, this corresponds to the popular triplet architecture

Another possibility proposed by Berlemont et al. [BLDG15] is to use tuples $(\mathbf{x}_R, \mathbf{x}_+, \mathbf{x}_{-,1}, \mathbf{x}_{-,2}, \ldots, \mathbf{x}_{-,K})$, with a reference $\mathbf{x}_R$, a positive example $\mathbf{x}_+$ and K negative examples $\mathbf{x}_{-,k}$, as illustrated in Fig. 5.5. Here, the loss function $\mathcal{L}_W$ (Eq. 5.15) is defined on a modified triplet cosine similarity that tries to bring closer the positive and separate all the negative examples from the reference. This objective essentially operates on $K + 1$ pairs of examples. Similarly, Song et al. [SXJS16] proposed a deep metric learning approach based on CNNs, where training batches are composed of several positive and negative pairs and the loss function (Eq. 5.17) brings closer similar pairs and separates each positive pair from its hardest negatives. Later, Yang et al. [YCS19] extended this approach defining a loss function on all positive and negative pairs in a tuple (Eq. 5.18) and a regularisation term preventing the variances of positive and negative distances from becoming too large. They also introduced a dynamically computed weight that focuses the loss on harder positive pairs.

Going beyond pair-wise loss formulations, Berlemont et al. [BLDG16, BLDG18] extended their previous approach by proposing a loss based on the polar sine function [LW09], a generalisation of the sine function to n dimensions (Eq. 5.22). The relation between all examples in the tuple is thus taken into account by maximising the hyper-volume spanned by the projected reference and the negative example vectors. By sampling each negative example in a tuple from a different class (different from the one of the reference example) the training becomes more balanced and the learnt similarity metric tends to better reflect the relationship between the classes.

The various SNN architectures and corresponding loss functions described here have been used and evaluated for different applications in the literature. It seems there is no universal strategy for the choice of these hyper-parameters, and their performance largely depends on the given application as well as the type and amount of training data.

5.2.3 Loss Functions

The loss function $\mathcal{L}_W$ that is to be minimised by the optimisation algorithm defines a global objective in terms of the distances d_W between the neural network output vectors of the training examples $\mathbf{o}_i$ $i \in 1..N$, i.e. the projections in the embedding. The loss should ensure that the distances between similar examples is small and between dissimilar examples large. As described above, this measure is usually defined over pairs, triplets or tuples in the loss function, and mostly based on the Euclidean or cosine distance. When minimising the empirical loss over the training examples by iteratively updating the shared weights W of the neural network, one essentially learns a projection function of inputs $\mathbf{o} = f(\mathbf{x})$ into an embedding that forms a similarity metric space where similar examples lie close (in terms of the defined distance d_W) and dissimilar examples are further apart.

In the following, we will briefly describe some of the most commonly used loss functions. For convenience and unless stated otherwise, we only note the loss on a given subset (pair, triplet, tuple etc.), and the overall loss is usually defined as the sum of these losses over all possible subsets.

5.2.3.1 Cosine Pair-Wise

Given a network with weights W and two examples $\mathbf{x}_1$ and $\mathbf{x}_2$ with their labels Y, a target $t(Y)$ is defined for the cosine value between the two respective output vectors $\mathbf{o}_1$ and $\mathbf{o}_2$ as "1" for similar pairs and "-1" (or "0") for dissimilar pairs [BGL+94]:

$$\mathcal{L}_W(\mathbf{x}_1, \mathbf{x}_2, Y) = (t(Y) - \cos(\mathbf{o}_1, \mathbf{o}_2))^2 . \tag{5.1}$$

A similar function is used in the Cosine Similarity Metric Learning (CSML) approach [HL11]:

$$\mathcal{L}_W(\mathbf{x}_1, \mathbf{x}_2, Y) = -t(Y)\cos(\mathbf{o}_1, \mathbf{o}_2) . \tag{5.2}$$

5.2.3.2 Triangular

Zheng et al. [ZDI+18] use the same targets for a pair-wise function and impose additional constraints on the norms of the output vectors: $\mathbf{o}_1$ and $\mathbf{o}_2$. Integrating a geometrical interpretation using the triangle inequality, the resulting loss function becomes:

$$\mathcal{L}_W(\mathbf{x}_1, \mathbf{x}_2, Y) = \frac{1}{2}\|\mathbf{o}_1\|^2 + \frac{1}{2}\|\mathbf{o}_2\|^2 - \|\mathbf{o}_1 + t(Y)\mathbf{o}_2\| + 1 . \tag{5.3}$$

5.2.3.3 Norm-Based

Several works [CHL05, HCL06, SCWT14, MBBS14] propose to use the norm, e.g. ℓ_2-norm, between the output vectors as a similarity measure:

$$d_W(\mathbf{x}_1, \mathbf{x}_2) = \|\mathbf{o}_1 - \mathbf{o}_2\|_2 \ . \tag{5.4}$$

For example, Chopra et al. [CHL05] define an objective composed of an "impostor" I (t(Y)=1) and a "genuine" G term (t(Y)=0):

$$\mathcal{L}_W(\mathbf{x}_1, \mathbf{x}_2, Y) = (1 - t(Y))E_W^G(\mathbf{x}_1, \mathbf{x}_2) + t(Y).E_W^I(\mathbf{x}_1, \mathbf{x}_2) \tag{5.5}$$

$$\text{with } E_W^G(\mathbf{x}_1, \mathbf{x}_2) = \frac{2}{Q}(d_W)^2, \ E_W^I(\mathbf{x}_1, \mathbf{x}_2) = 2Qe^{\left(-\frac{2.77}{Q}d_W\right)}, \tag{5.6}$$

where Q is the upper bound of d_W.

Many recent works use the so-called *contrastive loss* [HCL06], where

$$E_W^G(\mathbf{x}_1, \mathbf{x}_2) = d_W^2 \quad \text{and} \quad E_W^I(\mathbf{x}_1, \mathbf{x}_2) = \max(m - d_W^2, 0) \ , \tag{5.7}$$

with m being a fixed margin parameter.

5.2.3.4 Triplet

Weinberger et al. [WBS06] introduced the triplet loss for a linear metric learning approach using simultaneously targets for genuine and impostor pairs by forming triplets of a reference $\mathbf{x}_R$, a positive $\mathbf{x}_+$ and a negative $\mathbf{x}_-$ example:

$$\mathcal{L}_W(\mathbf{x}_R, \mathbf{x}_+, \mathbf{x}_-) = \max(d_W(\mathbf{o}_R, \mathbf{o}_+)^2 - d_W(\mathbf{o}_R, \mathbf{o}_-)^2 + m, 0) \ . \tag{5.8}$$

This loss has been used by numerous deep learning approaches in the literature and for various different applications.

Later, Lefebvre et al. [LG13] proposed a triplet similarity measure based on the cosine distance. Here, the output of the positive pair $(\mathbf{o}_R, \mathbf{o}_+)$ is trained to be collinear, whereas the output of the negative pair $(\mathbf{o}_R, \mathbf{o}_-)$ is trained to be orthogonal. Thus:

$$\mathcal{L}_W(\mathbf{x}_R, \mathbf{x}_+, \mathbf{x}_-) = (1 - \cos(\mathbf{o}_R, \mathbf{o}_+))^2 + (0 - \cos(\mathbf{o}_R, \mathbf{o}_-))^2 \ . \tag{5.9}$$

Note that the margin m is a hyper-parameter that is usually empirically determined. It is however not trivial to find its optimal value as it depends mostly on the application, the used dataset and the dimension of the embedding space.

5.2.3.5   Angular

Wang et al. [WZW+17] introduced a loss function based on the angles formed by triplets, called the *angular loss*. Compared to the classical triplet loss, it not only takes into account the reference-positive and reference-negative pairs but also the negative-positive relationship, thus improving its stability during training. It is defined as:

$$.\mathcal{L}_W(\mathbf{x}_R, \mathbf{x}_+, \mathbf{x}_-) = \max(0, ||\mathbf{o}_R - \mathbf{o}_+||^2 - 4\tan^2\alpha||\mathbf{o}_- - \mathbf{o}_c||^2) , \qquad (5.10)$$

where $\mathbf{o}_c = (\mathbf{o}_R + \mathbf{o}_+)/2$, and $\alpha > 0$ is a margin parameter defining an upper bound of the angle between the $\mathbf{o}_-$ and $\mathbf{o}_c$.

5.2.3.6   Deviance

Yi et al. [YLLL14] use the binomial deviance to define their loss function:

$$\mathcal{L}_W(\mathbf{x}_1, \mathbf{x}_2, Y) = \ln\left(\exp^{-2t(Y)\cos(\mathbf{o}_1, \mathbf{o}_2)} + 1\right) . \qquad (5.11)$$

5.2.3.7   Quadruplets

Yih et al. [YTPM11] consider two pairs of vectors, $(\mathbf{x}_{p1}, \mathbf{x}_{q1})$ and $(\mathbf{x}_{p2}, \mathbf{x}_{q2})$, the first being known to have a higher similarity than the second. The main objective is then to maximise

$$\Delta = \cos(\mathbf{o}_{x_{p1}}, \mathbf{o}_{x_{q1}}) - \cos(\mathbf{o}_{x_{p2}}, \mathbf{o}_{r_{q2}}) \qquad (5.12)$$

in a logistic loss function

$$\mathcal{L}_W(\Delta) = \log(1 + \exp(-\gamma\Delta)) , \qquad (5.13)$$

with γ being a scaling factor.

Chen et al. [CCZH17b] propose a deep metric learning approach based on quadruplets extending the classical triplet loss (Eq. 5.8) by introducing an additional constraint that enforces the distance between two negative examples to be larger then between two positive examples but with a different reference (anchor). The overall loss for N training examples is defined as:

$$\mathcal{L}_W = \sum_{i,j,k}^{N} \max(d_W(\mathbf{o}_i, \mathbf{o}_j)^2 - d_W(\mathbf{o}_i, \mathbf{o}_k)^2 + m_1, 0)$$

$$+ \sum_{i,j,k,l]}^{N} \max(d_W(\mathbf{o}_i, \mathbf{o}_j)^2 - d_W(\mathbf{o}_l, \mathbf{o}_k)^2 + m_2, 0) , \qquad (5.14)$$

where m_1 and m_2 are two margins, and the input triplets $(\mathbf{x}_i, \mathbf{x}_j, \mathbf{x}_k)$ and quadruplets $(\mathbf{x}_i, \mathbf{x}_j, \mathbf{x}_k, \mathbf{x}_l)$ are chosen under label constraints: $y_i = y_j$, $y_l \neq y_k$, $y_i \neq y_l$, $y_i \neq y_k$.

5.2.3.8 Tuples: Pair-Wise

The approach from Berlemont et al. [BLDG15] is based on tuples $\mathbf{T}$ composed of a reference example $\mathbf{x}_R$, a positive example $\mathbf{x}_+$ and K negative examples $\mathbf{x}_{-,k}$ ($k = 1..K$). Similar to the triplet loss (c.f. Eq. 5.9), with a tuple $\mathbf{T}$, a positive pair $(\mathbf{x}_R, \mathbf{x}_+)$ and K negative pairs are formed, with respective target values "1" and "0" for the cosine distance. Thus, given the respective outputs of the neural network $\mathbf{o}_R, \mathbf{o}_+, \mathbf{o}_{-,1}, \ldots, \mathbf{o}_{-,K}$, the loss for a tuple $\mathbf{T}$ is defined as follows:

$$\mathcal{L}_W(\mathbf{T}) = (1 - \mathbf{o}_R \cdot \mathbf{o}_+)^2 + \sum_{k=1}^{K}(0 - \mathbf{o}_R \cdot \mathbf{o}_{-,k})^2 + \sum_{\mathbf{x}_p \in \mathbf{T}} (1 - \|\mathbf{o}_p\|)^2 , \qquad (5.15)$$

where the cosine distances for the positive pair and the K negative pairs have been replaced by the scalar product as $\cos(\mathbf{o}_1, \mathbf{o}_2) = \frac{\mathbf{o}_1 \cdot \mathbf{o}_2}{\|\mathbf{o}_1\| \cdot \|\mathbf{o}_2\|}$, and the norms are enforced to be "1" in the term of the last sum.

A very similar approach has been proposed by Son [Soh16] based on the classical triplet formulation of Weinberger et al. (Eq. 5.8) using the so-called *N-pair loss*:

$$\mathcal{L}_W(\mathbf{T}) = \log \left(1 + \sum_{i=1}^{K} \exp \left(\mathbf{o}_R^\top \mathbf{o}_{-,k} - \mathbf{o}_R^\top \mathbf{o}_+ \right) \right) . \qquad (5.16)$$

Song et al. [SXJS16] proposed another pair-wise formulation based on several positive and several negative pairs, $\mathcal{P}$ and $\mathcal{N}$, respectively. For each positive pair $(\mathbf{x}_i, \mathbf{x}_j)$, only the closest negative examples (for either) have an influence during the training. The initial loss function involves a discontinuous *max* function that is replaced by a smooth upper bound defined by the "log-exp-sum" of the two terms for the positive pair, which leads to the final loss:

$$\mathcal{L}_W = \frac{1}{2|\mathcal{P}|} \sum_{(\mathbf{x}_i, \mathbf{x}_j) \in \mathcal{P}} \max(0, \mathcal{L}_W^{(i,j)})^2$$

$$\mathcal{L}_W^{(i,j)} = \log \left(\sum_{(\mathbf{x}_i, \mathbf{x}_k) \in \mathcal{N}} \exp(m - d(\mathbf{x}_i, \mathbf{x}_k)) + \sum_{(\mathbf{x}_j, \mathbf{x}_l) \in \mathcal{N}} \exp(m - d(\mathbf{x}_j, \mathbf{x}_l)) \right)$$
$$+ d(\mathbf{x}_i, \mathbf{x}_j), \qquad (5.17)$$

where $d(\mathbf{x}_i, \mathbf{x}_j) = ||\mathbf{x}_i - \mathbf{x}||_2$ is the Euclidean distance, and m is a margin.

Yang et al. [YZW18] proposed a similar "structural" loss function based on all positive and negative pairs, $\mathcal{P}$ and $\mathcal{N}$, in a tuple (mini-batch) $\mathbf{T}$:

$$\mathcal{L}_W(\mathbf{T}) = \frac{1}{B} \sum_{(\mathbf{x}_i, \mathbf{x}_j) \in \mathcal{P}} \beta_{ij} \log \left(1 + \sum_{(\mathbf{x}_i, \mathbf{x}_k) \in \mathcal{N}} \exp \left(d(\mathbf{x}_i, \mathbf{x}_j)^2 - d(\mathbf{x}_i, \mathbf{x}_k)^2 + m \right) / \xi \right)$$
$$+ \frac{\lambda}{2} (|\sigma_p^2 - m_p]_+ + [\sigma_n^2 - m_n]_+), \qquad (5.18)$$

where $\xi < 1$ is a constant making the triplet constraint "softer", m, m_p and m_n are margin parameters, and $[x]_+ = \max(0, x)$ is a hinge loss minimising the second-order statistics σ_p and σ_n of positive and negative pair distances. This second term serves as a regularisation (controlled by factor λ) preventing the variances of positive and negative distances from becoming too large. The factor β_{ij} (with $B = \sum_{(\mathbf{x}_i, \mathbf{x}_j) \in \mathcal{P}} \beta_{ij}$) weights each positive pair $(\mathbf{x}_i, \mathbf{x}_j)$ according to its "hardness", i.e. the distance in the Euclidean space:

$$\beta_{ij} = \exp(d(\mathbf{x}_i, \mathbf{x}_j)^2 - \tau_c), \qquad (5.19)$$

with τ_c being a class-dependent threshold, defined as twice the mean minus the minimum of positive distances in $\mathcal{P}$.

5.2.3.9 Tuples: Polar Sine

Later, Berlemont et al. [BLDG16, BLDG18] proposed a loss function based on tuples that simultaneously takes into account the relationship between *all* examples in the tuple, i.e. also between the negatives and not only pair-wise w.r.t. to a reference.

Inspired by the 2D sine function, Lerman et al. [LW09] define the polar sine for a set $V_m = \{\mathbf{v}_1, \dots, \mathbf{v}_n\}$ of m-dimensional linearly independent vectors ($m > n$) as a normalised hyper-volume. Given $\mathbf{A} = \begin{bmatrix} \mathbf{v}_1 \ \mathbf{v}_2 \ \cdots \ \mathbf{v}_m \end{bmatrix}$ and its transpose $\mathbf{A}^\top$:

$$PolarSine(\mathbf{v}_1, \ldots, \mathbf{v}_n) = \frac{\sqrt{\det\left(\mathbf{A}^\top \mathbf{A}\right)}}{\prod_{i=1}^n \|\mathbf{v}_i\|}. \tag{5.20}$$

In the special case where $m = n$, the matrix product in the determinant is replaced by the square matrix $\mathbf{A}$.

Given the matrix $\mathbf{S}$ such that $\forall(i, j) \in [1, .., n]^2$, $\mathbf{S}_{i,j} = \cos(\mathbf{v}_i, \mathbf{v}_j)$, this measure can be rewritten as $PolarSine(\mathbf{v}_1, \ldots, \mathbf{v}_n) = \sqrt{\det(\mathbf{S})}$. For numerical stability reasons during the derivation process, and to make this value independent from the number of classes, the authors define a polar sine metric as follows:

$$psine(\mathbf{v}_1, \ldots, \mathbf{v}_n) = \sqrt[n]{\det(\mathbf{S})}. \tag{5.21}$$

This metric only depends on the angles between every vector of the set. It reaches its maximum value when all the vectors are orthogonal, and thus can be used as a measure for dissimilarity. The loss function for a tuple $\mathbf{T}$ is defined as:

$$\mathcal{L}_W(\mathbf{T}) = \mathcal{L}_W^p(\mathbf{T}) + \mathcal{L}_W^n(\mathbf{T}) \, ,$$

$$\mathcal{L}_W^p(\mathbf{T}) = (1 - \cos(\mathbf{o}_R, \mathbf{o}_+))^2, \tag{5.22}$$

$$\mathcal{L}_W^n(\mathbf{T}) = (1 - psine(\mathbf{o}_R, \mathbf{o}_{-,1}, \ldots, \mathbf{o}_{-,K}))^2 \, .$$

Optimising the polar sine corresponds to assigning a target of 0 to the cosine value of every pair of outputs from different vectors drawn in $\mathbf{T} \setminus \{\mathbf{x}_R\}$, i.e. a target for every pair of dissimilar examples.

5.2.3.10 SoftTriple

The approach proposed by Qian et al. [QSS+19] uses a loss formulation based on the SoftMax function. They showed that the SoftMax loss, commonly used in supervised learning, is essentially equivalent to a smoothed triplet loss, where each class has a single "centre", and they extended this to multiple centres allowing to cope with large intra-class variations and underlying multi-modal distributions. The similarity metric is learnt in a supervised way with C classes and an additional fully-connected layer l with weight vectors $[\mathbf{w}_1, \ldots, \mathbf{w}_C] \in \mathbb{R}^{d \times C}$ before the final SoftMax layer. The algorithm minimises the following loss defined on a single input $\mathbf{x}_i$ to the layer l and its label y_i over the entire training set:

$$\mathcal{L}_W = - \log \frac{\exp(\lambda(S_{i,y_i} - \delta))}{\exp(\lambda(S_{i,y_i} - \delta)) + \sum_{j \neq y_i} \exp(\lambda S_{i,j})} \tag{5.23}$$

where all weights $\mathbf{w}_i$ and $\mathbf{x}_i$ have unit length, λ is a regularisation constant, δ is a margin, and

$$S_{i,c} = \sum_k \frac{\exp(\frac{1}{\gamma}\mathbf{x}_i^\top w_c^k)}{\sum_k \exp(\frac{1}{\gamma}\mathbf{x}_i^\top w_c^k)} \mathbf{x}_i^\top w_c^k \tag{5.24}$$

is a relaxed (hence *soft*) version of

$$S'_{i,c} = \max_k \mathbf{x}_i^\top w_c^k , \tag{5.25}$$

a similarity between the input $\mathbf{x}_i$ and the (closest) k-th centre (i.e. weight vector) of class c. The final embedding representing the similarity metric is thus produced at the output of layer l.

5.2.3.11 Sphere Loss

Similar to the SoftTriple loss, Fan et al. [FJLF19] proposed a similarity metric learning approach based on the SoftMax function, where, again, the embeddings $\mathbf{o}_i$ and weights $\mathbf{w}_i$ $i = 1..C$ of the last layer are set to unit norm and thus lie on a hyper-sphere. Here, the similarity is expressed in terms of the angle θ_j between the weight ("centre") $\mathbf{w}_j$ and an embedding o. And the loss for one example $\mathbf{x}_i$ with label $\mathbf{y}_i$ is defined as:

$$\mathcal{L}_W = -\log \frac{e^{s \cos\theta_{y_i}}}{\sum_{j=1}^{C} e^{s \cos\theta_j}} , \tag{5.26}$$

where s is a scaling factor. Thus, after training, the embedding is formed by the hyper-sphere manifold described by the last layer outputs, and similarity is measured in terms of the angular distance.

5.2.3.12 Probability-Driven

Nair et al. [NH10] add a final unit to their neural network architecture whose activation function computes the probability P of two examples $\mathbf{x}_1, \mathbf{x}_2$ being from the same class:

$$P = \frac{1}{1 + \exp(-(w.\cos(\mathbf{o}_1, \mathbf{o}_2) + b))} , \tag{5.27}$$

with w and b being scalar parameters.

5.2.3.13 Statistical

Chen et al. [CS11] compute the first and second-order statistics, $\mu^{(i)}$ and $\Sigma^{(i)}$, over sliding windows on the SNN outputs of a speech sample i, and define the loss function as:

$$\mathcal{L}_W(\mathbf{x}_1, \mathbf{x}_2, Y) = (1 - t(Y))(D_m + D_S) + t(Y).(\exp(\frac{-D_m}{\lambda_m}) + \exp(\frac{-D_S}{\lambda_S})) ,$$

(5.28)

where

$$D_m = \left\| \mu^{(i)} - \mu^{(j)} \right\|_2^2 , \quad D_S = \left\| \Sigma^{(i)} - \Sigma^{(j)} \right\|_F^2$$

(5.29)

are incompatibility measures of these statistics between two samples i and j, λ_m and λ_s are tolerance bounds on these measures, and $\|.\|_F$ is the Frobenius norm.

5.2.4 Training Algorithms and Schemes

In principle, any optimisation algorithm for neural networks can be used to train a SNN. In most similarity learning approaches from the literature, the standard Stochastic Gradient Descent (SGD) with mini-batches has been used. The batch size is sometimes an important hyper-parameter because it may have an influence on the overall convergence. Furthermore, as mentioned in Sect. 5.2.2, the selection of examples used for one training iteration might be pre-determined by the SNN architecture or loss function. For instance, for an SNN with a pair-wise loss function, e.g. contrastive divergence, different pairs can be formed with the examples in the batch, and it is often recommended to balance the possible number of positive and negative pairs. For triplets, it may be appropriate to have at least one positive/similar pair for each example. For tuples, there may need to be at least one example for each class. In any case, at each training iteration, the loss function (c.f. Sect. 5.2.3) is generally evaluated on the set of examples in the batch. And the gradients are computed for each input example and summed. Then the (shared) weights are updated according to the standard error backpropagation algorithm.

To circumvent the problem of exhaustive triplet sampling, Movshovitz-Attias et al. [MATL+17] introduced the notion of *proxies*, where a representative point (proxy) from a much smaller subset is assigned to each positive and negative training example in the triplets as an approximation. The proxies are updated during training, and the algorithm thus requires much less memory and further improves the convergence.

Another original similarity metric learning algorithm proposed by Duan et al. [DZL+18] uses a Generative Adversarial Network (GAN) model to generate synthetic hard negative examples for efficient learning of the metric space. The generator loss encourages the creation of realistic negative examples that are as close as possible to the positive ones and that violate the triplet constraint. The algorithm jointly minimises this loss with the metric loss, and the authors experimented with various loss functions: contrastive, triplet, N-pair tuples etc., and obtained improved image retrieval results using the N-pair loss (c.f. Eq. 5.16).

For large embeddings and complex, deep SNN models, the convergence can be difficult when learning only with similarity and dissimilarity constraints. For this reason, some approaches suggest to either pre-train the model in a supervised way using the labels of the training dataset and a temporary final fully-connected SoftMax layer [CS11] or to simultaneously train identities and the metric space using the combination of two (or more) loss functions [MdRM17, CCZH17b, YWLY18, MMLJ20].

Finally, when the metric that is to be learnt is applied to a ranking problem, like image retrieval for example, additional constraints on this ranking can be imposed. For example, the approach proposed by Chen et al. [CDS+18] integrates the classical mean average precision and rank-1 measures as a weighting term in the loss function and specifically selects mis-ranked triplets in the training batches in order to correct ranking errors. Another approach for deep metric learning to rank, called *FastAP*, has been presented by Cakir et al. [CHX+19]. Here, instead of sampling pairs or triplets of training examples, the training is directly optimising the mean average precision within mini-batches by computing the distance histograms between each query and sets of gallery examples in the batch. Using a differentiable soft-binning technique, gradient descent can directly be performed on these histograms of positive and negative retrieval results.

5.3   Conclusion

In this chapter, we gave an overview on the principal models and training algorithms for similarity metric learning with neural networks. Numerous variants of loss functions, neural architectures and training sample selection have been proposed in the literature, and these choices mostly depend on the given training data, the task to solve and the application domain. The general trend seems to go away from simple pair-wise training and triplets to samples involving tuples of several similar and dissimilar examples allowing to model richer structural relations, but these strategies are nowadays mostly based on application-dependant heuristics involving empirical parameters. More research work needs to be done to study the generalisation capacity and the general applicability (beyond person re-identification, for example) for more complex state-of-the-art deep metric learning models. In this regard, the interplay between classification, feature extraction and the learnt metric needs to be better understood. Finally, the notion of similarity (and dissimilarity) is to be defined more formally and extended, possibly by introducing other (prior) knowledge, in order design new models and algorithms that learn a rich, abstract and semantic representation of training data.

References

[ASS20] Mohammad Adiban, Hossein Sameti, and Saeedreza Shehnepoor. Replay spoofing countermeasure using autoencoder and siamese networks on ASVspoof 2019 challenge. *Computer Speech & Language*, 64, 2020.

[BGL+94] Jane Bromley, Isabelle Guyon, Yann Lecun, Eduard Säckinger, and Roopak Shah. Signature Verification using a "Siamese" Time Delay Neural Network. In *Proceedings of NIPS*, 1994.

[BHS13] A. Bellet, A. Habrard, and M. Sebban. A survey on metric learning for feature vectors and structured data. *Computing Research Repository*, abs/1306.6709, 2013.

[BLDG15] Samuel Berlemont, Gregoire Lefebvre, Stefan Duffner, and Christophe Garcia. Siamese neural network based similarity metric for inertial gesture classification and rejection. In *Proceedings of the International Conference on Automatic Face and Gesture Recognition (FG)*, Ljubljana, Slovenia, May 2015.

[BLDG16] Samuel Berlemont, Grégoire Lefebvre, Stefan Duffner, and Christophe Garcia. Polar sine based siamese neural network for gesture recognition. In *Proceedings of the International Conference on International Conference on Artificial Neural Networks (ICANN)*, Barcelona, Spain, 2016.

[BLDG18] Samuel Berlemont, Grégoire Lefebvre, Stefan Duffner, and Christophe Garcia. Class-balanced siamese neural networks. *Neurocomputing*, 273:47–56, 2018.

[BWCB11] Antoine Bordes, Jason Weston, Ronan Collobert, and Yoshua Bengio. Learning structured embeddings of knowledge bases. In *Conference on Artificial Intelligence*, 2011.

[CBB19] Ushasi Chaudhuri, Biplab Banerjee, and Avik Bhattacharya. Siamese graph convolutional network for content based remote sensing image retrieval. *Computer Vision and Image Understanding*, 184:22–30, 2019.

[CCZH17a] Weihua Chen, Xiaotang Chen, Jianguo Zhang, and Kaiqi Huang. Beyond triplet loss: a deep quadruplet network for person re-identification. In *Proceedings of the International Conference on Computer Vision and Pattern Recognition (CVPR)*, 2017.

[CCZH17b] Weihua Chen, Xiaotang Chen, Jianguo Zhang, and Kaiqi Huang. A multi-task deep network for person re-identification. In *Proceedings of the AAAI Conference on Artificial Intelligence*, 2017.

[CDS+18] Yiqiang Chen, Stefan Duffner, Andrei Stoian, Jean-Yves Dufour, and Atilla Baskurt. Similarity learning with listwise ranking for person re-identification. In *Proceedings of the International Conference on Image Processing (ICIP)*, 2018.

[CHL05] S. Chopra, R. Hadsell, and Y. LeCun. Learning a similarity metric discriminatively, with application to face verification. In *Proceedings of the International Conference on Computer Vision and Pattern Recognition (CVPR)*, volume 1, pages 539–546. IEEE, 2005.

[CHX+19] Fatih Cakır, Kun He, Xide Xia, Brian Kulis, and Stan Sclaroff. Deep metric learning to rank. In *Proceedings of the International Conference on Computer Vision and Pattern Recognition (CVPR)*, 2019.

[CLDG19] Paul Compagnon, Gregoire Lefebvre, Stefan Duffner, and Christophe Garcia. Routine modeling with time series metric learning. In *Proceedings of the International Conference on International Conference on Artificial Neural Networks (ICANN)*, September 2019.

[CS11] Ke Chen and Ahmad Salman. Extracting Speaker-Specific Information with a Regularized Siamese Deep Network. In *Proceedings of Advances in Neural Information Processing Systems (NIPS)*, pages 298–306, 2011.

[CSSB10] G. Chechik, V. Sharma, U. Shalit, and S. Bengio. Large scale online learning of image similarity through ranking. *Journal of Machine Learning Research*, 11:1109–1135, 2010.

[DZL+18] Yueqi Duan, Wenzhao Zheng, Xudong Lin, Jiwen Lu, and Jie Zhou. Deep adversarial metric learning. In *Proceedings of the International Conference on Computer Vision and Pattern Recognition (CVPR)*, 2018.

[FJLF19] Xing Fan, Wei Jiang, Hao Luo, and Mengjuan Fei. SphereReID: Deep hypersphere manifold embedding for person re-identification. *Journal of Visual Communication and Image Representation*, 60:51–58, 2019.

[GHDS18] Weifeng Ge, Weilin Huang, Dengke Dong, and Matthew R. Scott. Deep metric learning with hierarchical triplet loss. In *Proceedings of the European Conference on Computer Vision (ECCV)*, 2018.

[HBL17] Alexander Hermans, Lucas Beyer, and Bastian Leibe. In Defense of the Triplet Loss for Person Re-Identification. *arXiv preprint arXiv:1703.07737*, 2017.

[HCL06] Raia Hadsell, Sumit Chopra, and Yann LeCun. Dimensionality reduction by learning an invariant mapping. In *Proceedings of the International Conference on Computer Vision and Pattern Recognition (CVPR)*, pages 1735–1742, 2006.

[HL11] N. V. Hieu and B. Li. Cosine similarity metric learning for face verification. In *Proceedings of the Asian Conference on Computer Vision (ACCV)*, pages 709–720. Springer, 2011.

[HLT14] J. Hu, J. Lu, and Y.-P. Tan. Discriminative deep metric learning for face verification in the wild. In *Proceedings of the International Conference on Computer Vision and Pattern Recognition (CVPR)*, pages 1875–1882, 2014.

[HSW89] Kurt Hornik, Maxwell Stinchcombe, and Halbert White. Multilayer feedforward networks are universal approximators. *Neural Networks*, 2(5):359–366, 1989.

[KBR16] Theofanis Karaletsos, Serge Belongie, and Gunnar Rätsch. Bayesian representation learning with oracle constraints. In *International Conference on Learning Representations (ICLR)*, 2016.

[KTS+12] D. Kedem, S. Tyree, F. Sha, G. R. Lanckriet, and K. Q. Weinberger. Non-linear metric learning. In *Proceedings of Advances in Neural Information Processing Systems (NIPS)*, pages 2573–2581, 2012.

[LBOM12] Y. A. LeCun, L. Bottou, G. B. Orr, and K.-R. Müller. Efficient backprop. In *Neural networks: Tricks of the trade*, pages 9–48. Springer, 2012.

[LG13] Grégoire Lefebvre and Christophe Garcia. Learning a bag of features based nonlinear metric for facial similarity. In *Proceedings of the International Conference on Advanced Video and Signal-Based Surveillance (AVSS)*, pages 238–243, 2013.

[LGD+19] Yujia Li, Chenjie Gu, Thomas Dullien, Oriol Vinyals, and Pushmeet Kohli. Graph matching networks for learning the similarity of graph structured objects. In *Proceedings of the International Conference on Machine Learning (ICML)*, 2019.

[LW09] Gilad Lerman and J. Tyler Whitehouse. On D-dimensional D-semimetrics and simplex-type inequalities for high-dimensional sine functions. *Journal of Approximation Theory*, 156(1):52–81, January 2009.

[MATL+17] Yair Movshovitz-Attias, Alexander Toshev, Thomas K. Leung, Sergey Ioffe, and Saurabh Singh. No fuss distance metric learning using proxies. In *Proceedings of the International Conference on Computer Vision (ICCV)*, 2017.

[MBBS14] Jonathan Masci, Michael M. Bronstein, Alexander M. Bronstein, and Jurgen Schmidhuber. Multimodal Similarity-Preserving Hashing. *IEEE Transactions on Pattern Analysis and Machine Intelligence*, 36(4):824–830, April 2014.

[MdRM17] Niall McLaughlin, Jesus Martinez del Rincon, and Paul C Miller. Person reidentification using deep convnets with multitask learning. *IEEE Transactions on Circuits and Systems for Video Technology*, 27(3):525–539, 2017.

[MLK16] Panagiotis Moutafis, Mengjun Leng, and Ioannis A Kakadiaris. An overview and empirical comparison of distance metric learning methods. *IEEE Transactions on Cybernetics*, 2016.

[MMLJ20] Weiqing Min, Shuhuan Mei, Zhuo Li, and Shuqiang Jiang. A two-stage triplet network training framework for image retrieval. *IEEE Transactions on Multimedia*, 2020.

[MT16] J. Mueller and A. Thyagarajan. Siamese recurrent architectures for learning sentence similarity. In *Proceedings of the AAAI Conference on Artificial Intelligence*, 2016.

[NH10] Vinod Nair and Geoffrey E. Hinton. Rectified Linear Units Improve Restricted Boltzmann Machines. In *Proceedings of the International Conference on Machine Learning (ICML)*, pages 807–814, 2010.

[NVR16] Paul Neculoiu, Maarten Versteegh, and Mihai Rotaru. Learning Text Similarity with Siamese Recurrent Networks. In *Proceedings of the 1st Workshop on Representation Learning for NLP*, pages 148–157, August 2016.

[QGCL08] A. M. Qamar, E. Gaussier, J. P. Chevallet, and J. H. Lim. Similarity learning for nearest neighbor classification. In *Proceedings of the International Conference on Data Mining (ICDM)*, pages 983–988. IEEE, 2008.

[QSS+19] Qi Qian, Lei Shang, Baigui Sun, Juhua Hu, Hao Li, and Rong Jin. SoftTriple loss: Deep metric learning without triplet sampling. In *Proceedings of the International Conference on Computer Vision (ICCV)*, 2019.

[SCWT14] Y. Sun, Y. Chen, X. Wang, and X. Tang. Deep learning face representation by joint identification-verification. In *Proceedings of Advances in Neural Information Processing Systems (NIPS)*, pages 1988–1996, 2014.

[SKP15] Florian Schroff, Dmitry Kalenichenko, and James Philbin. Facenet: A unified embedding for face recognition and clustering. In *Proceedings of the International Conference on Computer Vision and Pattern Recognition (CVPR)*, pages 815–823, 2015.

[SL20] Weijie Sheng and Xinde Li. Siamese denoising autoencoders for joints trajectories reconstruction and robust gait recognition. *Neurocomputing*, 395:86–94, 2020.

[Soh16] Kihyuk Sohn. Improved deep metric learning with multi-class N-pair loss objective. In *Proceedings of Advances in Neural Information Processing Systems (NIPS)*, 2016.

[SXJS16] Hyun Oh Song, Yu Xiang, Stefanie Jegelka, and Silvio Savarese. Deep metric learning via lifted structured feature embedding. In *Proceedings of the International Conference on Computer Vision and Pattern Recognition (CVPR)*, 2016.

[SYZ+16] Hailin Shi, Yang Yang, Xiangyu Zhu, Shengcai Liao, Zhen Lei, Weishi Zheng, and Stan Z. Li. Embedding deep metric for person re-identification: A study against large variations. In *Proceedings of the European Conference on Computer Vision (ECCV)*, 2016.

[WBS06] K. Weinberger, J. Blitzer, and L. Saul. Distance metric learning for large margin nearest neighbor classification. In *Proceedings of Advances in Neural Information Processing Systems (NIPS)*, volume 18, page 1473, 2006.

[WZL17] Chong Wang, Xue Zhang, and Xipeng Lan. How to train triplet networks with 100k identities? In *Proceedings of the International Conference on Computer Vision (ICCV)*, 2017.

[WZW+17] J. Wang, F. Zhou, S. Wen, X. Liu, and Y. Lin. Deep metric learning with angular loss. In *Proceedings of the International Conference on Computer Vision (ICCV)*, 2017.

[XNJR03] E. P. Xing, A. Y. Ng, M. I. Jordan, and S. Russell. Distance metric learning with application to clustering with side-information. In *Proceedings of Advances in Neural Information Processing Systems (NIPS)*, pages 521–528. MIT; 1998, 2003.

[YCS19] Yao Yang, Haoran Chen, and Junming Shao. Triplet enhanced autoencoder: Model-free discriminative network embedding. In *Proceedings of the International Joint Conference on Artificial Intelligence (IJCAI)*, 2019.

[YLLL14] Dong Yi, Zhen Lei, Shengcai Liao, and Stan Z. Li. Deep metric learning for person re-identification. In *Proceedings of International Conference on Pattern Recognition (ICPR)*, pages 34–39, 2014.

[YT19]   Baosheng Yu and Dacheng Tao. Deep metric learning with tuplet margin loss. In *Proceedings of the International Conference on Computer Vision (ICCV)*, 2019.

[YTPM11] Wen-tau Yih, Kristina Toutanova, John C. Platt, and Christopher Meek. Learning discriminative projections for text similarity measures. In *Proceedings of the Fifteenth Conference on Computational Natural Language Learning*, pages 247–256. Association for Computational Linguistics, 2011.

[YWLY18] Mang Ye, Zheng Wang, Xiangyuan Lan, and Pong C. Yuen. Visible thermal person re-identification via dual-constrained top-ranking. In *Proceedings of the International Joint Conference on Artificial Intelligence (IJCAI)*, 2018.

[YYGT16] Jun Yu, Xiaokang Yang, Fei Gao, and Dacheng Tao. Deep multimodal distance metric learning using click constraints for image ranking. *IEEE Transactions on Cybernetics*, 2016.

[YZW18] Xun Yang, Peicheng Zhou, and Meng Wang. Person reidentification via structural deep metric learning. *IEEE Transactions on Neural Networks and Learning Systems*, 30(10), 2018.

[ZDI+18] Lilei Zheng, Stefan Duffner, Khalid Idrissi, Christophe Garcia, and Atilla Baskurt. Pairwise identity verification via linear concentrative metric learning. *IEEE Transactions on Cybernetics*, 48(1):324–335, 2018.

[ZIG+15] Lilei Zheng, Khalid Idrissi, Christophe Garcia, Stefan Duffner, and Atilla Baskurt. Logistic similarity metric learning for face verification. In *Proceedings of the International Conference on Acoustics, Speech, and Signal Processing (ICASSP)*. IEEE, 2015.

Chapter 6
Zero-Shot Learning with Deep Neural Networks for Object Recognition

Yannick Le Cacheux, Hervé Le Borgne, and Michel Crucianu

6.1 Introduction

The core problem of supervised learning lies in the ability to generalize the prediction of a model learned on some samples *seen* in the training set to other *unseen* samples in the test set. A key hypothesis is that the samples of the training set allow a fair estimation of the distribution of the test set, since both result from the same independent and identically distributed random variables. Beyond the practical issues linked to the exhaustiveness of the training samples, such a paradigm is not adequate for all needs, nor reflects the way humans seem to learn and generalize. Despite the fact that, to our knowledge, nobody has seen a real dragon, unicorn or any beast of the classical fantasy, one could easily recognize some of them if met. Actually, from the single textual description of these creatures, and inferring from the knowledge of the real wildlife, there exist many drawings and other visual representations of them in the entertainment industry.

Zero-shot learning (ZSL) addresses the problem of recognizing categories of the test set that are not present in the training set [LEB08, LNH09, PPHM09, FEHF09]. The categories used at training time are called *seen* and those at testing time are *unseen*, and contrary to classical supervised learning, not any sample of unseen categories is available during training. To compensate this lack of information, each category is nevertheless described semantically either with a list of attributes, a set of words or sentences in natural language. The general idea of ZSL is thus to learn some intermediate features from training data, that can be used during the test to

Y. Le Cacheux · H. Le Borgne (✉)
Université Paris-Saclay, CEA,List, F-91120, Palaiseau, France
e-mail: yannick.lecacheux@cea.fr; herve.le-borgne@cea.fr

M. Crucianu
CNAM Case courrier 2D4P30, Paris Cedex 03, France
e-mail: michel.crucianu@cnam.fr

© Springer Nature Switzerland AG 2021
J. Benois-Pineau, A. Zemmari (eds.), *Multi-faceted Deep Learning*,
https://doi.org/10.1007/978-3-030-74478-6_6

map the sample to the unseen classes. These intermediate features can reflect the colors or textures (*fur, feathers, snow, sand...*) or even some part of objects (*paws, claws, eyes, ears, trunk, leaf...*). Since such features are likely to be present in both seen and unseen categories, and one can expect to infer a discriminative description of more complex concepts from them (e.g. some types of animals, trees, flowers...), the problem becomes tractable.

6.2 Formalism, Settings and Evaluation

6.2.1 Standard ZSL Setting

Formally, let us note the set of seen classes $\mathcal{C}^S$ and that of unseen classes $\mathcal{C}^{\mathcal{U}}$. The set of all classes is $\mathcal{C} = \mathcal{C}^S \cup \mathcal{C}^{\mathcal{U}}$, with $\mathcal{C}^S \cap \mathcal{C}^{\mathcal{U}} = \emptyset$.

For each class $c \in \mathcal{C}$, *semantic information* is provided. It can consist of binary attributes, such as "has stripes", "is orange" and "has hooves". With this example, the semantic representation of the class *tiger* would be $(1\ 1\ 0)^\top$, while the representation of class *zebra* would be $(1\ 0\ 1)^\top$. For a given class c, we write its corresponding semantic representation vector $\mathbf{s}_c$; such a vector is also called the *class prototype*. The prototypes of all classes have the same dimension K, and represent the same attributes. More generally, the semantic information does not have to consist of binary attributes, and may not correspond to attributes at all. More details on the most common types of prototypes are provided in Sect. 6.4.

For each class c, a set of images is available. One can extract a feature vector $\mathbf{x}_i \in \mathbb{R}^D$ from an image, usually using a pre-trained deep neural network, for example based on the VGG [SZ14] or ResNet [HZRS16] architectures mentioned in Chap. 2. In the latter case, the feature vector of an image corresponds to the internal representation in the network after the last max-pooling layer, before the last fully-connected layer. It is of course also possible to train a deep network from scratch on the available training images. In the following, we will refer to an image, a sample of a class or its feature vector with this unique notation $\mathbf{x}$.

During the training phase, the model has only access to the semantic representations of seen classes $\{\mathbf{s}_c\}_{c \in \mathcal{C}^S}$ and to N images belonging to these classes. Hence, the training dataset is $\mathcal{D}^{\text{tr}} = (\{(\mathbf{x}_n, y_n)\}_{n \in [\![1,N]\!]}, \{\mathbf{s}_c\}_{c \in \mathcal{C}^S})$, where $y_n \in \mathcal{C}^S$ is the label of the nth training sample. During the testing phase, the model has access to the semantic representations of unseen classes $\{\mathbf{s}_{c'}\}_{c' \in \mathcal{C}^{\mathcal{U}}}$, and to the N' unlabeled images belonging to unseen classes $\{\mathbf{x}_{n'}\}_{n' \in [\![1,N']\!]}$. The objective for the model is to make a prediction $\hat{y}_{n'} \in \mathcal{C}^{\mathcal{U}}$ for each test image $\mathbf{x}_{n'}^{\text{te}}$, assigning it to the most likely unseen class.

As a first basic example, a simple ZSL model may consist in simply predicting attributes $\hat{\mathbf{s}}$ corresponding to an image $\mathbf{x}$ such that $\hat{\mathbf{s}} = \mathbf{w}^\top \mathbf{x}$; the parameters $\mathbf{w}$ can be estimated on the training set $\mathcal{D}^{\text{tr}}$ using a least square loss. To make predictions, we can simply predict the unseen class c whose attributes $\mathbf{s}_c$ are closest to the estimated

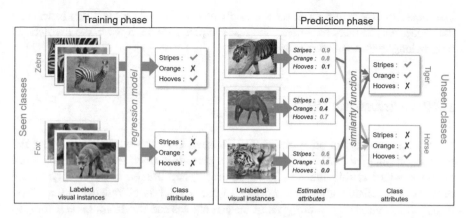

Fig. 6.1 Illustration of a basic ZSL model, with two seen classes *fox* and *zebra*, and two unseen classes *horse* and *tiger*. Each class is represented by a 3-dimensional semantic prototype corresponding to attributes "has stripes", "is orange" and "has hooves". During the training phase, the model learns the relations between the visual features and the attributes using the seen classes. During the evaluation phase, the model estimates the attributes for each test image and predicts the unseen class having the closest prototype

attributes $\hat{\mathbf{s}}$ as measured by a euclidean distance. Such a model is illustrated in Fig. 6.1, and will be presented in more details in Sect. 6.3.1.

More generally, most ZSL methods in the literature are based on a *compatibility function* $f : \mathbb{R}^D \times \mathbb{R}^K \rightarrow \mathbb{R}$ assigning a "compatibility" score $f(\mathbf{x}, \mathbf{s})$ to a pair composed of a visual sample $\mathbf{x} \in \mathbb{R}^D$ and a semantic prototype $\mathbf{s} \in \mathbb{R}^K$, that reflects the likelihood that $\mathbf{x}$ belongs to class c (if $\mathbf{s}$ is $\mathbf{s}_c$). This function may be parameterized by a vector $\mathbf{w}$ or a matrix $\mathbf{W}$, or by a set of parameters $\{\mathbf{w}_i\}_i$, leading to the notation $f_\mathbf{w}(\mathbf{x}, \mathbf{s})$ or $f(\mathbf{x}, \mathbf{s}; \{\mathbf{w}_i\}_i)$ in the following. These parameters are generally learned by selecting a suitable loss function $\mathcal{L}$ and minimizing the total training loss $\mathcal{L}_{\text{tr}}$ over the training dataset $\mathcal{D}^{\text{tr}}$ with respect to the parameters $\mathbf{w}$:

$$\mathcal{L}_{\text{tr}}(\mathcal{D}^{\text{tr}}) = \frac{1}{N} \sum_{n=1}^{N} \sum_{c \in \mathcal{C}^S} \mathcal{L}[(\mathbf{x}_n, y_n, \mathbf{s}_c), f_\mathbf{w}] + \lambda \Omega[f_\mathbf{w}] \tag{6.1}$$

where $\Omega[f]$ is a regularization penalty based on f and weighted by λ. Once the model learned, the predicted label $\hat{y}$ of a test image $\mathbf{x}$ can be selected among candidate testing classes based on their semantic representations $\{\mathbf{s}_c\}_{c \in \mathcal{C}^U}$:

$$\hat{y} = \underset{c \in \mathcal{C}^U}{\text{argmax}} \ f_\mathbf{w}(\mathbf{x}, \mathbf{s}_c) \tag{6.2}$$

In the standard setting of ZSL, the only data available during the training phase consists of the class prototypes of the *seen* classes and the corresponding labeled visual samples. The class prototypes of *unseen* classes, as well as the unlabeled instances from these unseen classes, are only provided during the testing phase, after the model was trained. Moreover, the test samples for which we make predictions only belong to these unseen classes.

6.2.2 Alternative ZSL Settings

When class prototypes of both seen *and* unseen classes are available during the training phase, [WZYM19] considers it as a *class-transductive* setting, as opposed to the standard setting that is *class-inductive*, when unseen class prototypes are only made available after the training of the model is completed. In a class-transductive setting, the prototypes of unseen classes can for example be leveraged by a generative model, which attempts to synthesize images of objects from unseen classes based on their semantic description (Sect. 6.3.3). They can also simply be used during training to ensure that the model does not misclassify a sample from a seen class as a sample from an unseen class. An access to this information as early as the training phase may be legitimate for some use-cases but new classes cannot be added as seamlessly as in a class-inductive setting, in which a new class can be introduced by simply providing its semantic representation (without any retraining).

A more permissive setting allows to consider that *unlabeled* instances of unseen classes are available during training. Such a setting is called *instance-transductive* in [WZYM19], as opposed to the *instance-inductive* setting. These two settings are often simply referred to as respectively *transductive* and *inductive*, even though there is some ambiguity on whether the (instance-)inductive setting designates a class-inductive or a class-transductive setting. Some methods use approaches which specifically take advantage of the availability of these unlabeled images, for example by extracting additional information on the geometry of the visual manifold [FYH+15]. Even though models operating in and taking advantage of a transductive setting can often achieve better accuracy than models designed for an inductive setting, one can argue that such a setting is not suitable for many real-life use cases. With a few exceptions [LG15], most transductive approaches consider that the actual (unlabeled) testing instances are available during the training phase, which excludes many practical applications. Even without this strong assumption, it is not always reasonable to expect to have access to unlabeled samples from many unseen classes during the training phase. One may further argue that this is all the more unrealistic as there is some evidence [XSSA19] that labeling even a single instance per class (in a "one-shot learning" scenario) can lead to a significant improvement in accuracy over a standard ZSL scenario.

In some settings, the available information itself can be different from the default setting. For example, in addition to the semantic prototypes, some methods make use of relations between classes defined with a graph [WYG18, KCL+19] or a hierarchical structure [RSS11]. Others make use of information regarding the environment of the object, for example by detecting surrounding objects [ZBS+19] or by computing co-occurrence statistics using an additional multilabel dataset [MGS14].

Other methods consider that instead of a semantic representation per class, a semantic representation *per image* is available, for example in the form of text descriptions [RALS16] or human gaze information [KASB17].

Another classification of ZSL settings is concerned with which classes have to be recognized during the testing phase. Indeed, one may legitimately want to recognize both seen and unseen classes. The setting in which testing instances may belong to both seen and unseen classes is usually called *generalized zero-shot learning* (GZSL) and has been introduced by [CCGS16b]. Approaches to extend ZSL to GZSL can be divided into roughly two categories: (1) approaches which explicitly try to identify when a sample does not belong to a seen class, and use either a standard classifier or a ZSL method depending on the result, and (2) approaches that employ a unified framework for both seen and unseen classes.

In [SGMN13], the authors explicitly estimate the probability $g_u(\mathbf{x}) = P(y \in \mathcal{C}^{\mathcal{U}}|\mathbf{x})$ that a test instance $\mathbf{x}$ belongs to an unseen class $c \in \mathcal{C}^{\mathcal{U}}$. They first estimate the class-conditional probability density $p(\mathbf{x}|c)$ for all seen classes $c \in \mathcal{C}^{\mathcal{S}}$, by assuming the projections $\hat{\mathbf{s}}(\mathbf{x})$ of visual features in the semantic space are normally distributed around the semantic prototype $\mathbf{s}_c$. We can then consider that an instance $\mathbf{x}$ does not belong to a seen class if its class-conditional probability is below a threshold γ for all seen classes:

$$g_u(\mathbf{x}) = \mathbb{1}[\forall c \in \mathcal{C}^{\mathcal{S}}, \ p(\mathbf{x}|c) < \gamma] \tag{6.3}$$

If one sees the compatibility $f(\mathbf{x}, \mathbf{s}_c)$ as the probability that the label of visual instance $\mathbf{x}$ is c, i.e. $P(y = c|\mathbf{x}) \propto f(\mathbf{x}, \mathbf{s}_c)$, the compatibilities of seen and unseen classes can be weighted by the estimated probabilities that $\mathbf{x}$ belongs to a seen or unseen class.

Most recent GZSL methods [VAMR18, XLSA18b, CCGS20] adopt a more direct approach: the unweighted compatibility function f is used to directly estimate compatibilities of seen and unseen classes, so that we simply have

$$\hat{y} = \underset{c \in \mathcal{C}^{\mathcal{S}} \cup \mathcal{C}^{\mathcal{U}}}{\mathrm{argmax}} \ f(\mathbf{x}, \mathbf{s}_c) \tag{6.4}$$

This approach has the advantage that using a trained ZSL model in a GZSL setting is straightforward, as all there is to do is adding the seen class prototypes to the list of prototypes whose compatibility with $\mathbf{x}$ needs to be evaluated. However, it has been empirically demonstrated [CCGS16b, XSA17] that many ZSL models suffer from a bias towards seen classes. With the example of Fig. 6.1, many models would thus tend to consider zebras as "weird" horses rather than members of a new, unseen class. To address this problem, a straightforward solution consists in penalizing seen classes to the benefit of unseen classes by decreasing the compatibility of the former by a constant value γ, similarly to Eq. 6.5. In [LCLBC19b] was put forward a simple method to select a suitable value of γ based on a training-validation-testing split specific to GZSL, which enabled a slight reduction in the accuracy on seen classes to result in a large improvement of the accuracy on unseen classes, thus significantly improving the GZSL score of any model.

Other even less restrictive tasks may be considered during the testing or application phase. For instance, one may want a model able to answer that a visual instance matches neither a seen nor an unseen class. Or one may aim to recognize entities that belong to several non-exclusive categories, a setting known as multilabel ZSL [MGS14, FYH+15, LFYFW18]. Other works are interested in the ZSL setting applied to other tasks such as object detection [BSS+18] or semantic segmentation [XCH+19].

6.2.3   ZSL Evaluation

Most of the (G)ZSL works to date address a classification task on mutually exclusive classes, thus the performance is evaluated with a classification rate. The standard accuracy nevertheless computes the score *per sample* (micro-average accuracy). Although many publicly available ZSL datasets [WBW+11, LNH14] have well-balanced classes, other datasets or use cases do not necessarily exhibit this property. [XSA17] therefore proposed to compute the score *per class* (macro-average accuracy) and most recent works adopted this metric.

For GZSL the performance measure is a more subtle issue. Of course, using $y_n \in \mathcal{C}^S \cup \mathcal{C}^U$ for each of the N testing instances, the micro and macro average accuracy can still be employed. However, this does not always provide the full picture regarding the performance of a (G)ZSL model: assuming per class accuracy is used and 80% of classes are seen classes, a perfect supervised model could achieve 80% accuracy with absolutely no ZSL abilities. This is all the more important as many GZSL models suffer from a bias towards seen classes, as mentioned previously.

To take the trade-off between seen and unseen classes into account, performance is often measured separately on each type of classes. Chao et al. [CCGS16b] defined $\mathcal{A}_{\mathcal{U} \rightarrow \mathcal{U}}$ as the (per class) accuracy evaluated only on test instances of unseen classes when candidate classes are the unseen classes $\mathcal{C}^U$. Also, $\mathcal{A}_{\mathcal{U} \rightarrow \mathcal{C}}$ is the accuracy evaluated on test instances of unseen classes when candidate classes are *all* classes $\mathcal{C}$, seen and unseen. Then $\mathcal{A}_{\mathcal{S} \rightarrow \mathcal{S}}$ and $\mathcal{A}_{\mathcal{S} \rightarrow \mathcal{C}}$ are defined correspondingly. Before the GZSL setting, test classes were all unseen classes so the (per-class) accuracy was $\mathcal{A}_{\mathcal{U} \rightarrow \mathcal{U}}$. $\mathcal{A}_{\mathcal{S} \rightarrow \mathcal{S}}$ corresponds to what is measured in a standard supervised learning setting. $\mathcal{A}_{\mathcal{C} \rightarrow \mathcal{C}}$ would correspond to the standard per class accuracy in a GZSL setting. $\mathcal{A}_{\mathcal{U} \rightarrow \mathcal{C}}$ and $\mathcal{A}_{\mathcal{S} \rightarrow \mathcal{C}}$ respectively measure how well a GZSL model is performing on respectively seen and unseen classes. [XSA17] proposes to use the harmonic mean as a trade-off between the two, to penalize models with a high score in one of these two sub-tasks but low performance in the other. This measure is the most commonly employed in the recent GZSL literature [CZX+18, VAMR18, LCLBC19a, MYX+20]. It can be noted that this metric requires to keep some instances from seen classes for the testing phase for a given ZSL benchmark dataset. When this is not convenient, for instance if the number of training samples per class is really small or datasets suffer from biases (see Sect. 6.2.4), sometimes only $\mathcal{A}_{\mathcal{U}\rightarrow\mathcal{C}}$ is evaluated [HAT19] in order to still provide some measure of GZSL performance. Alternatively, [CCGS16b] introduced *calibrated stacking*, where a weight γ is used as a trade-off between favoring $\mathcal{A}_{\mathcal{U}\rightarrow\mathcal{C}}$ (when $\gamma > 0$) and $\mathcal{A}_{\mathcal{S}\rightarrow\mathcal{C}}$ (when $\gamma < 0$):

$$\hat{y} = \underset{c\in\mathcal{C}}{\arg\max}\; f(\mathbf{x}, \mathbf{s}_c) - \gamma \mathbb{1}[c \in \mathcal{C}^{\mathcal{S}}] \tag{6.5}$$

[CCGS16b] defined the Area Under Seen-Unseen accuracy Curve (AUSUC) as the area under the curve of the plot with $\mathcal{A}_{\mathcal{U}\rightarrow\mathcal{C}}$ on the x-axis and $\mathcal{A}_{\mathcal{S}\rightarrow\mathcal{C}}$ on the y-axis, when γ goes from $-\infty$ to $+\infty$. Similarly to the area under a receiver operating characteristic curve, the AUSUC can be used as a metric to evaluate the performance of a GZSL model.

6.2.4 Standard ZSL Datasets and Evaluation Biases

We briefly describe a few datasets commonly used to benchmark ZSL models, provide the rough accuracy obtained on these datasets by mid-2020 and mention a few common biases to avoid when measuring ZSL accuracy on such benchmarks. Some examples of typical images from these datasets are shown in Fig. 6.2. The dataset list is by no means exhaustive, as many other ZSL evaluation datasets can be found in the literature.

Animals with Attributes or **AwA** [LNH14] is one of the first proposed benchmarks for ZSL [LNH09]; it has recently been replaced by the very similar **AwA2** [XLSA18a] due to copyright issues on some images. It consists of 37,322 images of 50 animal species such as *antelope, grizzly bear* or *dolphin*, 10 of which

Fig. 6.2 Images from the standard ZSL benchmarks AwA *(top)*, CUB *(middle)* and ImageNet *(bottom)*

are being used as unseen test classes, the rest being seen training classes. Class prototypes have 85 binary attributes such as *brown*, *stripes*, *hairless* or *claws*. As mentioned in Sect. 6.2.1, visual features are typically extracted from images using a deep network such as ResNet pre-trained on a generic dataset like ImageNet. As evidenced in [XSA17], this can induce an important bias on the AwA2 dataset. Indeed, 6 of the 10 unseen test classes are among the 1000 classes of ImageNet used to train the ResNet model; thus, such classes cannot be considered as truly "unseen". In [XSA17] it is therefore proposed to employ a different train/test split, called the *proposed split*, such that no unseen (test) class is present among the 1000 ResNet training classes. This setting has been widely adopted by the ZSL community. Recent ZSL models in a standard ZSL setting can reach an accuracy of around 71% [XSSA19] on the 10 test classes of this proposed split.

Caltech UCSD Birds 200–2011 or **CUB** [WBW+11] is referred to as a "fine-grained" dataset, as its 200 classes all correspond to bird species (*black footed albatross*, *rusty blackbird*, *eastern towhee*. . .) and are considered to be fairly similar (Fig. 6.2). Fifty classes are used as unseen testing classes; similarly to AwA2, the standard train/test split has been proposed in [XSA17]. The class prototypes consist of 312 usually continuous attributes with values between 0 and 1. Examples of attributes include "has crown color blue", "has nape color white" or "has bill shape cone". Recent models can reach a ZSL accuracy of around 64% [LCLBC19a] on the 50 test classes.

The **ImageNet** [DDS+09] dataset has also been used as a large-scale ZSL benchmark [RSS11]. Contrary to AwA or CUB, the usual semantic prototypes do not consist of attributes but rather of word embeddings of the class names—more details are provided in Sect. 6.4. This dataset contains classes as diverse as *coyote*, *goldfish*, *lipstick* or *speedboat*. The training classes usually consist of the 1000 classes of the ILSVRC challenge [RDS+15]. In the past, the approximately 20,000 remaining classes were used as unseen test classes. However, [HAT19] recently showed that this induces a bias, in part due to the fact that unseen classes are often subcategories or supercategories of seen classes. The authors suggested instead to use only a subset of 500 of the total unseen classes such that they do not exhibit this problem. The best ZSL models in [HAT19] can reach an accuracy of around 14% on these 500 test classes; the fact that this accuracy is significantly lower than on the other two datasets can be attributed to the much larger number of classes, but also to the lower quality of the semantic prototypes (Sect. 6.4).

6.3 Methods

There exist several surveys of the ZSL literature, each with its own classification of existing approaches [XLSA18a, FXJ+18, WZYM19]. Here we separate the state of the art into three main categories: *regression methods* (Sect. 6.3.1), *ranking methods* (Sect. 6.3.2) and *generative methods* (Sect. 6.3.3). We start by presenting the most simple methods which can be considered as baselines. Sometimes, the methods described below are slightly different from their initial formulation in the original articles, for the sake of brevity and simplicity. We aim at giving a general overview with the strengths, weaknesses and underlying hypotheses of these types of methods, not to dive deep into specific implementation details. As well, we mainly address the GZSL and standard ZSL settings, since they are the most easily applicable to real use-cases.

The **Direct Attribute Prediction** or **DAP** [LNH09] approach consists in training K standard classifiers which provide the probability $P(a_k|\mathbf{x})$ that attribute a_k is present in visual input $\mathbf{x}$. At test time, we predict the class c which maximizes the probability to have attributes corresponding to its class prototype $\mathbf{s}_c$. Assuming deterministic binary attributes, identical class priors $P(c)$, uniform attribute priors and independence of attributes, we have:

$$\underset{c\in\mathcal{C}^{\mathcal{U}}}{\operatorname{argmax}} P(c|\mathbf{x}) = \underset{c\in\mathcal{C}^{\mathcal{U}}}{\operatorname{argmax}} \prod_{k=1}^{K} P(a_k = (\mathbf{s}_c)_k|\mathbf{x}) \qquad (6.6)$$

Similar results may be obtained with continuous attributes by using probability density functions and regressors instead of classifiers. The **Indirect Attribute Prediction** or **IAP** was also proposed in [LNH09] and is very close to DAP. A notable difference is that it does not require any model training beyond a standard multi-class classifier on seen classes, and in particular does not require any training related to the attributes. As such, it enables to seamlessly convert any pre-trained standard supervised classification model to a ZSL setting provided a semantic representation is available for each seen and unseen class. In Eq. (6.6), writing $P(a_k|\mathbf{x})$ as $P(a_k|c)P(c|\mathbf{x})$ and considering that $P(c|\mathbf{x})$ for seen classes can be obtained using any supervised classifier trained on the training dataset on the one hand, and $P(a_k|c)$ is 1 if $a_k = (\mathbf{s}_c)_k$ and 0 otherwise on the other hand, we finally have:

$$\hat{y} = \underset{c\in\mathcal{C}^{\mathcal{U}}}{\operatorname{argmax}} \prod_{k=1}^{K} \sum_{\substack{c'\in\mathcal{C}^{\mathcal{S}} \\ (\mathbf{s}_c)_k=(\mathbf{s}_{c'})_k}} P(c'|\mathbf{x}) \qquad (6.7)$$

Similarly to IAP, a method based on Convex Semantic Embeddings, or **ConSE** [NMB+14], relies only on standard classifiers and can be used to adapt pre-trained standard models to a ZSL setting without any further training. Given a visual sample $\mathbf{x}$, we estimate its semantic representation $\hat{\mathbf{s}}(\mathbf{x}) \in \mathbb{R}^K$ as a convex combination of the semantic prototypes $\mathbf{s}_{\hat{c}}$ of the best predictions $\hat{c}_t(\mathbf{x})$ for $\mathbf{x}$, each prototype being weighted by its classification score. For a test instance $\mathbf{x}$, we can then simply predict $\hat{y}$ as the class whose class prototype is the closest to the estimated semantic representation as measured with cosine similarity. We can notice that contrary to DAP and IAP, ConSE does not make any implicit assumption regarding the nature of the class prototypes, and can be used with semantic representations having binary or continuous components. It is also interesting to note that if the convex combination is restricted to one prototype, this method is equivalent to simply finding the best matching seen class to the (unseen) test instance $\mathbf{x}$ and predicting the unseen class whose prototype is closest to the prototype of the best matching seen class.

6.3.1 Ridge Regression Approaches

One simple approach to ZSL is to view this task as a regression problem, where we aim to predict continuous attributes from a visual instance. Linear regression is a straightforward baseline. Given a visual sample $\mathbf{x}$ and the corresponding semantic representation $\mathbf{s}$, we aim to predict each semantic component s_k of $\mathbf{s}$ as $\hat{s}_k = \mathbf{w}_k^\top \mathbf{x}$, so as to minimize the squared difference between the prediction and the true value $(\hat{s}_k - s_k)^2$, $\mathbf{w}_k \in \mathbb{R}^D$ being the parameters of the model. If we write $\mathbf{W} = (\mathbf{w}_1, \ldots, \mathbf{w}_K)^\top \in \mathbb{R}^{K \times D}$, we can directly estimate the entire prototype with $\hat{\mathbf{s}} = \mathbf{W}\mathbf{x}$. We can also directly compare how close $\hat{\mathbf{s}}$ is to $\mathbf{s}$ with $\|\hat{\mathbf{s}} - \mathbf{s}\|_2^2 = \sum_k (\hat{s}_k - s_k)^2$. As with a standard linear regression, we determine the optimal parameters $\mathbf{W}$ by minimizing the error over the training dataset $\mathcal{D}^{\mathrm{tr}}$. Let us note $\mathbf{X} = (\mathbf{x}_1, \ldots, \mathbf{x}_N)^\top \in \mathbb{R}^{N \times D}$ the matrix whose N lines correspond to the visual features of training samples, and $\mathbf{T} = (\mathbf{t}_1, \ldots, \mathbf{t}_N)^\top \in \mathbb{R}^{N \times K}$ that containing the class prototypes associated to each image so that $\mathbf{t}_n = \mathbf{s}_{y_n}$. To simplify notations, we denote $\|\cdot\|_2$ both the $\ell 2$ norm when applied to a vector and the Frobenius norm when applied to a matrix. The loss can then be expressed in matrix form and regularized with an $\ell 2$ penalty on the model parameters weighted by a hyperparameter λ:

$$\frac{1}{N}\|\mathbf{X}\mathbf{W}^\top - \mathbf{T}\|_2^2 + \lambda\|\mathbf{W}\|_2^2 \tag{6.8}$$

Such a loss has a closed-form solution, which directly gives the value of the optimal parameters:

$$\mathbf{W} = \mathbf{T}^\top \mathbf{X}(\mathbf{X}^\top \mathbf{X} + \lambda N \mathbf{I}_D)^{-1} \tag{6.9}$$

At test time, given an image $\mathbf{x}$ belonging to an unseen class, we estimate its corresponding semantic representation $\hat{\mathbf{s}} = \mathbf{W}\mathbf{x}$ and predict the class with the closest semantic prototype. Note that it's also possible to use other distances or similarity measures such as a cosine similarity during the prediction phase.

The **Embarrassingly Simple approach to Zero-Shot Learning** [RPT15], often abbreviated **ESZSL**, makes use of a similar idea with a few additional steps. Similarly, the projection $\hat{\mathbf{t}}_n = \mathbf{W}\mathbf{x}_n$ of an image $\mathbf{x}_n$ should be close to the expected semantic representations. This last similarity is nevertheless estimated by a dot product $\hat{\mathbf{t}}_n^\top \mathbf{t}_n$ that should be close to 1 for the ground truth $t_n = s_{y_n}$ and to -1 for the prototypes of other classes. Considering the matrix $\mathbf{Y} \in \{-1, 1\}^{N \times C}$ that is 1 on line n and column y_n and -1 everywhere else, and $\mathbf{S} = (\mathbf{s}_1, \ldots, \mathbf{s}_C)^\top \in \mathbb{R}^{|\mathcal{C}^S| \times K}$ the matrix that contains the prototypes of seen classes, the loss to minimize is $\frac{1}{N} \|\mathbf{X}\mathbf{W}^\top \mathbf{S}^\top - \mathbf{Y}\|_2^2$. In [RPT15], it is further regularized such that visual features projected on the semantic space, $\mathbf{X}\mathbf{W}^\top$, have similar norms to allow for fair comparison, and similarly for the semantic prototypes projected on the visual space $\mathbf{W}^\top \mathbf{S}^\top$. Adding an $\ell2$ penalty on the model parameters, we have:

$$\frac{1}{N} \|\mathbf{X}\mathbf{W}^\top \mathbf{S}^\top - \mathbf{Y}\|_2^2 + \gamma \|\mathbf{W}^\top \mathbf{S}^\top\|_2^2 + \frac{\lambda}{N} \|\mathbf{X}\mathbf{W}^\top\|_2^2 + \gamma\lambda \|\mathbf{W}\|_2^2 \tag{6.10}$$

λ and γ being hyperparameters controlling the weights of the different regularization terms. The minimization of this expression also leads to a closed-form solution:

$$\mathbf{W} = (\mathbf{S}^\top \mathbf{S} + \lambda N \mathbf{I}_K)^{-1} \mathbf{S}^\top \mathbf{Y}^\top \mathbf{X} (\mathbf{X}^\top \mathbf{X} + \gamma N \mathbf{I}_D)^{-1} \tag{6.11}$$

Instead of aiming to predict the class prototypes $\mathbf{s}$ from the visual features $\mathbf{x}$, we can consider **predicting the visual features** from the class prototypes' features. Each visual dimension is expressed as a linear combination of prototypes such that we can estimate the "average" visual representation associated with prototype $\mathbf{s}$ with $\hat{\mathbf{x}} = \mathbf{W}^\top \mathbf{s}$. The distances between the observations and our predictions are then computed in the visual space by minimizing the distance $\|\mathbf{x}_n - \hat{\mathbf{x}}_n\|^2$ between the sample $\mathbf{x}_n$ and the predicted visual features $\hat{\mathbf{x}}_n = \mathbf{W}^\top \mathbf{s}_{y_n}$ of the corresponding semantic prototype $\mathbf{s}_{y_n}$. The resulting regularized loss $\frac{1}{N} \|\mathbf{X} - \mathbf{T}\mathbf{W}\|_2^2 + \lambda \|\mathbf{W}\|_2^2$ also has a closed-form solution:

$$\mathbf{W} = (\mathbf{T}^\top \mathbf{T} + \lambda N \mathbf{I}_K)^{-1} \mathbf{T}^\top \mathbf{X} \tag{6.12}$$

The label of a test image $\mathbf{x}$ is then predicted through a nearest-neighbor search in the visual space. Although this approach is very similar to previous ones, it turns out that projecting semantic objects to the visual space has an advantage. Like other machine learning methods, ZSL methods can be subject to the *hubness problem* [RNI10], which describes a situation where certain objects, referred to as *hubs*, are the nearest neighbors of many other objects. In the case of ZSL, if a semantic prototype is too centrally located, if may be the nearest neighbor of many projections of visual samples into the semantic space even if these samples belong to other classes, thus leading to incorrect predictions and decreasing the performance of the model. When using ridge regression for ZSL, it has been verified experimentally [LDB15, SSH+15] that this situation tends to happen. However, [SSH+15] shows that this effect is mitigated when projecting from the semantic to the visual space, compared to the opposite situation. It should be noted that the hubness problem does not occur exclusively when using ridge regression, and more complex ZSL methods such as [ZXG17] make use of the findings of [SSH+15].

The semantic autoencoder (**SAE**) [KXG17] approach can be seen as a combination of the two ridge regression projections, from the semantic space to the visual one and the opposite. The idea consists in first *encoding* a visual sample by linearly projecting it onto the semantic space and then *decoding* it by projecting the result into the visual space again. Contrary to the previous proposals, there is no immediate closed-form solution to this problem. However, it can be expressed as a Sylvester equation and a numerical solution can be computed efficiently using the Bartels-Stewart algorithm [BS72]. During the testing phase, predictions can be made either in the semantic space or in the visual space.

All previous methods project linearly from one modality (visual or semantic) to the other, but they can be adapted to non-linear regression methods, as proposed by **Cross-Modal Transfer** or **CMT** [SGMN13]. It consists in a simple fully-connected, 1-hidden layer neural network with hyperbolic tangent non-linearity, which is used to predict semantic prototypes from visual features. Equation (6.8) can therefore be re-written as

$$\frac{1}{N} \sum_n \|\mathbf{t_n} - \mathbf{W}_2 \tanh(\mathbf{W}_1 \mathbf{x}_n)\|_2^2 \tag{6.13}$$

$\mathbf{W}_1 \in \mathbb{R}^{H \times D}$ and $\mathbf{W}_2 \in \mathbb{R}^{K \times H}$ being the parameters of the model, and H the dimension of the hidden layer which is a hyperparameter. Similar or more complex adaptations can easily be made for other methods. The main drawback of such non linear projections compared to the linear methods presented earlier is that there is no general closed-form solution, and iterative numerical algorithms must be used to determine suitable values for the parameters.

6.3.2 Triplet-Loss Approaches

Triplet loss methods make a more direct use of the compatibility function f. The main idea behind these methods is that the compatibility of matching pairs should be much higher than the compatibility of non-matching pairs. More specifically, given a visual sample $\mathbf{x}$ with label y, we expect that its compatibility with the corresponding prototype $\mathbf{s}_y$ should be much higher than its compatibility with $\mathbf{s}_c$, the prototype of a different class $c \neq y$. How "much higher" can be more precisely defined through the introduction of a margin m, such that $f(\mathbf{x}, \mathbf{s}_y) \geq m + f(\mathbf{x}, \mathbf{s}_c)$. To enforce this constraint, we can penalize triplets $(\mathbf{x}, \mathbf{s}_y, \mathbf{s}_c)$, $c \neq y$, for which this inequality is not true, using the triplet loss

$$\mathcal{L}_{\text{triplet}}(\mathbf{x}, \mathbf{s}_c, \mathbf{s}_y; f) = [m + f(\mathbf{x}, \mathbf{s}_c) - f(\mathbf{x}, \mathbf{s}_y)]_+ \tag{6.14}$$

where $[\cdot]_+ = \max(0, \cdot)$. This way, for a given triplet $(\mathbf{x}, \mathbf{s}_y, \mathbf{s}_c)$, $c \neq y$, the loss is 0 if $f(\mathbf{x}, \mathbf{s}_c)$ is much smaller than $f(\mathbf{x}, \mathbf{s}_y)$, and is all the higher as $f(\mathbf{x}, \mathbf{s}_c)$ gets close to, or surpasses $f(\mathbf{x}, \mathbf{s}_y)$. In general, it is not possible to derive a solution analytically for methods based on a triplet loss, so we must resort to the use of numerical optimization.

In many triplet loss approaches to ZSL, the compatibility function f is simply defined as a bilinear mapping between the visual and semantic spaces parameterized by a matrix $\mathbf{W} \in \mathbb{R}^{D \times K}$, so that $f(\mathbf{x}, \mathbf{s}) = \mathbf{x}^\top \mathbf{W} \mathbf{s}$. This compatibility function is actually the same as with ESZSL, even though the loss function used to learn its parameters $\mathbf{W}$ is different. The **Deep Visual-Semantic Embedding** model or **DeViSE** [FCS+13] is one of the most direct applications of a triplet loss with a linear compatibility function to ZSL: the total loss is simply the sum of the triplet loss over all training triplets $(\mathbf{x}_n, \mathbf{s}_{y_n}, \mathbf{s}_c)$, $c \neq y$:

$$\mathcal{L}_{\mathrm{tr}}(\mathcal{D}^{tr}) = \frac{1}{N} \sum_{n=1}^{N} \sum_{\substack{c \in \mathcal{C}^S \\ c \neq y_n}} [m + f(\mathbf{x}_n, \mathbf{s}_c) - f(\mathbf{x}_n, \mathbf{s}_{y_n})]_+ \qquad (6.15)$$

DeViSE can also be viewed as a direct application of the Weston-Watkins loss [WW+99] to ZSL. It can be noted that the link with the generic loss framework in Eq. (6.1) is this time quite straightforward, as with many triplet loss methods. Although no explicit regularization Ω on f is mentioned in the original publication—even though the authors make use of early stopping in the gradient descent—it is again straightforward to add an $\ell 2$ penalty. The **Structured Joint Embedding** approach, or **SJE** [ARW+15], is fairly similar to DeViSE. It is inspired by works on structured SVMs [THJA04, TJHA05], and makes use of the Cramer-Singer loss [CS01] for multi-class SVM. Applied to ZSL, this means that only the class which is violating the triplet-loss constraint the most is taken into account for each sample. In our case, this results in:

$$\mathcal{L}_{\mathrm{tr}}(\mathcal{D}^{\mathrm{tr}}) = \frac{1}{N} \sum_{n=1}^{N} \max_{\substack{c \in \mathcal{C}^S \\ c \neq y_n}} \left([m + f(\mathbf{x}_n, \mathbf{s}_c) - f(\mathbf{x}_n, \mathbf{s}_{y_n})]_+ \right) \qquad (6.16)$$

The **Attribute Label Embedding** approach or **ALE** [APHS13, APHS15] considers the ZSL task as a ranking problem, where the objective is to rank the correct class c as high as possible on the list of candidate unseen classes. From this perspective, we can consider that SJE only takes into account the top element of the ranking list provided the margin m is close to 0. By contrast, DeViSE penalizes all ranking mistakes: given labeled sample $(\mathbf{x}, y)$, for all classes c mistakenly ranked higher than y, we have $f(\mathbf{x}, \mathbf{s}_c) > f(\mathbf{x}, \mathbf{s}_y)$ which contributes to the loss. The ALE approach aims to be somewhere in between these two proposals, so that a mistake on the rank when the true class is close to the top of the ranking list weighs more than a mistake when the true class is lower on the list.

Similarly to CMT in the previous section, all triplet-loss models can be extended to the nonlinear case. Such an extension is even more straightforward as this time, having no closed-form solution, all models require the use of numerical optimization. One such example of a nonlinear model worth describing due to its historical significance and still fair performance is the **Synthesized Classifiers** approach, or **SynC** [CCGS16a, CCGS20]. Based on a manifold learning framework, it aims to learn *phantom classes* in both the semantic and visual spaces, so that linear classifiers for seen and unseen classes can be synthesized as a combination of such phantom classes. More precisely, the goal is to synthesize linear classifiers $\mathbf{w}_c$ in the visual space such that the compatibility between image $\mathbf{x}$ and class c can be computed with $f(\mathbf{x}, \mathbf{s}_c) = \mathbf{w}_c^\top \mathbf{x}$. The prediction is then $\hat{y} = \arg\max_c \mathbf{w}_c^\top \mathbf{x}$.

Let us note respectively $\{\overset{*}{\mathbf{x}}_p\}_{p \in [\![1,P]\!]}$ and $\{\overset{*}{\mathbf{s}}_p\}_{p \in [\![1,P]\!]}$ the P phantom classes in the respective visual and semantic spaces. These phantom classes are learned and constitute the parameters of the model.[1] Each visual classifier is synthesized as a linear combination of visual phantom classes $\mathbf{w}_c = \sum_{p=1}^{P} v_{c,p} \overset{*}{\mathbf{x}}_p$. The value of each coefficient $v_{c,p}$ is set so as to correspond to the conditional probability of observing phantom class $\overset{*}{\mathbf{s}}_p$ in the neighborhood of real class $\mathbf{s}_c$ in the semantic space. Following works on manifold learning [HR03, MH08], this can be expressed according to $\mathbf{s}_c$ and $\overset{*}{\mathbf{s}}_p$. The parameters of the model, i.e. the phantom classes $\{(\overset{*}{\mathbf{x}}_p, \overset{*}{\mathbf{s}}_p)\}_p$, can be estimated by making use of the Crammer-Singer loss, with adequate regularization to obtain the following objective:

$$
\underset{\{(\overset{*}{\mathbf{x}}_p, \overset{*}{\mathbf{s}}_p)\}_p}{\text{minimize}} \frac{1}{N} \sum_{n=1}^{N} \max_{\substack{c \in \mathcal{C}^S \\ c \neq y_n}} \left([m + \mathbf{w}_c^\top \mathbf{x}_n - \mathbf{w}_{y_n}^\top \mathbf{x}_n]_+ \right) +
$$

$$
\lambda \sum_{c \in \mathcal{C}^S} \|\mathbf{w}_c\|^2 + \gamma \sum_{p=1}^{P} \|\overset{*}{\mathbf{s}}_p\|^2 \tag{6.17}
$$

where λ and γ are hyperparameters. It is interesting to note that ALE can actually be considered as a special case of SynC, where the classifiers are simply a linear combination of semantic prototypes.

Recently, [LCLBC19a] showed that modifications to the triplet loss could enable models obtained with this loss to reach (G)ZSL accuracy competitive with generative models (Sect. 6.3.3). Such modifications include a margin that depends on the similarity between $\mathbf{s}_y$ and $\mathbf{s}_c$ in Eq. (6.14) so that confusions between very similar classes are not penalized as much as confusions between dissimilar classes during training, as well as a weighting scheme that makes "representative" training samples have more impact than outliers.

---

[1]A number of simplifications were made for the sake of clarity and brevity: in the original article [CCGS16a], phantom classes are actually sparse linear combinations of semantic prototypes, $v_{c,p}$ can further use Mahalanobis distance, other losses such as squared hinge loss can be employed instead of the Crammer-Singer loss, Euclidean distances between semantic prototypes can be used instead of a fixed margin in the triplet loss, additional regularization terms and hyperparameters are introduced, and optimization between $\{\overset{*}{\mathbf{x}}_p\}_p$ and $\{\overset{*}{\mathbf{s}}_p\}_p$ is performed alternatingly.

6.3.3   Generative Approaches

Generative methods applied to ZSL aim to produce visual samples belonging to unseen classes based on their semantic description; these samples can then be used to train standard classifiers. Partly for this reason, most generative methods directly produce high-level visual features, as opposed to raw pixels—another reason being that generating raw images is usually not as effective [XLSA18b]. Generative methods have gained a lot of attention in the last few years: many if not most recent high-visibility ZSL approaches [VAMR18, XLSA18b, XSSA19] rely on generative models. This is partly because such approaches have interesting properties, which make them particularly suitable to certain settings such as GZSL. However, a disadvantage is that they can only operate in a class-transductive setting, since the class prototypes of unseen classes are needed to generate samples belonging to these classes; contrary to methods based on regression or explicit compatibility functions, at least some additional training is necessary to integrate novel classes to the model. We divide generative approaches into two main categories: methods generating a parametric distribution, which consider visual samples follow a standard probability distribution such as a multivariate Gaussian and attempt to estimate its parameters so that visual features can be sampled from this distribution, and non-parametric methods, where visual samples are directly generated by the model.

Methods based on parametric distributions assume that visual features for each class follow a standard parametric distribution. For example, one may consider that for each class c, visual features are samples from a multivariate Gaussian with mean $\boldsymbol{\mu}_c \in \mathbb{R}^D$ and covariance $\boldsymbol{\Sigma}_c \in \mathbb{R}^{D \times D}$, such that for samples $\mathbf{x}$ from class c we have $p(\mathbf{x}; \boldsymbol{\mu}_c, \boldsymbol{\Sigma}_c) = \mathcal{N}(\mathbf{x}|\boldsymbol{\mu}_c, \boldsymbol{\Sigma}_c)$. If one can estimate $\boldsymbol{\mu}$ and $\boldsymbol{\Sigma}$ for unseen classes, it is possible to generate samples belonging to these classes. Zero-shot recognition can then be performed by training a standard multi-class classifier on the labeled generated samples.

Alternatively, knowing the (estimated) distribution of samples from unseen classes, one may determine the class of a test visual sample $\mathbf{x}$ using maximum likelihood or similar methods [VR17]:

$$\hat{y} = \underset{c \in \mathcal{C}^{\mathcal{U}}}{\mathrm{argmax}} \; p(\mathbf{x}; \boldsymbol{\mu}_c, \boldsymbol{\Sigma}_c) \tag{6.18}$$

Other approaches [XLSA18b] also propose to further train a ZSL model based on an explicit compatibility function using the generated samples and the corresponding class prototypes, and then perform zero-shot recognition as usually with Eq. (6.2).

The **Generative Framework for Zero-Shot Learning** [VR17] or **GFZSL** assumes that visual features are normally distributed given their class. The parameters of the distribution $(\boldsymbol{\mu}_c, \sigma_c^2)$ (to simplify, we assume that $\boldsymbol{\Sigma}_c = \mathrm{diag}(\sigma_c^2)$, with $\sigma_c^2 \in \mathbb{R}^{+D}$) are easy to obtain for seen classes $c \in \mathcal{C}^{\mathcal{S}}$ using e.g. maximum likelihood estimators, but are unknown for unseen classes. Since the only information available about unseen classes consists of class prototypes, one can assume that the parameters $\boldsymbol{\mu}_c$ and σ_c^2 of class c depend on class prototype $\mathbf{s}_c$. [VR17] further assumes a linear dependency, such that $\boldsymbol{\mu}_c = \mathbf{W}_\mu^\top \mathbf{s}_c$ and $\boldsymbol{\rho}_c = \log(\sigma_c^2) = \mathbf{W}_\sigma^\top \mathbf{s}_c$. The models' parameters $\mathbf{W}_\mu \in \mathbb{R}^{K \times D}$ and $\mathbf{W}_\sigma \in \mathbb{R}^{K \times D}$ can then be obtained with ridge regression, using the class distribution parameters $\{(\hat{\boldsymbol{\mu}}_c, \hat{\boldsymbol{\rho}}_c)\}_{c \in \mathcal{CS}}$ estimated on seen classes as training samples. Similarly to previous approaches, this consists in minimizing $\ell2$-regularized losses, with closed-form solutions. Noting $\mathbf{M} = (\hat{\boldsymbol{\mu}}_1, \ldots, \hat{\boldsymbol{\mu}}_C)^\top \in \mathbb{R}^{C \times D}$ and $\mathbf{R} = (\hat{\boldsymbol{\rho}}_1, \ldots, \hat{\boldsymbol{\rho}}_C)^\top \in \mathbb{R}^{C \times D}$, we have:

$$\mathbf{W}_\mu = (\mathbf{S}^\top \mathbf{S} + \lambda_\mu \mathbf{I}_K)^{-1} \mathbf{S}^\top \mathbf{M} \tag{6.19}$$

$$\mathbf{W}_\sigma = (\mathbf{S}^\top \mathbf{S} + \lambda_\sigma \mathbf{I}_K)^{-1} \mathbf{S}^\top \mathbf{R} \tag{6.20}$$

We can thus predict parameters $(\hat{\boldsymbol{\mu}}_c, \hat{\boldsymbol{\rho}}_c)$ for all unseen classes $c \in \mathcal{C}^\mathcal{U}$, and sample visual features of unseen classes accordingly. Predictions can then be made using either a standard classifier or the estimated distributions themselves. [VR17] also extends this approach to include more generic distributions belonging to the exponential family and non-linear regressors.

The **Synthesized Samples for Zero-Shot Learning** [GDHG17] or **SSZSL** method similarly assumes that $p(\mathbf{x}|c)$ is Gaussian, estimates parameters $(\boldsymbol{\mu}, \boldsymbol{\Sigma})$ for seen classes with techniques similar to GFZSL and aims to predict parameters $(\hat{\boldsymbol{\mu}}, \hat{\boldsymbol{\Sigma}})$ for unseen classes. In a way that reminds the ConSE method, the distributions parameters are estimated using a convex combination of parameters from seen classes d, such that $\hat{\boldsymbol{\mu}} = \frac{1}{Z} \sum_{d \in \mathcal{CS}} w_d \boldsymbol{\mu}_d$ and $\hat{\sigma}^2 = \frac{1}{Z} \sum_{d \in \mathcal{CS}} w_d \sigma_d^2$, with $Z = \mathbf{1}^\top \mathbf{w} = \sum_d w_d$. The model therefore has one vector parameter $\mathbf{w}_c \in \mathbb{R}^{|\mathcal{C}^\mathcal{S}|}$ to determine per unseen class c. These last are set such that the semantic prototype $\mathbf{s}_c^{\text{te}}$ from unseen class c is approximately a convex combination of prototypes from seen classes, i.e. $\mathbf{s}_c^{\text{te}} \simeq \mathbf{S}^\top \mathbf{w}_c / Z_c$, while preventing classes dissimilar to $\mathbf{s}_c^{\text{te}}$ from being assigned a large weight. This results in the following loss for unseen class c:

$$\|\mathbf{s}_c^{\text{te}} - \mathbf{S}^\top \mathbf{w}_c\|_2^2 + \lambda \mathbf{w}_c^\top \mathbf{d}_c \tag{6.21}$$

where each element $(\mathbf{d}_c)_i$ of $\mathbf{d}_c$ is a measure of how dissimilar unseen class c is to seen class i, and λ is a hyperparameter. Minimizing the second term in Eq. (6.21) naturally leads to assigning smaller weights to classes dissimilar to c.

The **non parametric approaches** do not explicitly make simplifying assumptions about the shape of the distribution of visual features, and use powerful generative methods such as variational auto-encoders (VAEs) [KW14] or generative adversarial networks (GANs) [GPAM+14] to directly synthesize samples. Although these models are in principle able to capture complex data distribution, they can prove to be hard to train [AB17].

The **Synthesized Examples for GZSL** method [VAMR18], or **SE-GZSL**, is based on a conditional VAE [SLY15] architecture. It consists of two main parts: an *encoder* $\mathcal{E}(\cdot)$ which maps an input $\mathbf{x}$ to an R-dimensional internal representation or latent code $\mathbf{z} \in \mathbb{R}^R$, and a *decoder* $\mathcal{D}(\cdot)$ which tries to reconstruct the input $\mathbf{x}$ from the internal representation. An optional third part can be added to the model: a *regressor* $\mathcal{R}(\cdot)$ which estimates the semantic representation $\mathbf{t}$ of the visual input $\mathbf{x}$. See Chap. 2 for more details on the VAE architecture. To help the decoder to produce class-dependant reconstructed outputs, the corresponding class prototype $\mathbf{t}_n = \mathbf{s}_{y_n}$ is concatenated to the representation $\mathbf{z}_n$ for input $\mathbf{x}_n$.

Other approaches such as [MKRMM18] consider that the encoder outputs a probability distribution, assuming that the true distribution of visual samples is an isotropic Gaussian given the latent representation, i.e. $p(\mathbf{x}|\mathbf{z}, \mathbf{t}) = \mathcal{N}(\mathbf{x}|\boldsymbol{\mu}(\mathbf{z}, \mathbf{t}), \sigma^2 \mathbf{I})$. In this case, the output of the decoder should be $\hat{\mathbf{x}} = \boldsymbol{\mu}(\mathbf{z}, \mathbf{t})$, and it can be shown that minimizing $-\log(p(\mathbf{x}|\mathbf{z}, \mathbf{t}))$ is equivalent to minimizing $\|\mathbf{x} - \hat{\mathbf{x}}\|^2$. Furthermore, in [MKRMM18], the class prototype is appended to the visual sample as opposed to the latent code.

The authors of [VAMR18] further propose to use the regressor $\mathcal{R}$ to encourage the decoder to generate discriminative visual samples. An example of such components consists in evaluating the quality of predicted attributes from synthesized samples, and takes the form $\mathcal{L} = -\mathbb{E}_{p(\hat{\mathbf{x}}|\mathbf{z}, \mathbf{t})p(\mathbf{z})p(\mathbf{t})}[\log(p(\mathbf{a}|\hat{\mathbf{x}}))]$. The regressor itself is trained on both labeled training samples and generated samples, and the parameters of the encoder/decoder and the regressor are optimized alternatingly.

f-GAN [XLSA18b] is based on a similar approach, but makes use of conditional GANs [MO14] to generate visual features. It consists of two parts: a *discriminator* $\mathcal{D}$ which tries to distinguish real images from synthesized images, and a *generator* $\mathcal{G}$ which tries to generate images that $\mathcal{D}$ cannot distinguish from real images. Both encoder and decoder are multilayer perceptrons. The generator is similar to the decoder from the previous approach in that it takes as input a latent code $\mathbf{z} \in \mathbb{R}^R$ and the semantic representation $\mathbf{s}_c$ of a class c, and attempts to generate a visual sample $\hat{\mathbf{x}}$ of class c: $\mathcal{G} : \mathbb{R}^R \times \mathbb{R}^K \to \mathbb{R}^D$. The key difference is that the latent code is not the output of an encoder but consists of random Gaussian noise. The discriminator takes as input a visual sample, either real or generated, of a class c as well as the prototype $\mathbf{s}_c$, and predicts the probability that the visual sample was generated: $\mathcal{D} : \mathbb{R}^D \times \mathbb{R}^K \to [0, 1]$. $\mathcal{G}$ and $\mathcal{D}$ compete in a two-player minimax game, such that the optimization objective is:

$$\min_{\mathcal{G}} \max_{\mathcal{D}} \mathbb{E}_{p(\mathbf{x}, y), p(\mathbf{z})}[\log(\mathcal{D}(\mathbf{x}, \mathbf{s}_y)) + \log(1 - \mathcal{D}(\mathcal{G}(\mathbf{z}, \mathbf{s}_y), \mathbf{s}_y))] \qquad (6.22)$$

The authors of [XLSA18b] further propose to train an improved Wasserstein GAN [ACB17, GAA+17], and similarly to [VAMR18], they add another component to the loss to ensure that generated features are discriminative, using a classification loss instead of a regression loss. They call this extended approach f-CLSWGAN.

6.4 Semantic Features for Large Scale ZSL

In most of the work on ZSL, the semantic features s_c were usually assumed to be vectors of attributes such as "is red", "has stripes" or "has wings". Such attributes can either be binary or continuous with values in $[0, 1]$. In the latter case, a value of 0.8 associated with the attribute "is red" could mean that the animal or object is mostly red. But for a large-scale dataset with hundreds or even thousands of classes, or an open dataset where novel classes are expected to appear over time, it is impractical or even impossible to define a priori all the useful attributes and manually provide semantic prototypes based on these attributes for all the classes. It is all the more time-consuming as fine-grained datasets may require hundreds of different attributes to reach a satisfactory accuracy [LCPLB20].

For large-scale or open datasets it is therefore necessary to identify appropriate sources of semantic information and means to extract this information in order to obtain relevant semantic prototypes. In the case of ImageNet, readily available sources are the word embeddings of the class names and the relations between them according to WordNet [Mil95], a large lexical database of English that has been developed for many years by human efforts, but is now openly available. Word embeddings are obtained in an unsupervised way and such embeddings of the class names have been employed in ZSL as semantic class representations since [RSS11]. The word embedding model is typically trained on a large text corpus, such as Wikipedia, where a neural architecture is assigned the task to predict the context of a given word. For instance, the skip-gram objective [MSC+13] aims to find word representations containing predictive information about the words occurring in the neighborhood of a given word. Given a sequence $\{w_1, \ldots, w_T\}$ of T training words and a context window of size S, the goal is to maximize

$$\sum_{t=1}^{T} \sum_{\substack{-S \leq i \leq S \\ i \neq 0}} \log p(w_{t+i}|w_t) \tag{6.23}$$

Although deep neural architectures could be used for this task, it is much more common to use a shallow network with a single hidden layer. In this case, each unique word w is associated with an "input" vector $\mathbf{v}_w$ and an "output" vector $\mathbf{v}'_w$, and $p(w_i|w_t)$ is computed such that

$$p(w_i|w_t) = \frac{\exp(\mathbf{v}'^{\top}_{w_i} \mathbf{v}_{w_t})}{\sum_w \exp(\mathbf{v}'^{\top}_w \mathbf{v}_{w_t})} \tag{6.24}$$

The internal representation corresponding to the hidden layer, i.e. the input vector representation $\mathbf{v}_w$, can then be used as the word embedding. Other approaches such as [BGJM17] or [PSM14] have also been proposed.

Semantic information regarding ImageNet classes can also be provided by Word-Net subsumption relationships between the classes (or IS-A relations). They were obtained by several methods, e.g. with graph convolutional neural networks, and employed for ZSL on ImageNet with state-of-the-art results [WYG18, KCL+19]. However, as shown in [HAT19], these results are biased by the fact that in the traditional ImageNet ZSL split between seen and unseen classes, the unseen classes are often subcategories or supercategories of seen classes. When the ZSL split is modified so as to remove this bias, the WordNet graph-based embeddings lead to an accuracy of about 14% according to [HAT19]. However, although using word or graph embeddings can reduce the additional human effort required to obtain class prototypes to virtually zero—pre-trained word embeddings can easily be found online—there still exists a large performance gap between the use of such embeddings and manually crafted attributes [LCPLB20].

The use of complementary sources to produce semantic class representations for ZSL relies on the assumption that the information these sources provide reflects visual similarity relations between classes. However, the word embeddings are typically developed from generic text corpora, like Wikipedia, that do not focus on the visual aspect. Also, the subsumption relationships issued from WordNet and supporting the graph-based class embeddings represent hierarchical conceptual relations. In both cases, the sought-after visual relations are at best represented in a very indirect and incomplete way.

To address this limitation and include more visual information to build the semantic prototypes, it was recently proposed to employ text corpora with a more visual nature, by constructing such a corpus from Flickr tags [LCPLB20]. Following [PG11], the authors of [LCPLB20] further suggested to address the problem of *bulk tagging* [OM13]—users attributing the exact same tags to numerous photos—by ensuring that a tuple of words (w_i, w_j) can only appear once for each user during training, thus preventing a single user from having a disproportionate weight on the final embedding. Also, [LCLBC20] suggested to exploit the sentence descriptions of WordNet concepts, in addition to the class name embedding, to produce semantic representations better reflecting visual relations. Any of these two proposals allow to reach an accuracy between 17.2 and 17.8 on the 500 test classes of the ImageNet ZSL benchmark with the linear model from the semantic to the visual space (Sect. 6.3.1), compared to 14.4 with semantic prototypes based on standard embeddings.

6.5 Conclusion and Current Challenges

Zero-shot learning addresses the problem of recognizing categories that are missing from the training set. ZSL has grown from an endeavor of some machine learning and computer vision researchers to find approaches that come closer to how humans learn to identify object classes. It now aims to become a radical answer to the concern that the amount of labeled data grows much slower than the volume of data in general, so supervised learning alone cannot produce satisfactory solutions for many real-world applications.

Key to the possibility of recognizing previously unseen categories is the availability for all categories, both seen and unseen, of more than just conventional labels. For each category we should have complementary information (or features) reflecting the characteristics of the modality used for recognition (visual if recognition is directed to images). The relation between these features and the target modality can thus be learned from the seen categories and then employed for recognizing unseen categories.

Most of the work on ZSL took advantage of the existence of some small or medium-size datasets for which the complementary information, under the form of attributes, was devised and manually provided to support the development of ZSL methods. However, in general applications one has to deal with large and even open sets of categories, so other approaches should be found for identifying associated sources of complementary information and exploiting them in ZSL.

For the large ImageNet dataset several readily available complementary sources were found, including word embeddings of class names, WordNet-based concept hierarchies including the classes as nodes, and short textual definitions from WordNet. While this allowed to extend ZSL methods to such large-scale datasets, the state-of-the-art accuracy obtained on the unseen categories of ImageNet is yet disappointing. This is because the information provided by these sources reflects mostly the conceptual relations and not so much the visual characteristics of the categories. To go beyond this level of performance we consider that two important steps should be taken. First, it is necessary to assemble large corpora including rather detailed textual descriptions of the visual characteristics of a large number of object categories. Partial corpora do exist in various domains (e.g. flora descriptions) and different languages. Second, zero-shot recognition should rely on a deeper, compositional analysis of an image as well as on visual reasoning.

References

[AB17]   Martin Arjovsky and Léon Bottou.   Towards principled methods for training generative adversarial networks. *arXiv preprint arXiv:1701.04862*, 2017.

[ACB17] Martin Arjovsky, Soumith Chintala, and Léon Bottou. Wasserstein GAN. *arXiv preprint arXiv:1701.07875*, 2017.

[APHS13] Zeynep Akata, Florent Perronnin, Zaid Harchaoui, and Cordelia Schmid. Label-embedding for attribute-based classification.   In *Computer Vision and Pattern Recognition*, pages 819–826, 2013.

[APHS15] Zeynep Akata, Florent Perronnin, Zaid Harchaoui, and Cordelia Schmid. Label-embedding for image classification.   *IEEE T. Pattern Analysis and Machine Intelligence*, 38(7):1425–1438, 2015.

[ARW+15] Zeynep Akata, Scott Reed, Daniel Walter, Honglak Lee, and Bernt Schiele. Evaluation of output embeddings for fine-grained image classification. In *Computer Vision and Pattern Recognition*, 2015.

[BGJM17] Piotr Bojanowski, Edouard Grave, Armand Joulin, and Tomas Mikolov. Enriching word vectors with subword information. *Transactions of the Association for Computational Linguistics*, 5:135–146, 2017.

[BS72] Richard H. Bartels and George W Stewart. Solution of the matrix equation AX+ XB= C [F4]. *Communications of the ACM*, 15(9):820–826, 1972.

[BSS+18] Ankan Bansal, Karan Sikka, Gaurav Sharma, Rama Chellappa, and Ajay Divakaran. Zero-shot object detection. In *European Conference on Computer Vision*, 2018.

[CCGS16a] Soravit Changpinyo, Wei-Lun Chao, Boqing Gong, and Fei Sha. Synthesized classifiers for zero-shot learning. In *Computer Vision and Pattern Recognition*, pages 5327–5336, 2016.

[CCGS16b] Wei-Lun Chao, Soravit Changpinyo, Boqing Gong, and Fei Sha. An empirical study and analysis of generalized zero-shot learning for object recognition in the wild. In *European Conference on Computer Vision*, pages 52–68. Springer, 2016.

[CCGS20] Soravit Changpinyo, Wei-Lun Chao, Boqing Gong, and Fei Sha. Classifier and exemplar synthesis for zero-shot learning. *International Journal of Computer Vision*, 128(1):166–201, 2020.

[CS01] Koby Crammer and Yoram Singer. On the algorithmic implementation of multiclass kernel-based vector machines. *Journal of Machine Learning Research*, 2(Dec):265–292, 2001.

[CZX+18] Long Chen, Hanwang Zhang, Jun Xiao, Wei Liu, and Shih-Fu Chang. Zero-shot visual recognition using semantics-preserving adversarial embedding networks. In *Computer Vision and Pattern Recognition*, pages 1043–1052, 2018.

[DDS+09] Jia Deng, Wei Dong, Richard Socher, Li-Jia Li, Kai Li, and Li Fei-Fei. Imagenet: A large-scale hierarchical image database. In *2009 IEEE conference on computer vision and pattern recognition*, pages 248–255. Ieee, 2009.

[FCS+13] Andrea Frome, Greg S. Corrado, Jon Shlens, Samy Bengio, Jeff Dean, Tomas Mikolov, et al. Devise: A deep visual-semantic embedding model. In *Advances in Neural Information Processing Systems*, 2013.

[FEHF09] Ali Farhadi, Ian Endres, Derek Hoiem, and David Forsyth. Describing objects by their attributes. In *Computer Vision and Pattern Recognition*, pages 1778–1785. IEEE, 2009.

[FXJ+18] Yanwei Fu, Tao Xiang, Yu-Gang Jiang, Xiangyang Xue, Leonid Sigal, and Shao-gang Gong. Recent advances in zero-shot recognition: Toward data-efficient understanding of visual content. *IEEE Signal Processing Magazine*, 35(1):112–125, 2018.

[FYH+15] Yanwei Fu, Yongxin Yang, Tim Hospedales, Tao Xiang, and Shaogang Gong. Transductive multi-label zero-shot learning. *arXiv preprint arXiv:1503.07790*, 2015.

[GAA+17] Ishaan Gulrajani, Faruk Ahmed, Martin Arjovsky, Vincent Dumoulin, and Aaron C Courville. Improved training of Wasserstein GANs. In *Advances in neural information processing systems*, pages 5767–5777, 2017.

[GDHG17] Yuchen Guo, Guiguang Ding, Jungong Han, and Yue Gao. Synthesizing samples fro zero-shot learning. In *IJCAI*. IJCAI, 2017.

[GPAM+14] Ian Goodfellow, Jean Pouget-Abadie, Mehdi Mirza, Bing Xu, David Warde-Farley, Sherjil Ozair, Aaron Courville, and Yoshua Bengio. Generative adversarial nets. In *Advances in neural information processing systems*, pages 2672–2680, 2014.

[HAT19] Tristan Hascoet, Yasuo Ariki, and Tetsuya Takiguchi. On zero-shot recognition of generic objects. In *Computer Vision and Pattern Recognition*, pages 9553–9561, 2019.

[HR03] Geoffrey E Hinton and Sam T Roweis. Stochastic neighbor embedding. In *Advances in Neural Information Processing Systems*, pages 857–864, 2003.

[HZRS16] Kaiming He, Xiangyu Zhang, Shaoqing Ren, and Jian Sun. Deep residual learning for image recognition. In *Computer Vision and Pattern Recognition*, pages 770–778, 2016.

[KASB17] Nour Karessli, Zeynep Akata, Bernt Schiele, and Andreas Bulling. Gaze embeddings for zero-shot image classification. In *Computer Vision and Pattern Recognition*, pages 4525–4534, 2017.

[KCL+19] Michael Kampffmeyer, Yinbo Chen, Xiaodan Liang, Hao Wang, Yujia Zhang, and Eric P. Xing. Rethinking knowledge graph propagation for zero-shot learning. In *Computer Vision and Pattern Recognition*, pages 11487–11496, 2019.

[KW14] Diederik P Kingma and Max Welling. Auto-encoding variational bayes. In *International Conference on Learning Representations*, 2014.

[KXG17] Elyor Kodirov, Tao Xiang, and Shaogang Gong. Semantic autoencoder for zero-shot learning. In *Computer Vision and Pattern Recognition*, pages 4447–4456. IEEE, 2017.

[LCLBC19b] Yannick Le Cacheux, Hervé Le Borgne, and Michel Crucianu. From classical to generalized zero-shot learning: A simple adaptation process. In *International Conference on Multimedia Modeling*, pages 465–477. Springer, 2019.

[LCLBC19a] Yannick Le Cacheux, Hervé Le Borgne, and Michel Crucianu. Modeling inter and intra-class relations in the triplet loss for zero-shot learning. In *Computer Vision and Pattern Recognition*, pages 10333–10342, 2019.

[LCLBC20] Yannick Le Cacheux, Hervé Le Borgne, and Michel Crucianu. Using sentences as semantic embeddings for large scale zero-shot learning. In *ECCV 2020 Workshop: Transferring and Adapting Source Knowledge in Computer Vision*. Springer, 2020.

[LCPLB20] Yannick Le Cacheux, Adrian Popescu, and Hervé Le Borgne. Webly supervised semantic embeddings for large scale zero-shot learning. *arXiv preprint arXiv:2008.02880*, 2020.

[LDB15] Angeliki Lazaridou, Georgiana Dinu, and Marco Baroni. Hubness and pollution: Delving into cross-space mapping for zero-shot learning. In *Proceedings of the 53rd Annual Meeting of the Association for Computational Linguistics and the 7th International Joint Conference on Natural Language Processing (Volume 1: Long Papers)*, pages 270–280, 2015.

[LEB08] Hugo Larochelle, Dumitru Erhan, and Yoshua Bengio. Zero-data learning of new tasks. In *AAAI Conference on Artificial Intelligence*, volume 2, pages 646–651, 2008.

[LFYFW18] Chung-Wei Lee, Wei Fang, Chih-Kuan Yeh, and Yu-Chiang Frank Wang. Multi-label zero-shot learning with structured knowledge graphs. In *Computer Vision and Pattern Recognition*, pages 1576–1585, 2018.

[LG15] Xin Li and Yuhong Guo. Max-margin zero-shot learning for multi-class classification. In *Artificial Intelligence and Statistics*, pages 626–634, 2015.

[LNH09] Christoph H Lampert, Hannes Nickisch, and Stefan Harmeling. Learning to detect unseen object classes by between-class attribute transfer. In *Computer Vision and Pattern Recognition*, pages 951–958. IEEE, 2009.

[LNH14] Christoph H Lampert, Hannes Nickisch, and Stefan Harmeling. Attribute-based classification for zero-shot visual object categorization. *IEEE T. Pattern Analysis and Machine Intelligence*, 36(3):453–465, 2014.

[MGS14] Thomas Mensink, Efstratios Gavves, and Cees GM Snoek. Costa: Co-occurrence statistics for zero-shot classification. In *Computer Vision and Pattern Recognition*, pages 2441–2448, 2014.

[MH08] Laurens van der Maaten and Geoffrey Hinton. Visualizing data using t-SNE. *Journal of machine learning research*, 9(Nov):2579–2605, 2008.

[Mil95] George A. Miller. Wordnet: A lexical database for english. *Commun. ACM*, 38(11):39–41, November 1995.

[MKRMM18] Ashish Mishra, Shiva Krishna Reddy, Anurag Mittal, and Hema A Murthy. A generative model for zero shot learning using conditional variational autoencoders. In *Proceedings of the IEEE Conference on Computer Vision and Pattern Recognition Workshops*, pages 2188–2196, 2018.

[MO14]   Mehdi Mirza and Simon Osindero. Conditional generative adversarial nets. *arXiv preprint arXiv:1411.1784*, 2014.

[MSC+13] Tomas Mikolov, Ilya Sutskever, Kai Chen, Greg S. Corrado, and Jeff Dean. Distributed representations of words and phrases and their compositionality. In *Advances in Neural Information Processing Systems*, pages 3111–3119, 2013.

[MYX+20] Shaobo Min, Hantao Yao, Hongtao Xie, Chaoqun Wang, Zheng-Jun Zha, and Yongdong Zhang. Domain-aware visual bias eliminating for generalized zero-shot learning. In *Computer Vision and Pattern Recognition*, pages 12664–12673, 2020.

[NMB+14] Mohammad Norouzi, Tomas Mikolov, Samy Bengio, Yoram Singer, Jonathon Shlens, Andrea Frome, Greg S Corrado, and Jeffrey Dean. Zero-shot learning by convex combination of semantic embeddings. In *International Conference on Learning Representations*, pages 488–501, 2014.

[OM13]   Neil O'Hare and Vanessa Murdock. Modeling locations with social media. *Information retrieval*, 16(1):30–62, 2013.

[PG11]   Adrian Popescu and Gregory Grefenstette. Social media driven image retrieval. In *Proceedings of the 1st ACM International Conference on Multimedia Retrieval*, pages 1–8, 2011.

[PPHM09] Mark Palatucci, Dean Pomerleau, Geoffrey E Hinton, and Tom M Mitchell. Zero-shot learning with semantic output codes. In *Advances in Neural Information Processing Systems*, pages 1410–1418, 2009.

[PSM14]   Jeffrey Pennington, Richard Socher, and Christopher D Manning. Glove: Global vectors for word representation. In *Proceedings of the conference on empirical methods in natural language processing*, pages 1532–1543, 2014.

[RALS16] Scott Reed, Zeynep Akata, Honglak Lee, and Bernt Schiele. Learning deep representations of fine-grained visual descriptions. In *Computer Vision and Pattern Recognition*, pages 49–58, 2016.

[RDS+15] Olga Russakovsky, Jia Deng, Hao Su, Jonathan Krause, Sanjeev Satheesh, Sean Ma, Zhiheng Huang, Andrej Karpathy, Aditya Khosla, Michael Bernstein, Alexander C. Berg, and Li Fei-Fei. ImageNet Large Scale Visual Recognition Challenge. *International Journal of Computer Vision (IJCV)*, 115(3):211–252, 2015.

[RNI10]   Miloš Radovanović, Alexandros Nanopoulos, and Mirjana Ivanović. Hubs in space: Popular nearest neighbors in high-dimensional data. *Journal of Machine Learning Research*, 11(Sep):2487–2531, 2010.

[RPT15]   Bernardino Romera-Paredes and Philip Torr. An embarrassingly simple approach to zero-shot learning. In *International Conference on Machine Learning*, pages 2152–2161, 2015.

[RSS11]   Marcus Rohrbach, Michael Stark, and Bernt Schiele. Evaluating knowledge transfer and zero-shot learning in a large-scale setting. In *Computer Vision and Pattern Recognition*, pages 1641–1648, 2011.

[SGMN13] Richard Socher, Milind Ganjoo, Christopher D Manning, and Andrew Ng. Zero-shot learning through cross-modal transfer. In *Advances in Neural Information Processing Systems*, pages 935–943, 2013.

[SLY15]   Kihyuk Sohn, Honglak Lee, and Xinchen Yan. Learning structured output representation using deep conditional generative models. In *Advances in neural information processing systems*, pages 3483–3491, 2015.

[SSH+15] Yutaro Shigeto, Ikumi Suzuki, Kazuo Hara, Masashi Shimbo, and Yuji Matsumoto. Ridge regression, hubness, and zero-shot learning. In *Joint European Conference on Machine Learning and Knowledge Discovery in Databases*, pages 135–151. Springer, 2015.

[SZ14]   Karen Simonyan and Andrew Zisserman. Very deep convolutional networks for large-scale image recognition. *arXiv preprint arXiv:1409.1556*, 2014.

[THJA04] Ioannis Tsochantaridis, Thomas Hofmann, Thorsten Joachims, and Yasemin Altun. Support vector machine learning for interdependent and structured output spaces. In *Proceedings of the twenty-first international conference on Machine learning*, page 104, 2004.

[TJHA05] Ioannis Tsochantaridis, Thorsten Joachims, Thomas Hofmann, and Yasemin Altun. Large margin methods for structured and interdependent output variables. *Journal of machine learning research*, 6(Sep):1453–1484, 2005.

[VAMR18] Vinay Kumar Verma, Gundeep Arora, Ashish Mishra, and Piyush Rai. Generalized zero-shot learning via synthesized examples. In *Computer Vision and Pattern Recognition*, pages 4281–4289, 2018.

[VR17] Vinay Kumar Verma and Piyush Rai. A simple exponential family framework for zero-shot learning. In *Joint European Conference on Machine Learning and Knowledge Discovery in Databases*, pages 792–808. Springer, 2017.

[WBW+11] Catherine Wah, Steve Branson, Peter Welinder, Pietro Perona, and Serge Belongie. The Caltech-UCSD Birds-200-2011 dataset, 2011.

[WW+99] Jason Weston, Chris Watkins, et al. Support vector machines for multi-class pattern recognition. In *European Symposium on Artificial Neural Networks*, volume 99, pages 219–224, 1999.

[WYG18] Xiaolong Wang, Yufei Ye, and Abhinav Gupta. Zero-shot recognition via semantic embeddings and knowledge graphs. In *Computer Vision and Pattern Recognition*, pages 6857–6866, 2018.

[WZYM19] Wei Wang, Vincent W Zheng, Han Yu, and Chunyan Miao. A survey of zero-shot learning: Settings, methods, and applications. *ACM Transactions on Intelligent Systems and Technology (TIST)*, 10(2):1–37, 2019.

[XCH+19] Y. Xian, S. Choudhury, Y. He, B. Schiele, and Z. Akata. Semantic projection network for zero- and few-label semantic segmentation. In *Computer Vision and Pattern Recognition*, pages 8248–8257, 2019.

[XLSA18a] Yongqin Xian, Christoph H Lampert, Bernt Schiele, and Zeynep Akata. Zero-shot learning—a comprehensive evaluation of the good, the bad and the ugly. *IEEE T. Pattern Analysis and Machine Intelligence*, 41(9):2251–2265, 2018.

[XLSA18b] Yongqin Xian, Tobias Lorenz, Bernt Schiele, and Zeynep Akata. Feature generating networks for zero-shot learning. In *Computer Vision and Pattern Recognition*, pages 5542–5551, 2018.

[XSA17] Yongqin Xian, Bernt Schiele, and Zeynep Akata. Zero-shot learning-the good, the bad and the ugly. In *Computer Vision and Pattern Recognition*, pages 4582–4591, 2017.

[XSSA19] Yongqin Xian, Saurabh Sharma, Bernt Schiele, and Zeynep Akata. f-VAEGAN-D2: A feature generating framework for any-shot learning. In *Computer Vision and Pattern Recognition*, pages 10275–10284, 2019.

[ZBS+19] Eloi Zablocki, Patrick Bordes, Laure Soulier, Benjamin Piwowarski, and Patrick Gallinari. Context-aware zero-shot learning for object recognition. In *International Conference on Machine Learning*, pages 7292–7303, 2019.

[ZXG17] Li Zhang, Tao Xiang, and Shaogang Gong. Learning a deep embedding model for zero-shot learning. In *Computer Vision and Pattern Recognition*, pages 2021–2030, 2017.

Chapter 7
Image and Video Captioning Using Deep Architectures

Danny Francis and Benoit Huet

7.1 Introduction

The image captioning task and the video captioning task consist in automatically generating short textual descriptions for images and videos respectively, as represented on Fig. 7.1. Automatic image captioning can be useful for visually impaired people, to give automatically a textual description of images they cannot see, on websites for instance. Automatic video captioning can be used for instance to enrich TV programs with textual information on scenes, that can improve the user experience. Captioning tasks are challenging multimedia tasks as they require to grasp all information contained in a visual document, such as objects, persons, context, actions, location, and to translate this information into text. This task can be compared to a translation task, except that instead of translating a sequence of words in a source language into a sequence of words in a target language, the aim is to translate a photograph or a sequence of frames into a sequence of words. Therefore, most of recent works in captioning rely on the encoder-decoder framework proposed in [SVL14], initially for text translation. In image or video captioning, the encoder aims at deriving an image or video representation, respectively. Recent advances in deep learning have shown to fit very well to that task. In particular, Convolutional Neural Networks (CNNs) have proved to give excellent results in producing highly descriptive image representations or video representations. The decoder part aims at generating a sentence based on the representation produced by the encoder. Recurrent Neural Networks such as Long Short-Term Units (LSTMs) [HS97] and Gated Recurrent Units (GRUs) [CVMG+14] are usually chosen for that task,

D. Francis (✉) · B. Huet
EURECOM, Biot, France

Median Technologies, Valbonne, France
e-mail: Danny.Francis@mediantechnologies.com; Benoit.Huet@mediantechnologies.com

© Springer Nature Switzerland AG 2021
J. Benois-Pineau, A. Zemmari (eds.), *Multi-faceted Deep Learning*,
https://doi.org/10.1007/978-3-030-74478-6_7

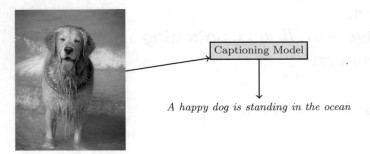

A happy dog is standing in the ocean

Fig. 7.1 Illustration of the captioning task: an image or a video is provided to a captioning model, which is expected to output a descriptive sentence for it

even though Transformer networks [VSP+17] have recently demonstrated high performances in captioning [LZLY19, HWCW19]. Image captioning [VTBE15] and video captioning can seem to be similar tasks, as both of them require to "translate" a visual object into a textual one. However, video captioning poses a problem that makes it more challenging than image captioning: it requires to take into account temporality.

In this chapter, we aim at giving insights on how to generate descriptive sentences from images and videos. In Sect. 7.2, we give an introduction to the basics of image and video captioning. Section 7.3 deals with the optimization of deep captioning models. In Sect. 7.4, we explain how to evaluate the quality of generated captions. Captioning datasets are introduced in Sect. 7.5, and experimental results that can be found in the literature are reported in Sect. 7.6. Section 7.7 briefly introduces related works that have not been mentioned before in this chapter. Section 7.8 concludes the chapter.

7.2 Basics of Visual Captioning

This section aims at explaining the basics of visual captioning. It explains how a captioning model can be built upon a deep neural network architecture.

7.2.1 From Neural Machine Translation to Visual Captioning

Captioning can be seen as a translation task: an image or a sequence of frames, which can be compared to a sequence of words in a source language, have to be translated in a target language. Some pioneering works in image captioning such as [FGI+15] and in video captioning such as [RQT+13] make use of Statistical Machine Translation techniques to generate captions from images or

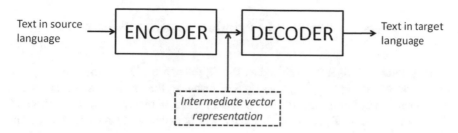

Fig. 7.2 The encoder-decoder scheme. A text in a source language is input to an encoder, which derives a vector representation of that text. That representation is then used by the decoder to generate a sentence in a target language

videos, respectively. However nowadays, most of recent works on image or video captioning rely on Deep Learning techniques, and more particularly on the encoder-decoder framework that has been developed in [SVL14] for text translation, and that we represented on Fig. 7.2. Moreover, attending to the hidden states of the encoder during the decoding phase has shown to give significant improvements in Neural Machine Translation [LPM15], which has been confirmed by [XBK+15] and [YTC+15] in the context of image captioning and video captioning, respectively. For more insights on how attention can improve captioning models, please refer to Sect. 7.2.4.

How exactly can captions be automatically generated? For that purpose, building a language model is necessary. In the next section, we will describe neural language models, which are language models based on neural networks.

7.2.2 Neural Language Models for Image and Video Captioning

A language model represents a probability distribution over sentences. These models are used for sentence generation tasks, including image and video captioning. More precisely, given an image or a video $\mathbf{V}$, and a caption $\mathbf{C} = (\mathbf{w}_1, \ldots, \mathbf{w}_L)$ of length L, a captioning language model outputs a probability $P_{\mathbf{V}}(\mathbf{w}_1, \ldots, \mathbf{w}_L)$—for convenience, in this chapter, $\mathbf{w}_i$ designates a word or its one-hot encoding. Our goal is then to find the most probable sentence with respect to that language model. How to find that sentence? As captions are not of fixed length, and the vocabulary upon which they are generated can be enormous, deriving the probability of all possible captions is not feasible. Therefore what is generally done is generating it word-by-word using a greedy search or a beam search algorithm (see Sect. 7.3.5.1) based on the following formula:

$$P_V(\mathbf{w}_1, \ldots, \mathbf{w}_L) = \left(\prod_{i=2}^{L} P_V(\mathbf{w}_i | \mathbf{w}_1, \ldots, \mathbf{w}_{i-1}) \right) \times P_V(\mathbf{w}_1).$$

In general, a ⟨START⟩ token and an ⟨END⟩ token are added at the beginning and at the end of the caption, respectively. For instance, the caption "a dog is playing with a cat" would be processed as "⟨START⟩ a dog is playing with a cat ⟨END⟩". Therefore, the term $P_V(\mathbf{w}_1) = P_V(⟨START⟩)$ is actually always equal to 1. The above formula then becomes:

$$P_V(\mathbf{w}_1, \ldots, \mathbf{w}_L) = \prod_{i=2}^{L} P_V(\mathbf{w}_i | \mathbf{w}_1, \ldots, \mathbf{w}_{i-1}).$$

Let us now dig into neural network architectures on which neural language models are built. Recurrent Neural Networks (RNNs) are an intuitive solution to model a language, as their outputs depend on previous inputs. More formally, an RNN is a neural network architectured to process sequences of inputs of variable length: if $(\mathbf{x}_1, \ldots, \mathbf{x}_L)$ is the sequence to be processed by the RNN, a sequence $(\mathbf{h}_1, \ldots, \mathbf{h}_L)$ can be output as follows:

$$\mathbf{h}_{i+1} = \text{RNN}(\mathbf{x}_i, \mathbf{h}_i).$$

Usually, $\mathbf{h}_i$ is called the i-th hidden state of the RNN. The initial hidden state $\mathbf{h}_1$ is generally set to zero.

Unlike some early works on captioning such as [KFF15], most recent works on captioning with RNNs are based on Long Short-Term Memory networks (LSTMs) [HS97] or on Gated Recurrent Units (GRUs) [CVMG+14], because these kinds of RNNs can grasp long-term dependencies and better avoid vanishing and exploding gradient issues. It should also be noted that recently, Transformer networks [VSP+17] have shown state-of-the-art performances on captioning tasks [HWCW19].

How do the inputs and the outputs of an RNN relate to images, videos and captions? In the next section, we will give a very generic method for generating captions. The reader will find other methods in references given in Sect. 7.6.

7.2.3 Building a Deep Caption Generation Model: A Generic Method

In this section we will propose a generic method for building a caption generation model. It is very similar to the method used in [VTBE15]. Since then, more efficient methods have been proposed in the literature. However, these methods are generally more or less complex derivations of [VTBE15]. For the sake of simplicity, we decided to introduce the most generic method. The interested reader can refer to Sect. 7.6 for more references. On top of that, a more efficient variant based on attention mechanisms will be discussed in Sect. 7.2.4.

7.2.3.1   Encoding Images and Videos

The first part to observe corresponds to the encoder of the encoder-decoder design. It consists in deriving vector representations of images—or videos in the case of video captioning. More specifically, the object to caption is transformed into a fixed-length vector of features. Since deep features have shown tremendous results at the ImageNet Large Scale Visual Recognition Challenge in 2012 [KSH12], a deep convolutional neural network is generally employed to derive features vectors for images and videos. More formally, if $\mathbf{V}$ is the image or the video to be captioned, our first step transforms $\mathbf{V}$ into a features vector of fixed-length CNN($\mathbf{V}$).

7.2.3.2   Decoding Images and Videos

The second part corresponds to the decoder of the encoder-decoder design. Even though other deep architectures like Transformer networks can be employed to build a model language, we will assume in this section that the model language is built upon an RNN—in practice, the RNN is an LSTM or a GRU. Our goal is to compute a probability for any sentence $(\mathbf{w}_1, \ldots, \mathbf{w}_L)$ of any length L. Words $\mathbf{w}_i$ with $i \in \{1, \ldots, L\}$—represented by one-hot vectors of dimension D where D is the size of the vocabulary—must be converted into vectors through a fully-connected layer— also called an embedding layer in the context of word vectors. More formally, a word $\mathbf{w}$ is converted into a word vector $\mathbf{v_w}$ through an embedding layer Emb($\bullet; \theta_e$) where θ_e is a set of trainable parameters:

$$\mathbf{v_w} = \text{Emb}(\mathbf{w}; \theta_e).$$

As the language model we want to build depends on the image or the video $\mathbf{V}$, inputs of the RNN must depend on words and on $\mathbf{V}$. One way to achieve that goal is to concatenate word embeddings with CNN($\mathbf{V}$). The $(i + 1)$-th hidden state of the RNN is then computed as follows:

$$\mathbf{h}_{i+1} = \text{RNN}(\mathbf{x}_i, \mathbf{h}_i; \theta_r),$$

where $\mathbf{x}_i = \text{Concat}(\mathbf{v_w}, \text{CNN}(\mathbf{V}))$ and θ_r is a set of trainable parameters.

Each hidden state $\mathbf{h}_i$ depends on the previous hidden state, which means that $\mathbf{h}_{i+1}$ depends on words $\mathbf{w}_1, \ldots, \mathbf{w}_i$. Therefore, we can make the assumption that the probability $P_{\mathbf{V}}(\mathbf{w}_{i+1}|\mathbf{w}_1, \ldots, \mathbf{w}_i)$ can be derived based on $\mathbf{h}_{i+1}$. This can be achieved through a fully-connected layer Pred($\bullet; \theta_p$) of output dimension D and trainable parameters θ_p followed by a softmax function, assigning a probability to each word of the vocabulary:

$$P_V(\mathbf{w}_{i+1}|\mathbf{w}_1, \ldots, \mathbf{w}_i) = \text{Softmax}_{\mathbf{w}_{i+1}}(\text{Pred}(\mathbf{h}_{i+1}; \theta_p)).$$

Altogether, we can then compute the probability of a sentence $(\mathbf{w}_1, \ldots, \mathbf{w}_L)$ as follows:

$$P_V(\mathbf{w}_1, \ldots, \mathbf{w}_L) = \prod_{i=2}^{L} \text{Softmax}_{\mathbf{w}_i}(\text{Pred}(\mathbf{h}_i; \theta_p)).$$

Parameters θ_e, θ_r and θ_p are optimized following one of the objectives mentioned in Sect. 7.3.

7.2.4 Improving Captioning Models with Attention

First works in image and video captioning were based on the simple Encoder-Decoder scheme we described in Sect. 7.2.1, with no attention mechanism. NeuralTalk by Karpathy et al. [KFF15] and Show and Tell by Vinyals et al. [VTBE15] fall into this category. After the image captioning model proposed by [XBK+15], using an attention mechanism to improve that Encoder-Decoder scheme has quickly become a standard in image captioning. From that Encoder-Decoder with attention baseline, different models for image captioning have been proposed, often taking into account local features [GCWC18] or object detections [AHB+18] instead of only global features. Some works such as [FH19] use attention mechanisms to deal with spatio-temporal features in video captioning. This section elaborates on how attention mechanisms can take advantage of multiple features.

Attention mechanisms have been proved to improve Neural Machine Translation (NMT) performances. An example has been depicted on Fig. 7.3: at each decoding step, the decoder attends to relevant parts of the source language sentence to decode

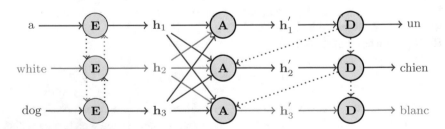

Fig. 7.3 Attention mechanism for neural machine translation. The attention module guides the prediction of words in the target language by inputting weighted sums of the encoder hidden states to the decoder, based on previous predictions

a corresponding sentence in the target language, based on previously decoded words.

The attention mechanisms that are usually employed in captioning tasks are inspired by attention mechanisms used in NMT. Unlike what has been presented in Sect. 7.2.3, they require that several features vectors are derived from the image or the video to be captioned, instead of a single one. They work as follows: the i-th hidden state from which the i-th word has been generated is used as a reference to help the decoder to generate the following $(i + 1)$-th word. More formally, if $\mathbf{h}_i$ is the hidden state from which the last word has been generated, and $\mathbf{v}_1, \ldots, \mathbf{v}_n$ are n visual features vectors (corresponding to image regions or video regions), then the features vector that will be input to the decoder to help it generate the next word is:

$$\mathbf{v} = \sum_{j=1}^{n} \alpha_j \mathbf{v}_j, \tag{7.1}$$

where the α_j are non-negative weights. These weights are derived through the following computations:

$$(\alpha_1, \ldots, \alpha_n) = \text{Softmax}(a_1, \ldots, a_n), \tag{7.2}$$

where a_j for $j \in \{1, \ldots, n\}$ are defined as:

$$a_j = f(\mathbf{w}_v \cdot \mathbf{v}_j + \mathbf{w}_h \cdot \mathbf{h}_i + b), \tag{7.3}$$

where f is an activation function. In the last equation, $\mathbf{w}_v$, $\mathbf{w}_h$ and b are optimized when the captioning model is trained, as explained in Sect. 7.3. The weighted sum of features vectors that is thus obtained emphasizes important regions for word prediction, and obliterates non-relevant ones. Attention mechanisms are nowadays widely used in captioning models because they improve performances a lot, as shown in Sect. 7.6.

7.3 Optimization of Visual Captioning Models

In the previous section, we saw that captioning models involved several parameters to optimize: word embedding layer, RNN decoder, attention mechanism if any, word prediction layer. The reader will have noticed that we did not mention trainable parameters regarding the encoder. The reason is that, as we will explain in Sect. 7.3.1, existing captioning datasets are not big enough to allow us to train efficient deep CNNs encoders: it leads to overfitting. In this section, we will elaborate on pretrained encoders and on objective functions for training decoders.

7.3.1 Pretraining Visual Features

Captioning models usually involve visual features extracted using pretrained models. As mentioned before, the reason is that captioning datasets are not big enough to allow training whole CNNs for features extraction. We will introduce three datasets that are commonly used to pretrain CNN encoders, with a view to use them in a captioning model.

The first dataset we introduce is ImageNet [DDS+09]. ImageNet contains images classified with respect to WordNet [Mil95] classes. Usually, 1000 classes are selected for training classifiers. These classifiers can then be used as features extractors by removing their last prediction layer. Figure 7.4 gives an example of four images of the "Chihuahua" class of ImageNet.

More recently, another big dataset of annotated images has been released. VisualGenome [KZG+17] is a dataset of more than 100,000 images, with very dense region annotations, as shown on Fig. 7.5. The interest of using a regional features extractor trained on VisualGenome is that local features can be used in the context of the attention mechanism we described in Sect. 7.2.4, which improves performances of captioning models. Features extractors that can be trained on that dataset include the widely used Faster R-CNN [RHGS16]: in Sect. 7.6, we show that image captioning models based on a Faster R-CNN trained on VisualGenome give state-of-the-art performance.

Video captioning can benefit from using features extracted through 3D-CNNs. An example of a widely used dataset for pretraining global video features extractors is Kinetics [CZ17]. In Kinetics, videos of different activities are classified in 700 different classes. An example of an annotated video from Kinetics is shown on Fig. 7.6.

Fig. 7.4 ImageNet: Four sample images from the "Chihuahua" class

Fig. 7.5 VisualGenome: an example of an image with several bounding boxes. These bounding boxes are annotated with classes ("dog", "shirt", "bottle", . . .), descriptions ("a white plate on a table", "a framed poster on a wall", . . .), attributes ("shirt is red", "dog is small", . . .) and relations ("woman WEARING shirt", "plate ON table", . . .)

Fig. 7.6   Kinetics: frames extracted from a video clip belonging to the "dog grooming" class

7.3.2 Optimizing the Language Model with a Cross-Entropy Loss

The goal of a captioning model is to output a descriptive sentence given an image or video input. That task is usually performed using a cross-entropy function as the loss function during training. Mathematically, given an image or video $\mathbf{V}$, and a ground-truth caption $\mathbf{C} = (\mathbf{w}_1, \ldots, \mathbf{w}_L)$ of length L, the cross-entropy loss is defined as:

$$\mathcal{L}(\theta) = -\sum_{i=2}^{L} \log\left(P_{\mathbf{V}}\left(\mathbf{w}_i | \mathbf{w}_1, \ldots, \mathbf{w}_{i-1}; \theta\right)\right) \tag{7.4}$$

where θ is the set of trainable weights of our captioning model.

The problem of the cross-entropy loss is that there is a discrepancy with the metrics that are used for evaluation—see Sect. 7.4 for the description of metrics used to evaluate captioning models. In the next section, we will introduce another loss function dealing with that problem.

7.3.3 Optimizing the Language Model by Reinforcement Learning

For the reason mentioned in the previous section, some works such as [RMM+17] introduced a reinforcement learning loss—or REINFORCE loss—aiming at directly optimizing the captioning model with respect to its evaluation metric. More formally, let $r(\mathbf{C})$ be the score assigned to a caption $\mathbf{C}$ for a given input image or video $\mathbf{V}$ ($r(\mathbf{C})$ is high when the caption is good and low otherwise). Then, the new loss function is:

$$\mathcal{L}(\theta) = -r(\mathbf{C}), \quad \mathbf{C} \sim P_{\mathbf{V}}(\bullet; \theta). \tag{7.5}$$

That loss function is not differentiable, therefore it is not usable directly for gradient descent. However the gradient of the expectation of L can be written as:

$$\begin{aligned}
\nabla_\theta \mathrm{E}\left[\mathcal{L}(\theta)\right] &= \nabla_\theta \mathrm{E}\left[-r(\mathbf{C})\right] \\
&= -\sum_{\mathbf{C}} r(\mathbf{C}) \nabla_\theta P_{\mathbf{V}}(\mathbf{C}; \theta) \\
&= -\sum_{\mathbf{C}} r(\mathbf{C}) P_{\mathbf{V}}(\mathbf{C}; \theta) \nabla_\theta \log(P_{\mathbf{V}}(\mathbf{C}; \theta)) \\
&= -\mathrm{E}\left[r(\mathbf{C}) \nabla_\theta \log(P_{\mathbf{V}}(\mathbf{C}; \theta))\right] \\
&= \nabla_\theta \mathrm{E}\left[-r(\mathbf{C}) \log(P_{\mathbf{V}}(\mathbf{C}; \theta))\right]
\end{aligned} \tag{7.6}$$

Therefore, it boils down to optimizing on the loss function $\mathcal{L}'$, defined as:

$$\mathcal{L}'(\theta) = -r(\mathbf{C})\log(P_V(\mathbf{C})), \quad \mathbf{C} \sim P_V(\bullet; \theta)$$
$$= -r(\mathbf{C}) \sum_{l=2}^{L} \log\left(P_V\left(\mathbf{w}_l | \mathbf{w}_1, \ldots, \mathbf{w}_{l-1}; \theta\right)\right). \tag{7.7}$$

As shown in Sect. 7.6, optimizing captioning models with the aforementioned reinforcement learning loss leads to high improvements with respect to models optimized based on the cross-entropy loss.

7.3.4 Regularizing Captioning Models

In addition to the loss functions we introduced in previous sections, some regularization terms can be added to the objective to increase the performances of captioning models. We will cover two types of regularization methods that can specifically enhance the quality of captions generated by a captioning model: matching regularization and attribute prediction.

7.3.4.1 Matching Regularization

Matching images or videos and corresponding texts is important for indexing and retrieval tasks. That matching task is different from the captioning task that we are discussing in this chapter. However, some works have shown that they can be complementary: jointly training one model for matching images or videos and captions and one model for generating captions improve results in both tasks [GCJ+18, LRY18, FH19].

Generally speaking, a vision-text matching model is composed of two embedding models aiming at projecting two modalities in a same vector space. The image or video embedding model is generally composed of a features extractor, followed by a fully-connected layer. The text embedding model is generally built upon an RNN—often an LSTM or a GRU. More formally, if $vEmb(\bullet; \theta_v)$ is the visual embedding model and $cEmb(\bullet; \theta_c)$ is the textual embedding model, where θ_v and θ_c are trainable parameters, the vision-text matching model is optimized to minimize the loss function $\mathcal{L}_m$, defined as follows:

$$\mathcal{L}_{m_1}(\theta_v, \theta_c) = \max(0, \alpha - S(\mathbf{v}, \mathbf{c}) + S(\mathbf{v}, \overline{\mathbf{c}}))$$

$$\mathcal{L}_{m_2}(\theta_v, \theta_c) = \max(0, \alpha - S(\mathbf{v}, \mathbf{c}) + S(\overline{\mathbf{v}}, \mathbf{c}))$$

$$\mathcal{L}_m(\theta_v, \theta_c) = \mathcal{L}_{m_1}(\theta_v, \theta_c) + \mathcal{L}_{m_2}(\theta_v, \theta_c),$$

where S is a similarity function, which is often a cosine similarity—but not always, see for example [VKFU15], $\mathbf{V}$ and $\mathbf{C}$ are an image or a video and a corresponding caption, $\overline{\mathbf{V}}$ is a non-corresponding image or video and $\overline{\mathbf{C}}$ is a non-corresponding caption, $\mathbf{v} = \text{vEmb}(\mathbf{V}; \theta_v)$, $\mathbf{c} = \text{cEmb}(\mathbf{C}; \theta_c)$, $\overline{\mathbf{v}} = \text{vEmb}(\overline{\mathbf{V}}; \theta_v)$ and $\overline{\mathbf{c}} = \text{cEmb}(\overline{\mathbf{C}}; \theta_c)$.

The interpretation of adding a matching regularization term can be the following: the additive matching term in the loss function enforces visual representations and textual representation to be close in a common space, and therefore enforces the captioning model to discriminate between similar images, videos or captions.

7.3.4.2 Attribute Prediction

Some works have shown that predicting image or video attributes could boost results in captioning tasks. More specifically, predicting which words can be used to describe an image or a video can boost captioning performances [YPL+17, PYLM17]. In that case, attributes are the words that we are trying to predict. On top of that, predicted attributes can be injected in the decoder of the captioning model to improve the quality of generated captions [YPL+17, PYLM17].

7.3.5 Improving Captions at Inference Time

Several tricks can help improving the quality of captions at inference time. We will cover two of them: beam search and reranking.

7.3.5.1 Greedy Search vs Beam Search

In Sect. 7.2.3, we saw that the probability that a caption corresponds to a given image or video could be computed iteratively based on the hidden states of the decoder. In particular, we saw that if $\mathbf{V}$ is an image or a video, $\mathbf{w}_i$ is the i-th word of the caption to evaluate, and $\mathbf{h}_i$ is the i-th hidden state of the decoder, we have the relation:

$$P_{\mathbf{V}}(\mathbf{w}_1, \ldots, \mathbf{w}_L) = \prod_{i=2}^{L} \text{Softmax}_{\mathbf{w}_i}(\text{Pred}(\mathbf{h}_i; \theta_p)).$$

At inference time, the captioning task boils down to finding the most probable caption for an input image or video. An easy way to achieve that goal is to perform a greedy search: at each step, we pick the most probable word, as follows:

$$\mathbf{w}_i = \text{argmax}_{\mathbf{w}}\left(\text{Softmax}_{\mathbf{w}}(\text{Pred}(\mathbf{h}_i; \theta_p))\right).$$

However, the greedy algorithm is biased: if the most probable caption has only one less probable word, it will never be selected by the greedy algorithm. To tackle this problem, many works use a beam search algorithm [MXY+14, FGI+15,

RMM+17, KFF15]. Instead of keeping track of only one possible sentence, the beam search keeps in memory B different possible sentences, where B is the beam size. At the end of the algorithm, the best caption among those that have been kept in memory is selected. It can be noticed that if $B = 1$, we obtain a greedy algorithm: greedy search is a particular case of beam search.

7.3.5.2 Captions Reranking

In addition to the beam search algorithm we mentioned in Sect. 7.3.5.1, some works have shown that taking the most probable caption was not always the best possible solution. Caption reranking can also boost results. Caption reranking consists in generating not only one, but several sentences, and then computing a score for each of these sentences to refine their ranking. For instance, it has been shown that using a matching model such as the ones we mentioned in Sect. 7.3.4.1 to evaluate how well sentences correspond to images or videos improved captioning results [DLL+16].

Let us now describe the evaluation metrics that are used for evaluating captioning models.

7.4 Evaluation of Captions Quality

Several metrics have been designed to evaluate translation tasks or captioning models. As we explained before, the captioning task can be seen as a translation task: evaluation metrics for translation tasks are actually widely used to evaluate captioning. Let us describe the ones we will use in this chapter.

7.4.1 *BLEU-n*

The BLEU-n metric has been defined in [PRWZ02]. It counts the proportion of n-grams in the reference sentences that appear in the candidate sentence, which is called the n-gram precision. If an n-gram is not more than p times in any reference sentence, we count it not more than p times in the candidate, even if there are more occurrences of that n-gram in the candidate sentence. Let us give an example:

Candidate sentence: "a dog is lying on a blue couch"
Reference sentence 1: "there is a black dog on a couch"
Reference sentence 2: "the dog is lying on the couch"
BLEU-1 = 0.875 (a dog is lying on a blue couch)
BLEU-2 = 0.429 (a dog, dog is, is lying, lying on, on a, a blue, blue couch).

7.4.2 ROUGE$_L$

The ROUGE$_L$ [Lin04] metric is based on the longest common subsequence (LCS) between the candidate sentence and a reference sentence. It is computed as follows:

$$\text{ROUGE}_L(c, r) = \frac{(1 + \beta^2)RP}{R + \beta^2 P}, \tag{7.8}$$

where c is the candidate sentence, r a reference, $P = \frac{\text{LCS}(c,r)}{\text{length}(c)}$, $R = \frac{\text{LCS}(c,r)}{\text{length}(r)}$ and $\beta = \frac{P}{R}$. In the previous example, the ROUGE$_L$ score of the candidate sentence with respect to the first reference sentence is 0.625, its ROUGE$_L$ score with respect to the second candidate sentence is 0.661. Therefore, its global ROUGE$_L$ score is 0.643.

For convenience, the ROUGE$_L$ metric will be designated as ROUGE in the rest of this chapter.

7.4.3 METEOR

The METEOR [BL05] metric is computed by first aligning candidate and reference sentences. Once the alignment done, a similarity score is computed between both sentences.

More precisely, the first step is to map unigrams in both sentences if they are exactly the same. Then, remaining unigrams are mapped together if they are equal after Porter stemming. The last matching step consists in matching unigrams if they are synonyms according to the WordNet hierarchy [Mil95].

Once all three mapping steps have been performed, a score F_{mean} is derived as follows:

$$P = \frac{\text{number of mapped unigrams in } c}{\text{length of } c},$$

$$R = \frac{\text{number of mapped unigrams in } c}{\text{length of } r},$$

$$F_{\text{mean}} = \frac{10PR}{9P + R},$$

where c is the candidate sentence and r is the reference sentence.

Then, a penalty is applied to that score based on the number of chunk-to-chunk maps in the candidate sentence: if the candidate sentence is mapped in one chunk to the reference sentence then the penalty is minimized, and it is maximized when chunks consist in unigrams. The penalty is defined as:

$$\text{Penalty} = 0.5 \times \left(\frac{\text{number of chunk-to-chunk maps in } c}{\text{number of mapped unigrams in } c} \right)^3.$$

The final METEOR score is given by the following formula:

$$\text{METEOR} = F_{\text{mean}} \times (1 - \text{Penalty})$$

7.4.4 CIDEr$_D$

The CIDEr metric [VLZP15] is derived as an average of cosine similarities of TF-IDF vectors of references and candidate captions, based on 1-gram, 2-gram, 3-gram and 4-gram tokenizations. The CIDEr$_D$ metric, also introduced in [VLZP15], is an improvement of CIDEr that makes gaming more difficult.

More specifically, if $\Omega^{(n)} = \{\omega_1^{(n)}, \ldots, \omega_{D_n}^{(n)}\}$ is the set of all n-grams, we compute the TF-IDF of a sentence s with respect to $\omega_k^{(n)}$ as follows:

$$g_k^{(n)}(s) = \frac{h_k(s)}{\sum_{\omega_l^{(n)} \in \Omega} h_l(s)} \times \log \left(\frac{|I|}{\sum_{I_p \in I} \min\left(1, \sum_q h_k(r_{pq})\right)} \right),$$

where $h_k(s)$ is the number of times $\omega_k^{(n)}$ appears in s, I is the set of all possible images or videos, and the r_{pq} are reference sentences. Then, the n-gram CIDER$_{D_n}$ score is derived as follows:

$$\text{CIDEr}_{D_n}(c, R_i) = \frac{10}{m} \sum_j e^{-\frac{(l(c) - l(r_{ij}))^2}{72}} \times \frac{\min(\mathbf{g}^{(n)}(c), \mathbf{g}^{(n)}(r_{ij})) \cdot \mathbf{g}^{(n)}(r_{ij})}{\|\mathbf{g}^{(n)}(c)\| \times \|\mathbf{g}^{(n)}(r_{ij})\|}$$

where $l(s)$ denotes the length of a sentence s, $\mathbf{g}^{(n)}(s)$ denotes the vector of all $g_k^{(n)}(s)$ and c is the candidate caption. The actual CIDEr$_D$ score is eventually derived as:

$$\text{CIDEr}_D = \frac{1}{4} \sum_{i=1}^4 \text{CIDEr}_{D_i}.$$

For convenience the CIDEr$_D$ metric will be designated as CIDEr in the rest of this chapter.

Reference captions

- a dog laying down on a fluffy carpet
- floor eye view of a supine dog hoping for some attention
- a brown and white dog laying on a carpet under a table
- a corgi dog resting on frizzy beige carpet
- a picture of a dog laying under a table on the rug

Fig. 7.7 An image captioning sample: a photograph of a dog is annotated with five captions

7.5 Captioning Datasets

In this section, we will introduce three image captioning datasets and two video captioning datasets. These captioning datasets are all composed of images or videos, with several human-written captions. A sample of an image captioning dataset is shown on Fig. 7.7.

7.5.1 Image Captioning Datasets

Flickr8k [HYH13] and Flickr30k [PWC+15] are two datasets based on Flickr images. Flickr30k contains 30,000 images with five captions each, and is an extension of Flickr8k, which contains 8000 images with five captions each. In the literature, each of these two datasets is commonly divided into 1000 images for validation, 1000 images for testing, and the rest for training. MSCOCO [LMB+14] is a much bigger dataset than the aforementioned ones, as it contains about 120,000 images, with five captions each. It is usually divided into 5000 images for validation, 5000 images for testing, and the rest for training. All these figures have been reported in Table 7.1.

7.5.2 Video Captioning Datasets

MSVD [CD11]—also called YouTube2Text—is a video captioning dataset based on YouTube videos. It contains 1970 videos, with about 35 captions each. In the

Table 7.1 Datasets for image captioning

	Flickr8k	Flickr30k	MSCOCO
Number of training images	6000	28,000	110,000
Number of validation images	1000	1000	5000
Number of test images	1000	1000	5000
Captions per image	5	5	5

Table 7.2 Datasets for video captioning

	MSVD (YouTube2Text)	MSR-VTT
Number of training images	1100	6500
Number of validation images	100	500
Number of test images	670	3000
Captions per image	35	20

literature, it is usually divided into 100 videos for validation, 670 videos for testing and the rest for training. MSR-VTT [XMYR16] is a much bigger dataset than MSVD, as it contains 10,000 videos with 20 captions each. It is usually divided into 500 images for validation, 3000 images for testing, and the rest for training. All these figures have been reported in Table 7.2.

7.6   Results Reported in Published Works

In this section, we report some experimental results that can be found in the literature.

7.6.1   Image Captioning

Some image captioning results from the literature on Flickr8k, Flickr30k and MSCOCO have been reported in Tables 7.3, 7.4, and 7.5, respectively. Table 7.6 gives some features of the models for which we reported results. Nowadays, most powerful image captioning models use Transformer networks as decoders, and are trained using the REINFORCE loss we introduced in Sect. 7.3.3.

7.6.2   Video Captioning

Some video captioning results from the literature on MSVD and MSR-VTT have been reported in Tables 7.7 and 7.8, respectively. Table 7.9 gives some features of the

Table 7.3 Results on
Flickr8k

Model	Bleu-4	METEOR	CIDEr
[KFF15]	16.0	16.7	31.8
[XBK+15]	21.3	20.3	–
[PGH+16]	25.0	–	–

Table 7.4 Results on
Flickr30k

Model	Bleu-4	METEOR	CIDEr
[KFF15]	15.7	15.3	24.7
[DAHG+15]	16.5	–	–
[XBK+15]	19.9	18.5	–
[PGH+16]	25	–	–

Table 7.5 Results on
MSCOCO

Model	Bleu-4	ROUGE	METEOR	CIDEr
[KFF15]	23.0	–	19.5	66.0
[DAHG+15]	24.9	–	–	–
[XBK+15]	25.0	–	23.0	–
[FGI+15]	–	–	23.6	–
[VTBE15]	27.7	–	23.7	85.5
[PGH+16]	28	–	24	90
[MXY+14]	30.4	51.9	23.9	93.8
[CZX+17]	31.1	–	25.0	–
[YPL+17]	32.6	54.0	25.4	100.2
[RMM+17]	35.4	56.6	27.1	117.5
[AHB+18]	36.3	56.9	27.7	120.1
[LZLY19]	39.9	59.0	28.9	127.6
[HWCW19]	40.2	59.4	29.3	132.0

models for which we reported results. Nowadays, most powerful video captioning models are trained using the REINFORCE loss we introduced in Sect. 7.3.3.

7.7 Other Related Works

In this section, we will briefly introduce some existing works on topics related to image and video captioning. It includes image dense captioning, video dense captioning and movie captioning.

7.7.1 Image Dense Captioning

Image dense captioning consists in generating not only one caption for a given image, but multiple ones, corresponding to different regions of interest [JKFF16].

Table 7.6 Some image captioning models. We report the type of encoder they use, the type of decoder, if they use an attention mechanism, if they use local features and the objective function. "BRNN" stands for "bidirectional RNN". "MELM" stands for "maximum entropy language model". "m-RNN" stands for "multimodal RNN". "ELBO" stands for "evidence lower bound"

Model	Encoder	Decoder	Attention	Local features	Text objective
[KFF15]	VGG16	BRNN	No	Yes	Cross-Entropy
[DAHG+15]	CaffeNet	LSTM	No	No	Cross-Entropy
[XBK+15]	VGG16	LSTM	Yes	Yes	Cross-Entropy
[FGI+15]	VGG16	MELM	No	Yes	Cross-Entropy
[VTBE15]	GoogLeNet	LSTM	No	No	Cross-Entropy
[PGH+16]	VAE	GRU	No	No	ELBO
[MXY+14]	VGG16	m-RNN	No	No	Cross-Entropy
[CZX+17]	ResNet-152	LSTM	Yes	Yes	Cross-Entropy
[YPL+17]	GoogleNet	LSTM	No	Yes	Cross-Entropy
[RMM+17]	ResNet-101	LSTM	Yes	Yes	REINFORCE
[AHB+18]	VG-ResNet-101	LSTM	Yes	Yes	REINFORCE
[LZLY19]	VG-VGG16	Transformer	Yes	Yes	REINFORCE
[HWCW19]	VG-ResNet-101	Transformer	Yes	Yes	REINFORCE

Table 7.7 Results on MSVD (YouTube2Text)

Model	Bleu-4	ROUGE	METEOR	CIDEr
[SGG+17]	53.0	–	33.6	73.8
[HHL+17]	53.9	–	32.2	67.4
[GGZ+17]	50.8	–	33.3	74.8
[PYLM17]	52.8	–	33.5	74.0
[GGH+17]	51.1	–	33.5	77.7
[PB17]	54.4	72.2	34.9	88.6
[SGG+18]	53.3	–	33.8	74.8
[WMZL18]	52.3	69.8	34.1	80.3
[PZW+19]	48.6	71.9	35.1	92.2
[FH19]	55.1	72.7	35.4	86.8

The recent dataset VisualGenome that we introduced in Sect. 7.3.1 contains dense region captions, that can be used to train dense captioning models.

7.7.2 Video Dense Captioning

Unlike image dense captioning, video dense captioning does not aim at generating captions for regions of interest. In video dense captioning, we aim at generating captions for interesting events that occur in a video. It involves capturing dependencies between different moments of a video. The dataset ActivityNet Captions

Table 7.8 Results on MSR-VTT. The first reported results (in italic characters) have been computed on the validation set of MSR-VTT

Model	Bleu-4	ROUGE	METEOR	CIDEr
[RDP+16]	*39.5*	*61.0*	*27.7*	*44.2*
[JCC+16]	40.8	60.9	28.2	44.8
[DLL+16]	38.7	58.7	26.9	45.9
[SGG+17]	38.3	–	26.3	–
[HHL+17]	39.7	–	25.5	40.0
[GGZ+17]	38.0	–	26.1	43.2
[PB17]	40.5	61.4	28.4	51.7
[SGG+18]	39.8	59.3	26.1	40.9
[WMZL18]	39.1	59.3	26.6	42.7
[WCW+18]	41.3	61.7	28.7	48.0
[PZW+19]	40.4	60.7	28.1	47.1
[FH19]	40.7	61.2	27.6	44.8

Table 7.9 Some video captioning models. We report the 2D encoder used if any, the 3D encoder used if any, the decoder, if an attention mechanism was used, if local features are used, if audio features are used and which objective function has been used. "XE" stands for "Cross-Entropy"

Model	2D encoder	3D encoder	Decoder	Att.	Local	Audio	Loss
[RDP+16]	ResNet	C3D	LSTM	No	No	Yes	XE
[JCC+16]	None	C3D	LSTM	No	No	Yes	XE
[DLL+16]	Googlenet-bu4k	C3D	LSTM	No	No	No	XE
[SGG+17]	ResNet-152	None	LSTM	Yes	No	No	XE
[HHL+17]	VGG16	C3D	LSTM	Yes	No	Yes	XE
[GGZ+17]	Inception-v3	None	LSTM	Yes	No	No	XE
[PYLM17]	VGG19	C3D	LSTM	No	No	No	XE
[GGH+17]	ResNet-152	C3D	LSTM	No	No	No	XE
[PB17]	Inception-v4	None	LSTM	Yes	No	No	REINFORCE
[SGG+18]	ResNet-152	None	LSTM	No	No	No	ELBO
[WMZL18]	Inception-V4	None	LSTM	Yes	No	No	XE
[WCW+18]	ResNet-152	None	LSTM	Yes	No	No	REINFORCE
[PZW+19]	ResNet-101	ResNeXt-101	LSTM	Yes	No	No	XE
[FH19]	ResNet-152	I3D	LSTM	Yes	Yes	No	XE

[KHR+17] contains 20,000 videos with 100,000 different captions corresponding to different events.

7.7.3 Movie Captioning

Movie captioning may seem very similar to video captioning, but is it not. Movie captioning models must capture the narrative of a movie clip to describe not only what is happening in the video, but also how events and characters relate to form a storyline. MPII-MD [RRTS15] and M-VAD [TPLC15] are examples of movie captioning datasets. More recently, the dataset M-VAD Names [PCB+19] extended M-VAD by adding characters names in the annotations: before that, character names were replaced by a SOMEONE token.

7.8 Conclusion

This chapter has given an overview of deep learning applied to image and video captioning, from first works to latest developments. In particular, we mentioned how to effectively build and train a deep captioning model with attention mechanisms, and discussed some improvements that have been established in the literature. Late works in captioning involve the use of local features, attention mechanisms, including especially the use of Transformer networks as an efficient replacement of the widely used LSTMs and GRUs.

Image and video captioning are hot topics, inducing every year many innovative scientific publications, and also giving birth to related and not less interesting research fields such as dense captioning or movie captioning.

References

[AHB+18] Peter Anderson, Xiaodong He, Chris Buehler, Damien Teney, Mark Johnson, Stephen Gould, and Lei Zhang. Bottom-up and top-down attention for image captioning and visual question answering. In *Proceedings of the IEEE conference on computer vision and pattern recognition*, pages 6077–6086, 2018.

[BL05] Satanjeev Banerjee and Alon Lavie. Meteor: An automatic metric for MT evaluation with improved correlation with human judgments. In *Proceedings of the ACL workshop on intrinsic and extrinsic evaluation measures for machine translation and/or summarization*, pages 65–72, 2005.

[CD11] David Chen and William B Dolan. Collecting highly parallel data for paraphrase evaluation. In *Proceedings of the 49th Annual Meeting of the Association for Computational Linguistics: Human Language Technologies*, pages 190–200, 2011.

[CVMG+14] Kyunghyun Cho, Bart Van Merriënboer, Caglar Gulcehre, Dzmitry Bahdanau, Fethi Bougares, Holger Schwenk, and Yoshua Bengio. Learning phrase representations using RNN encoder-decoder for statistical machine translation. *arXiv preprint arXiv:1406.1078*, 2014.

[CZ17] Joao Carreira and Andrew Zisserman. Quo vadis, action recognition? a new model and the kinetics dataset. In *proceedings of the IEEE Conference on Computer Vision and Pattern Recognition*, pages 6299–6308, 2017.

[CZX+17] Long Chen, Hanwang Zhang, Jun Xiao, Liqiang Nie, Jian Shao, Wei Liu, and Tat-Seng Chua. Sca-cnn: Spatial and channel-wise attention in convolutional networks for image captioning. In *Proceedings of the IEEE conference on computer vision and pattern recognition*, pages 5659–5667, 2017.

[DAHG+15] Jeffrey Donahue, Lisa Anne Hendricks, Sergio Guadarrama, Marcus Rohrbach, Subhashini Venugopalan, Kate Saenko, and Trevor Darrell. Long-term recurrent convolutional networks for visual recognition and description. In *Proceedings of the IEEE conference on computer vision and pattern recognition*, pages 2625–2634, 2015.

[DDS+09]   Jia Deng, Wei Dong, Richard Socher, Li-Jia Li, Kai Li, and Li Fei-Fei. Imagenet:
           A large-scale hierarchical image database. In *2009 IEEE conference on computer
           vision and pattern recognition*, pages 248–255. Ieee, 2009.

[DLL+16]   Jianfeng Dong, Xirong Li, Weiyu Lan, Yujia Huo, and Cees GM Snoek. Early
           embedding and late reranking for video captioning. In *Proceedings of the 24th ACM
           international conference on Multimedia*, pages 1082–1086, 2016.

[FGI+15]   Hao Fang, Saurabh Gupta, Forrest Iandola, Rupesh K Srivastava, Li Deng, Piotr
           Dollár, Jianfeng Gao, Xiaodong He, Margaret Mitchell, John C Platt, et al. From
           captions to visual concepts and back. In *Proceedings of the IEEE conference on
           computer vision and pattern recognition*, pages 1473–1482, 2015.

[FH19]   Danny Francis and Benoit Huet. L-STAP: Learned spatio-temporal adaptive pooling
           for video captioning. In *Proceedings of the 1st International Workshop on AI for
           Smart TV Content Production, Access and Delivery*, pages 33–41, 2019.

[GCJ+18]   Jiuxiang Gu, Jianfei Cai, Shafiq R Joty, Li Niu, and Gang Wang. Look, imagine and
           match: Improving textual-visual cross-modal retrieval with generative models. In
           Proceedings of the IEEE Conference on Computer Vision and Pattern Recognition,
           pages 7181–7189, 2018.

[GCWC18]   Jiuxiang Gu, Jianfei Cai, Gang Wang, and Tsuhan Chen. Stack-captioning: Coarse-
           to-fine learning for image captioning. In *Thirty-Second AAAI Conference on
           Artificial Intelligence*, 2018.

[GGH+17]   Zhe Gan, Chuang Gan, Xiaodong He, Yunchen Pu, Kenneth Tran, Jianfeng Gao,
           Lawrence Carin, and Li Deng. Semantic compositional networks for visual
           captioning. In *Proceedings of the IEEE conference on computer vision and pattern
           recognition*, pages 5630–5639, 2017.

[GGZ+17]   Lianli Gao, Zhao Guo, Hanwang Zhang, Xing Xu, and Heng Tao Shen. Video
           captioning with attention-based LSTM and semantic consistency. *IEEE Transactions
           on Multimedia*, 19(9):2045–2055, 2017.

[HHL+17]   Chiori Hori, Takaaki Hori, Teng-Yok Lee, Ziming Zhang, Bret Harsham, John R
           Hershey, Tim K Marks, and Kazuhiko Sumi. Attention-based multimodal fusion for
           video description. In *Proceedings of the IEEE international conference on computer
           vision*, pages 4193–4202, 2017.

[HS97]   Sepp Hochreiter and Jürgen Schmidhuber. Long short-term memory. *Neural
           computation*, 9(8):1735–1780, 1997.

[HWCW19]   Lun Huang, Wenmin Wang, Jie Chen, and Xiao-Yong Wei. Attention on attention
           for image captioning. In *Proceedings of the IEEE International Conference on
           Computer Vision*, pages 4634–4643, 2019.

[HYH13]   Micah Hodosh, Peter Young, and Julia Hockenmaier. Framing image description as a
           ranking task: Data, models and evaluation metrics. *Journal of Artificial Intelligence
           Research*, 47:853–899, 2013.

[JCC+16]   Qin Jin, Jia Chen, Shizhe Chen, Yifan Xiong, and Alexander Hauptmann. Describing
           videos using multi-modal fusion. In *Proceedings of the 24th ACM international
           conference on Multimedia*, pages 1087–1091, 2016.

[JKFF16]   Justin Johnson, Andrej Karpathy, and Li Fei-Fei. Densecap: Fully convolutional
           localization networks for dense captioning. In *Proceedings of the IEEE conference
           on computer vision and pattern recognition*, pages 4565–4574, 2016.

[KFF15]   Andrej Karpathy and Li Fei-Fei. Deep visual-semantic alignments for generating
           image descriptions. In *Proceedings of the IEEE conference on computer vision and
           pattern recognition*, pages 3128–3137, 2015.

[KHR+17]   Ranjay Krishna, Kenji Hata, Frederic Ren, Li Fei-Fei, and Juan Carlos Niebles.
           Dense-captioning events in videos. In *Proceedings of the IEEE international
           conference on computer vision*, pages 706–715, 2017.

[KSH12]   Alex Krizhevsky, Ilya Sutskever, and Geoffrey E Hinton. Imagenet classification
           with deep convolutional neural networks. In *NIPS*, 2012.

[KZG+17] Ranjay Krishna, Yuke Zhu, Oliver Groth, Justin Johnson, Kenji Hata, Joshua Kravitz, Stephanie Chen, Yannis Kalantidis, Li-Jia Li, David A Shamma, et al. Visual genome: Connecting language and vision using crowdsourced dense image annotations. *International journal of computer vision*, 123(1):32–73, 2017.

[Lin04] Chin-Yew Lin. Rouge: A package for automatic evaluation of summaries. In *Text summarization branches out*, pages 74–81, 2004.

[LMB+14] Tsung-Yi Lin, Michael Maire, Serge Belongie, James Hays, Pietro Perona, Deva Ramanan, Piotr Dollár, and C Lawrence Zitnick. Microsoft coco: Common objects in context. In *European conference on computer vision*, pages 740–755. Springer, 2014.

[LPM15] Minh-Thang Luong, Hieu Pham, and Christopher D Manning. Effective approaches to attention-based neural machine translation. *arXiv preprint arXiv:1508.04025*, 2015.

[LRY18] Sheng Liu, Zhou Ren, and Junsong Yuan. Sibnet: Sibling convolutional encoder for video captioning. In *Proceedings of the 26th ACM international conference on Multimedia*, pages 1425–1434, 2018.

[LZLY19] Guang Li, Linchao Zhu, Ping Liu, and Yi Yang. Entangled transformer for image captioning. In *Proceedings of the IEEE International Conference on Computer Vision*, pages 8928–8937, 2019.

[Mil95] George A Miller. Wordnet: a lexical database for English. *Communications of the ACM*, 38(11):39–41, 1995.

[MXY+14] Junhua Mao, Wei Xu, Yi Yang, Jiang Wang, Zhiheng Huang, and Alan Yuille. Deep captioning with multimodal recurrent neural networks (m-rnn). *arXiv preprint arXiv:1412.6632*, 2014.

[PB17] Ramakanth Pasunuru and Mohit Bansal. Reinforced video captioning with entailment rewards. In *Proceedings of the 2017 Conference on Empirical Methods in Natural Language Processing*, pages 979–985, 2017.

[PCB+19] Stefano Pini, Marcella Cornia, Federico Bolelli, Lorenzo Baraldi, and Rita Cucchiara. M-VAD names: A dataset for video captioning with naming. *Multimedia Tools and Applications*, 78(10):14007–14027, 2019.

[PGH+16] Yunchen Pu, Zhe Gan, Ricardo Henao, Xin Yuan, Chunyuan Li, Andrew Stevens, and Lawrence Carin. Variational autoencoder for deep learning of images, labels and captions. In *Advances in neural information processing systems*, pages 2352–2360, 2016.

[PRWZ02] Kishore Papineni, Salim Roukos, Todd Ward, and Wei-Jing Zhu. Bleu: a method for automatic evaluation of machine translation. In *Proceedings of the 40th annual meeting of the Association for Computational Linguistics*, pages 311–318, 2002.

[PWC+15] Bryan A Plummer, Liwei Wang, Chris M Cervantes, Juan C Caicedo, Julia Hockenmaier, and Svetlana Lazebnik. Flickr30k entities: Collecting region-to-phrase correspondences for richer image-to-sentence models. In *Proceedings of the IEEE international conference on computer vision*, pages 2641–2649, 2015.

[PYLM17] Yingwei Pan, Ting Yao, Houqiang Li, and Tao Mei. Video captioning with transferred semantic attributes. In *Proceedings of the IEEE conference on computer vision and pattern recognition*, pages 6504–6512, 2017.

[PZW+19] Wenjie Pei, Jiyuan Zhang, Xiangrong Wang, Lei Ke, Xiaoyong Shen, and Yu-Wing Tai. Memory-attended recurrent network for video captioning. In *Proceedings of the IEEE Conference on Computer Vision and Pattern Recognition*, pages 8347–8356, 2019.

[RDP+16] Vasili Ramanishka, Abir Das, Dong Huk Park, Subhashini Venugopalan, Lisa Anne Hendricks, Marcus Rohrbach, and Kate Saenko. Multimodal video description. In *Proceedings of the 24th ACM international conference on Multimedia*, pages 1092–1096, 2016.

[RHGS16] Shaoqing Ren, Kaiming He, Ross Girshick, and Jian Sun. Faster R-CNN: Towards real-time object detection with region proposal networks. *IEEE transactions on pattern analysis and machine intelligence*, 39(6):1137–1149, 2016.

[RMM+17] Steven J Rennie, Etienne Marcheret, Youssef Mroueh, Jerret Ross, and Vaibhava Goel. Self-critical sequence training for image captioning. In *Proceedings of the IEEE Conference on Computer Vision and Pattern Recognition*, pages 7008–7024, 2017.

[RQT+13] Marcus Rohrbach, Wei Qiu, Ivan Titov, Stefan Thater, Manfred Pinkal, and Bernt Schiele. Translating video content to natural language descriptions. In *Proceedings of the IEEE International Conference on Computer Vision*, pages 433–440, 2013.

[RRTS15] Anna Rohrbach, Marcus Rohrbach, Niket Tandon, and Bernt Schiele. A dataset for movie description. In *Proceedings of the IEEE Conference on Computer Vision and Pattern Recognition (CVPR)*, 2015.

[SGG+17] Jingkuan Song, Lianli Gao, Zhao Guo, Wu Liu, Dongxiang Zhang, and Heng Tao Shen. Hierarchical LSTM with adjusted temporal attention for video captioning. In *Proceedings of the 26th International Joint Conference on Artificial Intelligence*, pages 2737–2743, 2017.

[SGG+18] Jingkuan Song, Yuyu Guo, Lianli Gao, Xuelong Li, Alan Hanjalic, and Heng Tao Shen. From deterministic to generative: Multimodal stochastic RNNS for video captioning. *IEEE transactions on neural networks and learning systems*, 30(10):3047–3058, 2018.

[SVL14] Ilya Sutskever, Oriol Vinyals, and Quoc V Le. Sequence to sequence learning with neural networks. In *Advances in neural information processing systems*, pages 3104–3112, 2014.

[TPLC15] Atousa Torabi, Christopher Pal, Hugo Larochelle, and Aaron Courville. Using descriptive video services to create a large data source for video annotation research. *arXiv preprint arXiv:1503.01070*, 2015.

[VKFU15] Ivan Vendrov, Ryan Kiros, Sanja Fidler, and Raquel Urtasun. Order-embeddings of images and language. *arXiv preprint arXiv:1511.06361*, 2015.

[VLZP15] Ramakrishna Vedantam, C Lawrence Zitnick, and Devi Parikh. Cider: Consensus-based image description evaluation. In *Proceedings of the IEEE conference on computer vision and pattern recognition*, pages 4566–4575, 2015.

[VSP+17] Ashish Vaswani, Noam Shazeer, Niki Parmar, Jakob Uszkoreit, Llion Jones, Aidan N Gomez, Łukasz Kaiser, and Illia Polosukhin. Attention is all you need. In *Advances in neural information processing systems*, pages 5998–6008, 2017.

[VTBE15] Oriol Vinyals, Alexander Toshev, Samy Bengio, and Dumitru Erhan. Show and tell: A neural image caption generator. In *Proceedings of the IEEE conference on computer vision and pattern recognition*, pages 3156–3164, 2015.

[WCW+18] Xin Wang, Wenhu Chen, Jiawei Wu, Yuan-Fang Wang, and William Yang Wang. Video captioning via hierarchical reinforcement learning. In *Proceedings of the IEEE Conference on Computer Vision and Pattern Recognition*, pages 4213–4222, 2018.

[WMZL18] Bairui Wang, Lin Ma, Wei Zhang, and Wei Liu. Reconstruction network for video captioning. In *Proceedings of the IEEE Conference on Computer Vision and Pattern Recognition*, pages 7622–7631, 2018.

[XBK+15] Kelvin Xu, Jimmy Ba, Ryan Kiros, Kyunghyun Cho, Aaron Courville, Ruslan Salakhudinov, Rich Zemel, and Yoshua Bengio. Show, attend and tell: Neural image caption generation with visual attention. In *International conference on machine learning*, pages 2048–2057, 2015.

[XMYR16] Jun Xu, Tao Mei, Ting Yao, and Yong Rui. Msr-vtt: A large video description dataset for bridging video and language. In *Proceedings of the IEEE conference on computer vision and pattern recognition*, pages 5288–5296, 2016.

[YPL+17] Ting Yao, Yingwei Pan, Yehao Li, Zhaofan Qiu, and Tao Mei. Boosting image captioning with attributes. In *Proceedings of the IEEE International Conference on Computer Vision*, pages 4894–4902, 2017.

[YTC+15] Li Yao, Atousa Torabi, Kyunghyun Cho, Nicolas Ballas, Christopher Pal, Hugo Larochelle, and Aaron Courville. Describing videos by exploiting temporal structure. In *Proceedings of the IEEE international conference on computer vision*, pages 4507–4515, 2015.

Chapter 8
Deep Learning in Video Compression Algorithms

Ofer Hadar and Raz Birman

8.1 Introduction

Video content is becoming the primary means of consuming content over the Internet. Overwhelming usage of video content is constantly increasing due to availability of convenient edge-devices (such as smartphone and tablets) and the constantly increasingly available bandwidth of wireless and wireline networks that enable reasonable end-user experience when consuming the content. Various content providers have emerged, which provide rich content and further encourage its consumption, such as Netflix, YouTube, Facebook, Hulu, Amazon, and others. In addition, cameras are becoming more prevalent as means for ensuring public safety and security and as their penetration increases, the need to deliver and record large volumes of video content is also on the rise. The ever-increasing usage and shift of consumption patterns have fueled the need to obtain efficient and cost-effective video delivery over the channels. Vast research and standardization efforts have been invested during the last two decades in crafting innovative algorithms for compressing video content into a more compact representation, while retaining quality in the eye of the beholder, multiple metrics have been devised for assessing the results and comparing them against benchmarks in order to constantly demonstrate progress and achievements. The two most successful video coding standards so far, H.262/MPEG-2 Video and H.264/MPEG-4 AVC, were developed jointly by ITU-T and ISO/IEC. H.265/HEVC is now the third such joint project [OS13]. The evolution of video coding standards along the years is depicted in Fig. 8.1.

O. Hadar (✉) · R. Birman
Ben Gurion University of the Negev, School of Electrical and Computer Engineering, Beersheba, Israel
e-mail: hadar@bgu.ac.il; birmanr@post.bgu.ac.il

© Springer Nature Switzerland AG 2021
J. Benois-Pineau, A. Zemmari (eds.), *Multi-faceted Deep Learning*,
https://doi.org/10.1007/978-3-030-74478-6_8

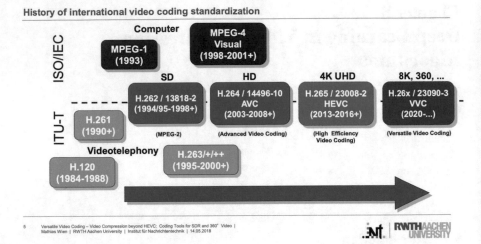

Fig. 8.1 Progression of video standards [OW]

The general concept that had been leading the evolution of video coding standards along the years is that a new standard is introduced when compression ratios, e.g., bit rate reduction, for the same quality reach an improvement of at least 50% over the standard that preceded it. The latest released standard is HEVC (or also known as H.265). The Joint Video Exploration Team (JVET) of ITU-T and ISO/IEC has been working on the Versatile Video Coding (VVC) standard, which has been finalized in July 2020 and will most likely be the next generation following HEVC. The codec was named Versatile Video Coding because it is "meant to be very versatile and address all of the video needs from low resolution and low bitrates to high resolution and high bitrates, HDR, 360 omnidirectional and so on. According to plans, the VVC standard should be released by end of 2020 and have 30–50% better compression rate for the same perceptual quality. This improvement has been accomplished using many small improvements of classic hybrid video coding design components.

The primary contributors that enable video compression are:

1. the inherent redundancy of data representations in the spatio-temporal domain;
2. the response of the Human Visual System (HVS), which has low sensitivity to high spatial frequencies; and
3. the statistical correlation between coding parameters and transform coefficients.

The foundation of all streaming video compression algorithms is built on two basic principles:

1. removing the data redundancy in all types of data by using transforms and entropy coding—this is a lossless process; and
2. removing the irrelevancy of the video signal by eliminating those features that are perceptually irrelevant for the HVS (e.g. high spatial frequencies and color resolution)—this is a lossy process.

Neural Networks have been around from the 1950s. Throughout the years they have seen some ups and downs, competing with other Artificial Intelligence (AI) methods for performing classification and prediction tasks. Recent years have seen a tremendous overflow of applications that take advantage of Deep Neural Networks (DNN), primarily due to major improvements in computing power and parallel processing using Graphical Processing Units (GPUs). Deep Learning (DL) has been gaining recognition in the last 5 years as an emerging best-of-breed candidate for replacing traditional analytics algorithms in a variety of applications, including identification, recognition, and classification. Considering their performance and applications, DNNs architectures seem like a natural choice for video compression algorithms, that primarily focus on predictions and filtering. This chapter explains the basic algorithms and methods that typically leverage Neural Networks and provides an overview of few neural network families/variants that are suitable for applying to various modules of video compression codecs.

8.2 Video Compression Standards

Traditional video coding standards have taken advantage of data redundancies in the video content as well as the characteristics of the Human Visual System (HVS) to reduce transmission rate requirements while retaining a satisfactory video quality at the receiving end. In the following paragraph, we review the different video compression steps/components, summarize their evolution between the different standards and thus lay the foundation to introducing neural network learning algorithms to improve them.

The general block diagram of video compression encoder is depicted in depicted in Fig. 8.2 [SBS14]. The various basic components, which contribute to compression, are indicated in this diagram.

The primary components are:

- Intra-Prediction (reducing spatial redundancies)
- Inter-Prediction (reducing temporal redundancies)
- DCT transform
- Quantization (lossy operation)
- Entropy coding (lossless operation)
- Handling of chroma components (color)

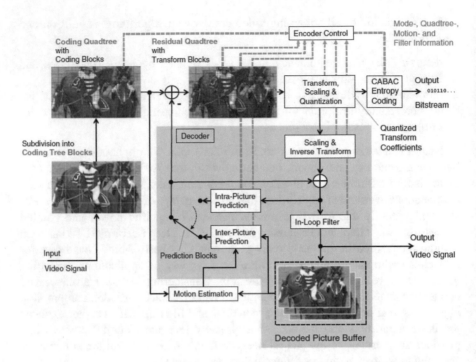

Fig. 8.2 Bock diagram of an HEVC encoder with built-in decoder (gray shaded) [SBS14]

Intra-Prediction   [Che14],[SPC13],[OY16],[LU11],[HSM+17],[Ric11]   is reducing spatial redundancies in the video I-frame content, which also corresponds to the Intra Random Access Point (IRAP) frame of the HEVC standard. These frames contain all the picture information in them and therefore are used to provide points in the bitstream where it is possible to start decoding. Intra-Prediction schemes predict one picture pixel values from their neighbors. The incentive to use prediction is to reduce transmitted values magnitudes. This is achieved by obtaining the residual errors between the predicted and the actual pixel values. The better the prediction, the lower the residual values are, and thus better compression rates can be obtained. Intra-Prediction follows the raster scan of the transmitted picture/frame thus allowing the decoder to utilize already known pixel values to predict new ones. The prediction is performed in blocks, which divide the video picture/frame. The dividing of pictures to individual prediction block units of different rectangular sizes varies between the different standards. Improvements of this division algorithm contribute to compression efficiency since the dependency between neighboring pixel values can be better optimized. However, the prediction concept remains the same. Prediction of block pixels is performed by interpolations of pixels surrounding the block. The different interpolation formulas are called 'Modes'. There is a total of 35 prediction modes in HEVC. 33 angular directional modes, which are good for predicting directional edges and the DC and Planar

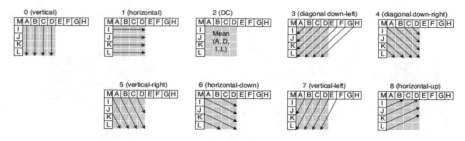

Fig. 8.3 4 × 4 Intra-Prediction modes used in H.264 [Ric11]

modes which provide good predictions for smooth image content. Figure 8.3 provides an illustration of the H.264 angular modes and the DC modes for 4 × 4 blocks [Ric11]. The arrows indicate the prediction direction of each one of the modes.

It is important to note that the selected mode should be signaled to the decoder. Therefore, the signaling bits should also be considered when calculating the compression efficiency.

Inter-Prediction [SBS14], [Ric11] is reducing temporal redundancy of pixel values between different video pictures/frames. Since the textual content of consecutive frames is, in many picture regions close or even identical, it is possible to predict pixel values of a particular picture from the values of similar regions in pictures that come before or after it in the frames sequence. Similarly to the Intra-Prediction, this is done at a block level. The picture is divided to blocks of different sizes and a similarity between blocks from different pictures is searched. This searching process is named Motion-Compensated Prediction (MCP). Once a similar block is discovered, a Motion Vector (MV) is calculated, which indicates the vertical and horizontal displacement between the predicted target block and the source block from which it is predicted. Since video frames are buffered, it is possible to detect such similar blocks in past as well as in future frames. The terms used to describe these frames are B-Frames and P-Frames respectively. It is also possible to perform bi-directional prediction that is a mathematical combination of blocks content from past and future frames. The concept is illustrated in Fig. 8.4.

After calculating the MV, the block pixel values are subtracted and only the residual is processes for transmission to the decoder, thus allowing the decoder to perform the same prediction, using the transmitted MV value and correct the actual block pixel values using the residual error, similarly to Intra-Prediction correction.pt

A 2-dimensional **Discrete Cosine Transform** is used to convert block residuals into the spatial frequencies domain. It has been observed that most of the energy of natural images resides in the low spatial frequencies and the HVS is less sensitive to high spatial frequencies. Therefore, the standards use subsequent **Quantization** of the transformed coefficients to zero-out the lower values and while thus reducing the rate, introducing a certain loss that impacts the restoration quality of the image at the receiving end. The extent of the loss is determined by the quantization parameter, which indicates the value resolution granularity factor.

Δt: Reference picture index:　　　　　　　　Δx Δy: Spatial displacement:

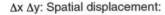
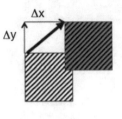
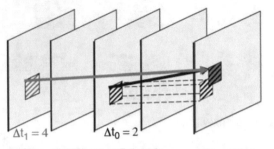

Prior Decoded Pictures as Reference　　　　　Current Picture

Fig. 8.4 Inter-prediction concept illustrated [SBS14]

The Red-Green-Blue (RGB) values of the video frame pixels are transformed into Luma and **Chroma** components, which represent the brightness and the color content, respectively. It was found that the HVS is less sensitive to chroma than to luma. Therefore the standards use lower resolution for transmitting the color, thus obtaining another compression gain.

The last stage of video encoding (and the first stage of video decoding) is **Entropy Coding**. The compressed video data is represented by a stream of numbers. Entropy coding are lossless compression schemes that leverage the statistic properties of these numbers to accomplish additional compression. In a simplistic way, each transmitted number is represented by a combination of bits. Entropy coding calculate how frequent each number is in the data stream and the more frequent the number is, the less bits are assigned for transmitting it. The number of bits used to represent the data is logarithmically proportional to the probability of the data. This minimizes the total number of bits required to transmit the complete data stream. The prevailing entropy coding method, which is used in H.264 as well as in HEVC is termed Context-Based Adaptive Binary Arithmetic Coding (CABAC) [MWS03] CABAC achieves good compression performance by [Ric11]:

1. selecting probability models for each syntax element according to the element's context;
2. adapting probability estimates based on local statistics; and
3. using arithmetic coding rather than variable-length coding.

8.3　Using Neural Networks for Video Compression

Neural Networks have been proven to excel in tasks of classification, regression, and prediction. Various architectures have been derived to provide suitable solutions that match different challenges. For example, Convolutional Neural Networks (CNN)

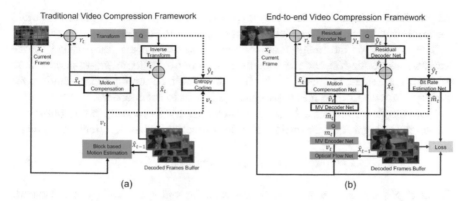

Fig. 8.5 (a) The predictive coding architecture used by the traditional video codec H.264 or H.265 compared to (b) the proposed Neural Network compression scheme [LOX+18]

are best at detecting textual patterns in images, such as visual objects and therefore assisting in classification tasks. Recurrent Neural Network (RNN) and their later evolution—Long Short Term Memory (LSTM) networks, provide a good learning architecture to handle time series, thus adapted by Natural Language Processing (NLP) applications. Given the capability of neural networks to learn complex data patterns and yield accurate, yet generalized results for similar samples from the same learned data probability distribution, they have been identified as having good potential for improving different components of the video compression chain. In a paper that we published [BSH20], we have provided an overview of the state-of-the-art research in the field of leveraging the power of neural networks to improve video compression algorithms.

An end-to-end optimization of a neural networks based video codec was proposed by G. Lu et. al. [LOX+18]. In this research work several networks have been used for mimicking the standard codec structure, by replacing complete functional blocks with Neural Networks [LOX+18]. They further replace the MV scheme by Optical Flow (OF),[1] which is expected to be more accurate, while avoiding the extensive data volumes for pixel-by-pixel movement representation by extracting distinctive features of the OF map using CNNs. The overall proposed codec architecture (b) alongside with the traditional architecture (a) are illustrated in Fig. 8.5.

The propose neural network architecture includes several networks which replace the distinctive analytic codec functions. The **Optical Flow Net** is a CNN that receives as inputs two consecutive video frames and extracts continues images representing the OF. The result is encoded using an auto-encoder style network, which compresses the OF. This operation is followed by quantization that feeds

---

[1] A heat map image that reflects the movement magnitude and direction of individual pixels between consecutive video frames.

the **Bit Rate Estimation Net** as well as dequantized in order to perform motion compensation by the **Motion Compensation Net** at the encoder.

The residual of the image after motion compensation are transformed by the **Residual Encoder Net** and then quantized in order to build an end-to-end training scheme. The residual values $\hat{y}_t$ as well as the quantized motion representation, $\hat{m}_t$ undergo Entropy Coding into bit stream that is sent to the decoder.

The complete network architecture, including all these networks is optimized end-to-end using a Rate-Distortion (RD) loss function that is described as follows:

$$\lambda D + R = \lambda d\left(x_t, \hat{x}_t\right) + \left(H\left(\hat{m}_t\right) + H\left(\hat{y}_t\right)\right) \tag{8.1}$$

Where $d\left(x_t, \hat{x}_t\right)$ denotes the distortion between x_t and $\hat{x}_t$, and $H\left(\cdot\right)$ represents the number of bits used for encoding the representations.

Different additional approaches have been proposed for improving codec individual components using neural networks as well as holistically performing non-standard conforming neural networks based video compression. In the following paragraphs of this chapter we describe improving the codec prediction algorithms as well as provide a short overview of some of the more predominant holistic approaches.

8.4 Improving Intra and Inter Predictions Using Neural Networks

Intra and Inter prediction are used to reduce redundancy of data pertaining to intra-frame pixel values and inter-frame pixel values, respectively. The general underlying concept is using already known pixel values, whether they are neighboring pixels in a single frame (intra) or already known pixels of previously transmitted frames (inter), to predict the value of yet unknown pixels. Due to intra/inter dependencies, the prediction allows us to reduce transmitted data by subtracting the predicted values from the true values and transmitting a residual. Since the coded residual determines the bit rate, the lower its value is, the lower the rate will be, which means that the better the prediction is, the lower the rate will be. Various human crafter analytic methods have been used in the video compression standards to improve the predictions. In recent years there have been efforts to use neural networks for that purpose.

Improvement of **Intra-Prediction** methods has taken a few possible directions:

1. Predicting the most suitable standard mode per block to prevent extensive MSE calculations at the encoder [LO16].
2. Using neighboring blocks for predicting block residual errors and correcting the values to accomplish an improved (lower) residual error [CZZ+18].

3. Using neighboring pixels to predict subsequent block pixels [LO16], [CZZ+18], [LLXX17].

While predicting the most suitable mode (first approach above) may increase encoder computational efficiency, it does not contribute to improved RD. Predicting next pixel values (second and third approaches above), if it is accurate enough, may replace existing modes altogether, or used as an additional mode that may be selected more frequently if it is better. This approach has a potential for improving RD, however, so far accomplished improvements are marginal at best. Due to the relatively small number of neurons (pixels) that are typically used for prediction, the networks are not too deep and usually use Fully Connected layers. The mode selection criteria are similar to the ones used by the standards.

In [LO16] the authors use a classification Convolution Neural Networks (CNN) and supervised learning to analyze image blocks and train the network to predict the most likely optimal HEVC mode. The considered modes are the 33 angular Intra-Prediction modes, the DC mode, and the Planar mode [LBH+12]. The network is trained on 32×32 blocks labeled according to the best selected mode. The general CNN architecture is provided in Fig. 8.6.

The results are evaluating the RD-Loss compared to a baseline of randomly selected modes. No RD-Loss (e.g. 0%) will indicate the optimal calculated mode selection whereas randomly selected mode represents the worst possible selection. The average obtained RD-Loss is 0.52% compared to an average of 6.46% for randomly selected mode, which means that the CNN networks perform much better than the worst option, yet they do not match the optimal mode selection results of the HEVC standard, thus introducing potential computation savings when selecting the most suitable mode per block, but not improving coding performance.

In [CZZ+18] the authors train CNNs for predicting residual error between HEVC predicted blocks and the original pixel values. The block prediction is performed for 8×8 Prediction Unit (PU) blocks. The network input is the three adjacent HEVC Intra-Prediction 8×8 blocks (top and left of the predicted block). The network is trained to predict the residual error between the target predicted block and the original values, thus providing additional prediction accuracy correction to

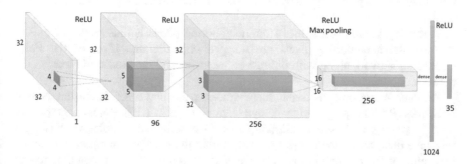

Fig. 8.6 CNN architecture for predicting the best mode [LO16]

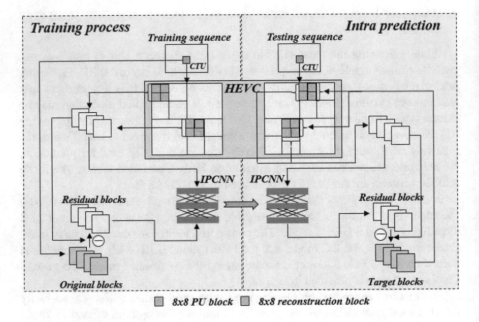

Fig. 8.7 CNN block prediction framework [CZZ+18]

standard calculated modes. The trained network is used for correcting prediction residual error. The overall framework is illustrated in Fig. 8.7. The image is divided to training and runtime parts on the left and on the right, respectively. The adjacent blocks are inserted into the IPCNN network with the purpose of predicting residual errors, which are the ground truth in this case. These predicted residual errors are then used to further reduce the actual error by subtracting the Residual blocks from the Original blocks.

The obtained lower residual errors contribute to coding performance. RD results were compared to standard HEVC and obtained some gains, as indicated in the results of Table 8.1. Note that these results indicate overall improvement whereas the actual gain is primarily relevant to the I-Frames (where only Intra-Prediction is used).

Predicting block pixel values from neighboring pixels is illustrated in Fig. 8.8, whereas the neighboring pixels are mapped onto the first layer of the Fully Connected network along with the rest of the layers and the predicted block as the network output.

Another interesting approach that harnesses the power of Recurrent Neural Networks (RNN) is proposed by Yueyu H. et al. in [HYLL19]. RNNs are typically used to predict time series, allowing to utilize weighted samples of historical sequence data, thus allowing to gradually forget the long past samples, which are less relevant for the prediction. The paper is proposing to use a combination of vertical and horizontal blocks for predicting the present block using RNN, thus

Table 8.1 Performance results of CNN based Intra-Prediction enhancements for HEVC [CZZ+18]

Sequences	BD-rate	Sequences	BD-rate
Traffic	−0.9%	PartyScene	−0.5%
PeopleOnStreet	−1.2%	RaceHorses	−0.7%
Kimono	−0.2%	BasketballPass	−0.4%
ParkScene	−0.8%	BQSquare	−0.1%
Cactus	−0.8%	BlowingBubbles	−0.7%
BasketballDrive	−0.6%	RaceHorces	−0.7%
BQTTerrace	−0.8%	FourPeople	−0.3%
BasketballDrill	−0.5%	Johnny	−1.0%
BQMall	−0.6%	KristenAndSara	−0.8%
All average	**−0.70%**		

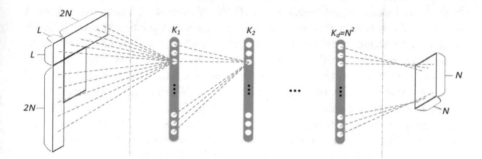

Fig. 8.8 Structure of Fully Connected network that uses neighboring pixels to predict block pixels [LLXX17]

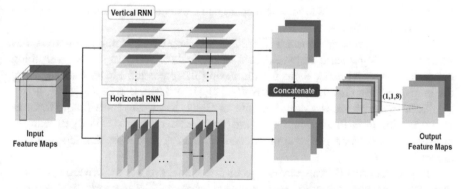

Fig. 8.9 Progressive Spatial Recurrent Neural Network (PS-RNN) based on spatial RNNs to learn to predict contents of PUs in intra prediction

learning features progression of the two directions. The used network is illustrated in Fig. 8.9. The first stage is using a CNN to extract block local feature map tensors. They are cascaded into RNN structures from the vertical and the horizontal directions respectively, the result is concatenated and converted to aggregate feature map that is then used to reconstruct the predicted block. The authors report on average 2.5% bit-rate reduction on variable-block-size settings under the same reconstruction quality compared with HEVC.

Due to prediction accuracy challenges when predicting full blocks, our research group has invested some efforts towards predicting the block pixel-by-pixel. In our paper [BSDMH18], we have used Fully-Connected Deep Neural Networks (FC-DNNs) to predict pixel blocks. The network was trained to predict either one pixel at a time or diagonal pixel groups based on pixels that preceded them in the frame scanning sequence. The method has utilized intra-block pixels for the prediction and due to quantization errors, we have shown that despite very promising Mean-Square-Error (MSE) improvement of up to 386% (prediction error before transform and quantization) compared to standard modes, further research has revealed that overall RD results do not match up to HEVC performance. Using this approach, it is possible to obtain very promising improvement of prediction error MSE, as compared to standard modes, however, this advantage was deemed mute when introducing quantization (lossy compression). The reason is that standard codecs perform predictions per block and introduce quantization correction loop per-block in the encoder to account for quantization errors. When using Intra-Block values during prediction, the quantization error cannot be fixed at the block level and thus the overall error is increased, and the overall RD results deteriorate. The accuracy advantage of Pixel-by-pixel prediction schemes can be leveraged to improve lossless compression, as indicated by I. Schiopu et al. [SLM18] where Convolutional Neural Networks (CNNs) are used to predict pixels from their neighbors in high-activity image regions. Due to ever improving computational efficiencies, some recent research work has been directed towards partial block predictions and/or pixel-by-pixel predictions. These will require a change in the codec design but hold a promise to improve prediction accuracy.

The most prevalent research direction for improving **Inter-prediction** has focused on using Neural Networks and in particular CNNs, to capture matching blocks features and using them for improving the predicted block, thus reducing the Inter-prediction residual error [ZWW+18], [HLWL18], [LKCK18], [WFJ+18]. Similarly to the previously described use of CNNs to capture Intra-Frame features for improving Intra-Prediction, CNNs are also used to capture similarity of features between frames when performing Inter-Prediction. A few examples are provided below.

The prediction of B-Frame is using a bi-directional average of forward and backward frame blocks. A method was proposed to use a CNN network that performs a more accurate weighing for bi-directional motion compensation [ZWW+18]. An illustration of the concept is provided in Fig. 8.10. Pair of past and future frames are used for training a CNN network to combine them and accomplish a more accurate predicted frame than the one traditionally calculated using simple averaging between them. The proposed method is claimed to have accomplished up to 10.5% BD-rate savings and an average of 3.1% BD-rate savings compared to HEVC.

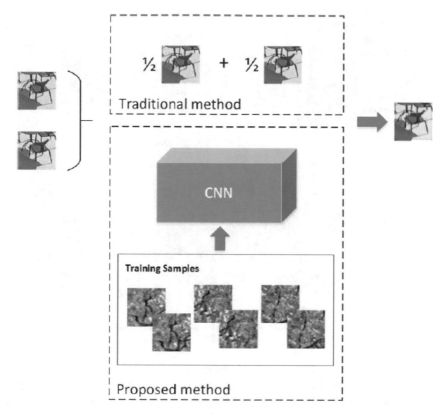

Fig. 8.10 Comparison between the proposed CNN based prediction and the traditional averaging method [ZWW+18]

Another method to improve Inter Prediction accuracy has combined temporal and spatial redundancies. CNN and FC networks were trained to predict any motion compensated (Inter-Prediction) block pixel values from the standard identified motion compensated block in a previous frame as well as neighboring block pixels of the same frame [ZWW+18], [LKCK18]. Combining the pixels from temporal and spatial domains into a network input layer and training the network, has yielded better prediction results than using the simple motion compensated block alone. The proposed method is reported to have accomplished an average of 5.2% BD-Rate reduction compared to HEVC.

In, [LKCK18], the authors propose to use Video Interpolation CNN (VI-CNN) to create a Virtual Reference (VR) frame from the forward and backward frames. The VR frame represents a higher temporal correlation with the current frame than any of the forward and back reference frames, thus allowing to calculate smaller MVs and lower block residual values. The idea is illustrated in Fig. 8.11. This approach has been shown to provide 1.4% BD-Rate reduction over HEVC.

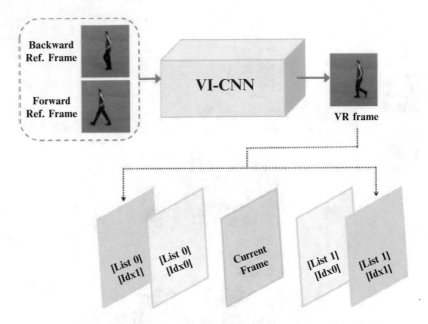

Fig. 8.11 VI-CNN using Adaptive Separable Convolution Neural Network [LKCK18]

Another research direction that has been proposed is improving the calculation efficiency of fractional/sub-pixel motion estimation. The conventional codecs improve temporal block predictions by calculating sub-pixel matches between blocks, as opposed to integer pixel matches. Neural Networks have been used for predicting the best matched Prediction Block (PB) in a reference frame with integer as well as with sub-pixel pixel precisions [IBA+18], [YLL+18], [ZSLY17],[LXY+19], thus eliminating the need to perform extensive interpolations that are required by conventional codecs for searching and finding the best sub-pixel match and obtaining better predictions despite the video signal not being low-passed nor stationary by nature.

We are presently focusing efforts on improving the Motion Vector Prediction (MVP) of the Ground Truth Motion Vector (GTMV), which is selected from a list of MVs belonging to neighboring spatial and previous frame temporal blocks [BSHBP20]. The selection of the most suitable MV at the decoder end requires signaling from the encoder. Being able to predict its value at the decoder more accurately and save the necessary signaling used by the HEVC standard, is expected to yield additional rate reduction. We perform this task using two different approaches:

1. Classification network that is train to predict the best MV from the list; and
2. Regression network that predicts the closest MV value from the list of candidate MVs.

The preliminary results obtained so far indicate a promising 34% reduction in bits required to transmit the best matching calculated Motion Vector Prediction (MVP) without reducing the quality, for fast forward movies with high motion.

8.5   Holistic Approaches

Video coding is a function applied to the original video stream at the encoder side in order to satisfy an optimization tradeoff between transmitted rate and quality of the reconstructed stream at the decoder side. The desirable function is not linear due to various non-linear optimization schemes that are used, such as quantization for example. Supervised neural networks excel in performing similar tasks. Therefore, instead of replacing specific human crafted components of the codec, it is possible to take more holistic approaches that implement this optimization function or large parts of it and accomplish performance improvements without aligning to the standard video coding schemes.

In this chapter we will briefly review the following proposed methods:

- End-to-end RD optimization scheme [TJZ+17], [LOX+18], [BLS17];
- Next frame prediction [CHJW20], [SCW+15], [CLS+17];
- Generative models [SBS18], [MCL16], [LSLW16];
- Content aware coding [SMPH18], [CBHA16],[ZLX19]; and
- Rate Control optimization scheme [LLLC17] .

End-to-end schemes model the codec complete block and optimize Rate-Distortion (RD) as the loss function to obtain best results. The model presented in [LOX+18] and described above is an end-to-end scheme that takes the approach of using different neural networks to replace standard codec functional blocks. However, there are other end-to-end schemes that deviate completely from the standard and introduce a whole new neural network complex to obtain results. One such distinctive method, which was applied to images, is proposed in [BLS17]. The general idea is illustrated in Fig. 8.12, whereas: (x) is a vector representing the input

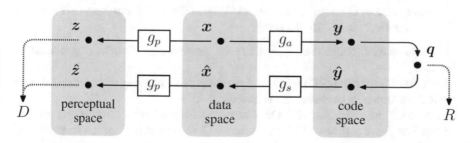

Fig. 8.12 General non-linear transform coding used for end-to-end image compression optimization [BLS17]

image, which resides in the Data Space; (y) is a mapped transformed Code Space that is quantized (represented by q); R is a calculated rate; converted back from Code Space to Data Space, at which point a transform is used to derive the distortion (D). The authors optimize the transform parameters $(\varnothing)$ and (θ) for a rated sum of the rate and distortion measures: $(R + \lambda D)$.

Whereas: $y = g_a(x; \varnothing)$, $\hat{x} = g_s(\hat{y}; \theta)$, $\hat{z} = g_p(\hat{x})$

The transforms used are Convolutional Neural Networks (CNN) with nonlinear activation functions.

The **next video frame prediction** method follows the same scheme used by standard codecs of leveraging spatio-temporal redundancies between pixel values to predict different parts of the next video frame and use these predictions to derive residuals, which require lower rates to transmit.

In [CHJW20] the researchers have implemented a neural network architecture to predict spatiotemporal relationship between blocks of reconstructed previous video frames along with adjacent blocks, to the left and top, of the same frame, in order to predict full blocks. The proposed codec utilizes two staged approach networks:

1. Network that takes as an input adjacent blocks from the present frame (Spatial redundancy) and block that has been derived using Motion Extension of the previous two frames (note the difference between Motion Estimation used in traditional codecs and the Motion Extension which is used here. The latest does not require extensive search calculations). This network is composed of Convolution layers (to extract spatial features) and ConvLSTM layers (to account for time series changes). The pixels used for prediction are based on the PixelCNN [vdOKK16] method.

2. Network that minimizes the residual block error in an iterative manner. Since the Decoder performs its predictions based on reconstructed blocks and therefore does not have access to the original pixel values, this network has to take into consideration the quantization process, such that the encoder can perform next block prediction with the de-quantized values (same as the ones that the Decoder uses). In order to account for quantization, which is a non-differentiable operation, the authors add a probabilistic quantization noise for the forward pass calculations and keep the gradients unchanged for the backward pass, thus allowing to include the quantization operation in the optimization. This network is composed of Convolution as well as ConvLSTM layers and processes residual error iteratively to reduce the error.

A schematic architecture of single block processing networks is provided in Fig. 8.13.

The Rate-Distortion (RD) results are promising but, in many cases, do not match the RD performance curves of the H.264 standard.

In [LT18] the authors use FC-DNN networks for predicting a frame one pixel at a time. The paper is divided to Intra-frame and Inter-frame predictions respectively. In the Intra-frame scheme multiple DNN networks are trained to predict each and every pixel of the frame from all the pixels that preceded it, thus the number of networks equals to the number of frame pixels.

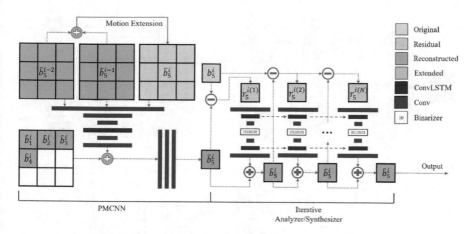

Fig. 8.13 Video compression scheme architecture [CHJW20]

Another promising approach for predicting full video frame was presented in [SCW+15] whereas the authors have used ConvLSTM network architecture to predict near term future rainfall intensity. Extending previous works of predicting future frames using Fully-Connected LSTM networks (FC-LSTM) [SMS15], the authors have realized the FC-LSTM networks are powerful enough to handle temporal correlation but contain too much redundancy for spatial data. Therefore, they have devised a multi-layered ConvLSTM network. Since the paper does not pretend to utilize the networks for video compression, the results are not stated in metrics that can be used to assess potential compression benefits. Yet, the approach can be applied to next video frame prediction.

Generative image and video compression models take advantage of Generative Adversarial Networks (GANs), proposed by Ian Goodfellow et al. in a key research work from 2014 [GPM+14]. The power of GANs was harnessed to compression. One of the most challenging obstacles is building complex and expressive models that are also tractable and scalable. This is achieved by conditioning the model on external information by using as an input, the actual image instance that we need to compress, thus ensuring that the decoded image is an accurate match of that same instance rather than an arbitrary instance from the same probability space.

In their Generative compression work [SBS18], the authors have used Deep Convolutional GANs to compress images. The architecture for images is depicted in the left of Fig. 8.14, while the right-hand side is extending the framework to video compression.

A Discriminative network $((d_v))$ is trained to detect images that are produced from the Generative network $((g_\varnothing (z)))$, which produces these images from a latent variable with a uniform distribution. A second network $((g_\varnothing (f_\theta (x)))$ is trained to predict the input image. By using the same parameters for the $(g_\varnothing)$ network, it is ensured that the predicted image is indeed matching the input image. In order to ensure proper prediction of the image, a subset of AlexNet [KSH12] is used, which

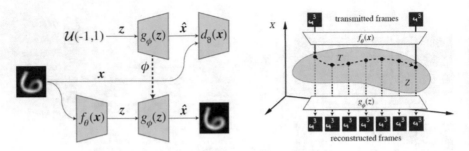

Fig. 8.14 Generative compression architectures: for (left) images and (right) video [SBS18]

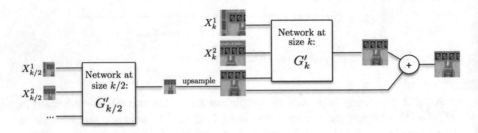

Fig. 8.15 Multi-scale prediction architecture [MCL16]

learns the features necessary for reconstruction. The extension to video is performed by predicting every Nth frame and interpolating between them in the latent domain (z), while also transmitting the differences between the interpolated result and the actual frame prediction.

The networks are run, and image compression results are compared with those of JPEG/2000. The proposed network yields higher quality reconstructions (in terms of SSIM and visual inspection) than JPEG/2000 at ~ fourfold higher compression levels. It also provides more recognizable images at compression ratios of up to ~190-fold and also provides a comparison to MPEG4 video compression. It indicates an average of 20–50% performance improvement. However, the results are not presented in a coherently enough manner such as to ascertain this improvement and report it here.

The researchers of [MCL16] take advantage of Generative Adversarial Networks (GAN) architecture [GPM+14] to predict future frames. The network architecture is depicted in Fig. 8.15.

The model utilizes CNNs for predicting next video frame at multiple resolutions and using a GAN model for each and every resolution. The multi scale resolutions allow the CNN networks to learn different scale features. This is the Generative part of the network (G). The Discriminative network is trained to distinguish between frames which are generated by G and real frames from the original sequence. By learning to distinguish between a generated frame and a real frame, the model is ensuring realistic frame samples by preventing observed blurriness phenomena that

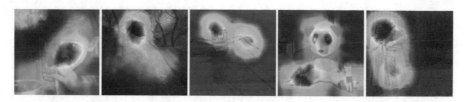

Fig. 8.16 ROI depicted by MS-ROI heatmaps [SMPH18]

occurs due to using L2 distance as the optimization loss function. The prediction results are promising in terms of visual outcomes and PSNR/SSIM metrics. In [LLLW18] the authors have explored a similar approach for extrapolating a next video frame from previous sequence of frames using GAN and employing a similar multi-resolution scheme. They have incorporated their prediction results in the HEVC coder algorithm and are reporting a 2.0% BD-rate reduction compared to the standard.

The exploitation of redundancies in the video content and the sensitivity of the HVS to accomplish compression are generalization of **Content aware coding** that applies to natural characteristic of video at large. It is possible to focus the implementation to local parts of the video stream and thus device new content aware schemes that rely on particular features of specific frame features. One such approach is based on Region of Interest (ROI). By detecting saliency of regions in the image, it is possible to apply better resolution in ROIs while reducing the resolution in other areas [SMPH18], [CBHA16]. An example of ROIs as detected by MS-ROI heatmaps is depicted in Fig. 8.16.

CNN networks excel at extracting specific features from images and classifying objects. The ROI detection challenge is simpler and can be performed by a degenerated network, which is able to detect objects without classifying them. Once identified, ROIs can be compressed with a higher quality, less lossy. This is accomplished by applying the quantization factor (Q) non-uniformly across the image. Compression is achieved by introducing higher Q for non-ROI regions, which will in most cases cover the majority of the image. The saliency concept was extended by S. Zhu et al. [ZLX19], where the authors propose a flexible QP selection algorithm based on saliency of the spatio-temporal domains, whereas the saliency maps are validated by comparing to a video-based-eye-tracking-dataset, which was produced with the participation of 100 subjects.

Rate Control (RC) means maximizing compressed video stream quality under a fixed rate constraint. RC has an important role in the transmission of video streams over Wide Area Networks (WAN) due to bandwidth limitations and utilization efficiencies. The typical method to apply RC is by building a model that characterizes the relationship between rate (R) and a quantization parameter ((Q_p)) or Lagrange multiplier ((λ)), which controls the Rate-Distortion tradeoff. This is easier to accomplish for inter frames, since it is possible to adjust RC parameters between frames. However, for intra frames, especially the first frame of a video

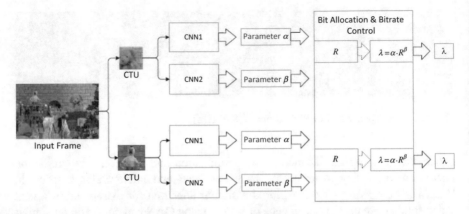

Fig. 8.17 A proposed CNN-based intra frame rate control method [LLLC17]

sequence (I-Frame), which contains much more bits than inter frames, there is no prior information to rely on. In order to address this challenge, a CNN network was introduced, that effectively analyzes the frame content at the Coding Tree Unit (CTU)[2] level and facilitates selecting a most suitable quantization parameter for each one [LLLC17]. The concept is illustrated in Fig. 8.17.

The model parameters are predicted by the CNN network for each CTU, thus providing a suitable quantization parameter per CTU, which is affected by the block pixels' texture, as it is learnt by the network.

8.6 Summary

A fairly bit of research has been invested over the last 5 years (2015–2020) in harnessing the power of Deep Learning to video compression. Despite the described research efforts, no breakthrough has been reported so far, that will suggest changing the traditional course of video compression standards.

Improvement of Intra/Inter-Prediction algorithms have yielded minor incremental codec performance gains, and therefore can become integrated as part of future codec and in-line with the standard scheme. The Inter-Prediction improvements described above address different parts of the existing temporal prediction algorithms. Further improvements can be achieved in reduced computation requirements due to more accurate original search area for the matching block, and in further reducing the residual error between the true matching block MV and the median of the neighboring blocks' MVs.

---

[2]A basic processing unit of HEVC that is the equivalent to block in previous standards (such as H.264).

Pure learning approaches, that do not align to the currently available codec standards, including next frame predictions, so far provide inferior or comparable results to standard video codecs. However, the direction will no doubt continue to be explored since the promise of creating a next generation holistic DNN based video compression scheme lies in this research direction. Due to the nature of neural networks, that can process large data structures while aggregating inputs using large number of tunable weights, it is possible to account for global as well as local frame/frames-sequence correlations and obtain improved compression rates. Along this line, we believe that further improvements can be obtained by designing networks that mix global with local scopes and are architecture properly to benefit from Intra and Inter pixels' correlations. Prevailing, human crafter video coding standards pay little attention to the content and for the most part (with some general exceptions, such as the different prediction block division of the I-Frame) do not adapt accurately to video content. It is reasonable to expect, as already demonstrated with the described ROI dependent quantization, that multiple neural networks could be potentially harnessed to perform a more accurate and efficient content adaptive coding scheme.

Another point to keep in mind when using neural networks for video compression is that while deeper networks in many cases provide better prediction results, they have a substantial computational cost and should be assessed carefully in the context of video coding due to the real-time impact of dealing with live video streaming.

References

[BLS17] Johannes Ballé, Valero Laparra, and Eero P. Simoncelli. End-to-end optimized image compression. In *5th International Conference on Learning Representations, ICLR 2017, Toulon, France, April 24–26, 2017, Conference Track Proceedings*. OpenReview.net, 2017.

[BSDMH18] Raz Birman, Yoram Segal, Avishay David-Malka, and Ofer Hadar. Intra prediction with deep learning. In *Applications of Digital Image Processing XLI*, volume 10752, page 1075214. International Society for Optics and Photonics, 2018.

[BSH20] Raz Birman, Yoram Segal, and Ofer Hadar. Overview of research in the field of video compression using deep neural networks. *Multim. Tools Appl.*, 79(17–18):11699–11722, 2020.

[BSHBP20] Raz Birman, Yoram Segal, Ofer Hadar, and Jenny Benois-Pineau. Improvements of motion estimation and coding using neural networks. *arXiv preprint arXiv:2002.10439*, 2020.

[CBHA16] Souad Chaabouni, Jenny Benois-Pineau, Ofer Hadar, and Chokri Ben Amar. Deep learning for saliency prediction in natural video. *CoRR*, abs/1604.08010, 2016.

[Che14] Roman I Chernyak. Analysis of the intra predictions in h. 265/hevc. *Applied Mathematical Sciences*, 8(148):7389–7408, 2014.

[CHJW20] Zhibo Chen, Tianyu He, Xin Jin, and Feng Wu. Learning for video compression. *IEEE Trans. Circuits Syst. Video Techn.*, 30(2):566–576, 2020.

[CLS+17] Tong Chen, Haojie Liu, Qiu Shen, Tao Yue, Xun Cao, and Zhan Ma. Deepcoder: A deep neural network based video compression. In *2017 IEEE Visual Communications and Image Processing, VCIP 2017, St. Petersburg, FL, USA, December 10–13, 2017*, pages 1–4. IEEE, 2017.

[CZZ+18] Wenxue Cui, Tao Zhang, Shengping Zhang, Feng Jiang, Wangmeng Zuo, and Debin Zhao. Convolutional neural networks based intra prediction for HEVC. *CoRR*, abs/1808.05734, 2018.

[GPM+14] Ian J. Goodfellow, Jean Pouget-Abadie, Mehdi Mirza, Bing Xu, David Warde-Farley, Sherjil Ozair, Aaron C. Courville, and Yoshua Bengio. Generative adversarial networks. *CoRR*, abs/1406.2661, 2014.

[HLWL18] Shuai Huo, Dong Liu, Feng Wu, and Houqiang Li. Convolutional neural network-based motion compensation refinement for video coding. In *IEEE International Symposium on Circuits and Systems, ISCAS 2018, 27–30 May 2018, Florence, Italy*, pages 1–4. IEEE, 2018.

[HSM+17] Ofer Hadar, Ariel Shleifer, Debargha Mukherjee, Urvang Joshi, Itai Mazar, Michael Yuzvinsky, Nitzan Tavor, Nati Itzhak, and Raz Birman. Novel modes and adaptive block scanning order for intra prediction in av1. In *Applications of Digital Image Processing XL*, volume 10396, page 103960G. International Society for Optics and Photonics, 2017.

[HYLL19] Yueyu Hu, Wenhan Yang, Mading Li, and Jiaying Liu. Progressive spatial recurrent neural network for intra prediction. *IEEE Trans. Multimedia*, 21(12):3024–3037, 2019.

[IBA+18] Ehab M. Ibrahim, Emad Badry, Ahmed M. Abdelsalam, Ibrahim L. Abdalla, Mohammed Sayed, and Hossam Shalaby. Neural networks based fractional pixel motion estimation for HEVC. In *2018 IEEE International Symposium on Multimedia, ISM 2018, Taichung, Taiwan, December 10–12, 2018*, pages 110–113. IEEE Computer Society, 2018.

[KSH12] Alex Krizhevsky, Ilya Sutskever, and Geoffrey E. Hinton. Imagenet classification with deep convolutional neural networks. In Peter L. Bartlett, Fernando C. N. Pereira, Christopher J. C. Burges, Léon Bottou, and Kilian Q. Weinberger, editors, *Advances in Neural Information Processing Systems 25: 26th Annual Conference on Neural Information Processing Systems 2012. Proceedings of a meeting held December 3– 6, 2012, Lake Tahoe, Nevada, United States*, pages 1106–1114, 2012.

[LBH+12] Jani Lainema, Frank Bossen, Woojin Han, Junghye Min, and Kemal Ugur. Intra coding of the HEVC standard. *IEEE Trans. Circuits Syst. Video Techn.*, 22(12):1792–1801, 2012.

[LKCK18] Jung Kyung Lee, Na-Young Kim, Seunghyun Cho, and Je-Won Kang. Convolution neural network based video coding technique using reference video synthesis. In *Asia-Pacific Signal and Information Processing Association Annual Summit and Conference, APSIPA ASC 2018, Honolulu, HI, USA, November 12–15, 2018*, pages 505–508. IEEE, 2018.

[LLLC17] Ye Li, Bin Li, Dong Liu, and Zhibo Chen. A convolutional neural network-based approach to rate control in HEVC intra coding. In *2017 IEEE Visual Communications and Image Processing, VCIP 2017, St. Petersburg, FL, USA, December 10–13, 2017*, pages 1–4. IEEE, 2017.

[LLLW18] Jianping Lin, Dong Liu, Houqiang Li, and Feng Wu. Generative adversarial network-based frame extrapolation for video coding. In *IEEE Visual Communications and Image Processing, VCIP 2018, Taichung, Taiwan, December 9–12, 2018*, pages 1–4. IEEE, 2018.

[LLXX17] Jiahao Li, Bin Li, Jizheng Xu, and Ruiqin Xiong. Intra prediction using fully connected network for video coding. In *2017 IEEE International Conference on Image Processing, ICIP 2017, Beijing, China, September 17–20, 2017*, pages 1–5. IEEE, 2017.

[LO16] Thorsten Laude and Jörn Ostermann. Deep learning-based intra prediction mode decision for HEVC. In *2016 Picture Coding Symposium, PCS 2016, Nuremberg, Germany, December 4–7, 2016*, pages 1–5. IEEE, 2016.

[LOX+18]   Guo Lu, Wanli Ouyang, Dong Xu, Xiaoyun Zhang, Chunlei Cai, and Zhiyong Gao. DVC: an end-to-end deep video compression framework. *CoRR*, abs/1812.00101, 2018.

[LSLW16]   Anders Boesen Lindbo Larsen, Søren Kaae Sønderby, Hugo Larochelle, and Ole Winther.   Autoencoding beyond pixels using a learned similarity metric.   In Maria-Florina Balcan and Kilian Q. Weinberger, editors, *Proceedings of the 33nd International Conference on Machine Learning, ICML 2016, New York City, NY, USA, June 19–24, 2016*, volume 48 of *JMLR Workshop and Conference Proceedings*, pages 1558–1566. JMLR.org, 2016.

[LT18]   Honggui Li and Maria Trocan.   Deep neural network based single pixel prediction for unified video coding. *Neurocomputing*, 272:558–570, 2018.

[LU11]   Jani Lainema and Kemal Ugur.   Angular intra prediction in high efficiency video coding (HEVC).   In *IEEE 13th International Workshop on Multimedia Signal Processing (MMSP 2011), Hangzhou, China, October 17–19, 2011*, pages 1–5. IEEE, 2011.

[LXY+19]   Jiaying Liu, Sifeng Xia, Wenhan Yang, Mading Li, and Dong Liu.   One-for-all: Grouped variation network-based fractional interpolation in video coding. *IEEE Trans. Image Process.*, 28(5):2140–2151, 2019.

[MCL16]   Michaël Mathieu, Camille Couprie, and Yann LeCun.   Deep multi-scale video prediction beyond mean square error. In Yoshua Bengio and Yann LeCun, editors, *4th International Conference on Learning Representations, ICLR 2016, San Juan, Puerto Rico, May 2–4, 2016, Conference Track Proceedings*, 2016.

[MWS03]   Detlev Marpe, Thomas Wiegand, and Heiko Schwarz. Context-based adaptive binary arithmetic coding in the H.264/AVC video compression standard. *IEEE Trans. Circuits Syst. Video Techn.*, 13(7):620–636, 2003.

[OS13]   Jens-Rainer Ohm and Gary J. Sullivan.   High efficiency video coding: The next frontier in video compression [standards in a nutshell]. *IEEE Signal Process. Mag.*, 30(1):152–158, 2013.

[OW]   From trends and recent developments in video coding standardization by J.-R. Ohm and M. Wien (via slideshare). https://www.slideshare.net/MathiasWien/trends-and-recent-developments-in-video-coding-standardization.

[OY16]   Carlo Noel Ochotorena and Yukihiko Yamashita. Regression-based intra-prediction for image and video coding. *CoRR*, abs/1605.03754, 2016.

[Ric11]   Iain E Richardson. *The H. 264 advanced video compression standard*. John Wiley & Sons, 2011.

[SBS14]   Vivienne Sze, Madhukar Budagavi, and Gary J. Sullivan, editors. *High Efficiency Video Coding (HEVC), Algorithms and Architectures*.   Integrated Circuits and Systems. Springer, 2014.

[SBS18]   Shibani Santurkar, David M. Budden, and Nir Shavit. Generative compression. In *2018 Picture Coding Symposium, PCS 2018, San Francisco, CA, USA, June 24–27, 2018*, pages 258–262. IEEE, 2018.

[SCW+15]   Xingjian Shi, Zhourong Chen, Hao Wang, Dit-Yan Yeung, Wai-Kin Wong, and Wang-chun Woo.   Convolutional LSTM network: A machine learning approach for precipitation nowcasting.   In Corinna Cortes, Neil D. Lawrence, Daniel D. Lee, Masashi Sugiyama, and Roman Garnett, editors, *Advances in Neural Information Processing Systems 28: Annual Conference on Neural Information Processing Systems 2015, December 7–12, 2015, Montreal, Quebec, Canada*, pages 802–810, 2015.

[SLM18]   Ionut Schiopu, Yu Liu, and Adrian Munteanu.   Cnn-based prediction for lossless coding of photographic images. In *2018 Picture Coding Symposium, PCS 2018, San Francisco, CA, USA, June 24–27, 2018*, pages 16–20. IEEE, 2018.

[SMPH18] Alena Selimovic, Blaz Meden, Peter Peer, and Ales Hladnik. Analysis of content-aware image compression with VGG16. In *IEEE International Work Conference on Bioinspired Intelligence, IWOBI 2018, San Carlos, Alajuela, Costa Rica, July 18–20, 2018*, pages 1–7. IEEE, 2018.

[SMS15] Nitish Srivastava, Elman Mansimov, and Ruslan Salakhutdinov. Unsupervised learning of video representations using lstms. In Francis R. Bach and David M. Blei, editors, *Proceedings of the 32nd International Conference on Machine Learning, ICML 2015, Lille, France, 6–11 July 2015*, volume 37 of *JMLR Workshop and Conference Proceedings*, pages 843–852. JMLR.org, 2015.

[SPC13] Maxim P Sharabayko, Oleg G Ponomarev, and Roman I Chernyak. Intra compression efficiency in VP9 and HEVC. *Applied Mathematical Sciences*, 7(137):6803–6824, 2013.

[TJZ+17] Wen Tao, Feng Jiang, Shengping Zhang, Jie Ren, Wuzhen Shi, Wangmeng Zuo, Xun Guo, and Debin Zhao. An end-to-end compression framework based on convolutional neural networks. In Ali Bilgin, Michael W. Marcellin, Joan Serra-Sagristà, and James A. Storer, editors, *2017 Data Compression Conference, DCC 2017, Snowbird, UT, USA, April 4–7, 2017*, page 463. IEEE, 2017.

[vdOKK16] Aäron van den Oord, Nal Kalchbrenner, and Koray Kavukcuoglu. Pixel recurrent neural networks. In Maria-Florina Balcan and Kilian Q. Weinberger, editors, *Proceedings of the 33nd International Conference on Machine Learning, ICML 2016, New York City, NY, USA, June 19–24, 2016*, volume 48 of *JMLR Workshop and Conference Proceedings*, pages 1747–1756. JMLR.org, 2016.

[WFJ+18] Yang Wang, Xiaopeng Fan, Chuanmin Jia, Debin Zhao, and Wen Gao. Neural network based inter prediction for HEVC. In *2018 IEEE International Conference on Multimedia and Expo, ICME 2018, San Diego, CA, USA, July 23–27, 2018*, pages 1–6. IEEE Computer Society, 2018.

[YLL+18] Ning Yan, Dong Liu, Houqiang Li, Tong Xu, Feng Wu, and Bin Li. Convolutional neural network-based invertible half-pixel interpolation filter for video coding. In *2018 IEEE International Conference on Image Processing, ICIP 2018, Athens, Greece, October 7–10, 2018*, pages 201–205. IEEE, 2018.

[ZLX19] Shiping Zhu, Chang Liu, and Ziyao Xu. High-definition video compression system based on perception guidance of salient information of a convolutional neural network and HEVC compression domain. *IEEE Transactions on Circuits and Systems for Video Technology*, 2019.

[ZSLY17] Han Zhang, Li Song, Zhengyi Luo, and Xiaokang Yang. Learning a convolutional neural network for fractional interpolation in HEVC inter coding. In *2017 IEEE Visual Communications and Image Processing, VCIP 2017, St. Petersburg, FL, USA, December 10–13, 2017*, pages 1–4. IEEE, 2017.

[ZWW+18] Zhenghui Zhao, Shiqi Wang, Shanshe Wang, Xinfeng Zhang, Siwei Ma, and Jiansheng Yang. CNN-based bi-directional motion compensation for high efficiency video coding. In *IEEE International Symposium on Circuits and Systems, ISCAS 2018, 27–30 May 2018, Florence, Italy*, pages 1–4. IEEE, 2018.

Chapter 9
3D Convolutional Networks for Action Recognition: Application to Sport Gesture Recognition

Pierre-Etienne Martin, Jenny Benois-Pineau, Renaud Péteri, Akka Zemmari, and Julien Morlier

9.1 Introduction

Movement is one of the most important aspects of visual perception, and stimuli associated with movement tend to be biologically significant [RMB+10]. Psycho-visual experiments show that a complete and exhaustive measurement of a scene is not always necessary to interpret its content. In the animal world, movement information, even partial, may be sufficient to recognize potential food: a frog can distinguish the flight of a fly from a falling leaf without needing to geometrically reconstruct the whole scene.

An illustration of recognition through movement by human beings is provided with the experiments of Johannsson (*Moving Light Display experiment* [Joh73]) in which the only source of information on a moving actor is given by bright spots attached to a few joints. People shown a static image can only see meaningless

P.-E. Martin (✉)
Department of Comparative Cultural Psychology, Max Planck Institute for Evolutionary Anthropology, Leipzig, Germany
e-mail: pierre_etienne_martin@eva.mpg.de

J. Benois-Pineau · A. Zemmari
LaBRI UMR 5800, University of Bordeaux, Talence Cedex, France
e-mail: jenny.benois-pineau@u-bordeaux.fr; akka.zemmari@u-bordeaux.fr

R. Péteri
MIA, University of La Rochelle, La Rochelle, France
e-mail: renaud.peteri@univ-lr.fr

J. Morlier
IMS, University of Bordeaux, Talence Cedex, France

© Springer Nature Switzerland AG 2021
J. Benois-Pineau, A. Zemmari (eds.), *Multi-faceted Deep Learning*,
https://doi.org/10.1007/978-3-030-74478-6_9

199

dot patterns. However, as soon as they are shown the entire sequence of images, they can recognize characteristic actions such as running or walking, and even the male or female gender of the actors. Such abilities suggest that it is possible to use movement as a means of recognition by itself.

In the field of computer vision, describing a video scene through natural language involves the narration of some key events or actions occurring in the scene. Movement analysis is obviously a central element in order to be able to take full advantage of the temporal coherence of the image sequences. Moreover, [Lap13] states that on average, on each video from the web, 35% of the pixels represent humans. Describing an image sequence from the human movements and activities performed is hence potentially discriminant and relevant for analyzing or indexing videos. These different elements, added to the complexity of the task, highlight why human action recognition in videos has been in past years a very active research topic in computer vision. Nevertheless, as shows the already large history of research for solving the problem of action recognition in video, when a real-world video scenes have to be analysed [SFBC16], motion characteristics alone are not sufficient. This is why the approaches using both temporal information expressed in terms of motion characteristics, such as velocity field, i.e. Optical Flow (OF) and spatial characteristics derived from colour pixel values in video frames have shown better performances. This is hold both for methods on the basis of hand-crafted features or the ever-winning Convolutional Neural Networks (CNNs).

Focusing on Deep Learning approaches, it is nevertheless interesting to show the problem in its historical perspective for better predicting the future. In this chapter, we will briefly present approaches for action recognition with different features going from handcrafted to produced by Deep NNs. We will speak about the evolution of datasets for development and testing action recognition methods, and introduce recently created dataset in our research for fine-grained classification of Table Tennis strokes. We will also present recent contributions in fine -grained action recognition with Twin Spatio-temporal networks.

9.2   Highlights on Action Recognition Problem

The problem of recognition of human actions in video has a wide-range of applications, this is why the history of research is quite long-term one. From historical perspective one can distinguish two main approaches with progressive combination of them. The first one consists in designing the so-called "handcrafted" features, expressing characteristics of video frames in a local or holistic manner thus forming a new description space. Classification of features in this new space with machine learning approaches brings the solution to the action recognition problem. The second approach consists in the "end-to-end" solution with Deep Neural Networks which extract features and then classify them.

9.2.1 Action Classification from Videos with Handcrafted Features

Handcrafted features extracted from videos development started to our best knowledge from feature extraction from images. Efforts were afterwards made for extracting information from the temporal domain. Such features were mainly used in the action recognition task, but also for other tasks such as scene and event recognition from videos. Most of the approaches using handcrafted features seek for their compact representation. The model of Bag of Words (BoW) or Bag-of-Visual Words (BoVW) [CDF+04] was introduced which allowed for quantizing a large amount of feature vectors into a set of classes—words of a dictionary of a predefined size. The classes-words of the dictionary were built by statistical clustering methods such as K-means [Mac67] or more sophisticated vector quantization techniques. This final descriptor of the images, areas in them or video frames were thus a statistical model—a histogram of class—occurrence of the "words" in the image. This model, for a long period was used for action recognition, and scene classification in video.

The use of temporal dimension of video with regard to static images was first introduced in 2003 with Spatio-Temporal Interest Points (STIP) features [SLC04]. They were an extension of the 2D corner detector to the temporal dimension for video. The equivalent to image corners in video are points which change direction over space and time. The authors of [SLC04] show that their descriptor matches with the action performed, meaning that STIPs will be located where and when the action happens. They apply their method on their newly created KTH dataset, which became one of the first widely used action datasets. Comparison is done using Support Vector Machine (SVM) classifier and Nearest Neighbor Classification (NNC) on BoW of their spatio-temporal local features. It results in a good classification score for actions which are not similar; but scores on similar actions such as "Walking", "Jogging" and "Running" remained low. The motion in those actions is very similar and the STIPs are concentrated on the same body parts.

The authors of [LP07] use jointly the Histogram of Oriented Gradients (HOG) descriptor and Motion Boundary Histogram (MBH) descriptors, using AdaBoost algorithm for recognition of two actions—smoking and drinking in their own dataset based on movies. At the same time, [GBS+07] introduce Space-Time Shapes (STS) features for action classification, along with the new Weizmann action dataset. Classification of STS is based on the same ideas as image shapes classification using Poisson equation. The authors classify the computed shape with Nearest Neighbor Clustering with euclidean distance. They reach an accuracy of 97.8% on their dataset with however a low confidence on similar actions. It is important to stress that their dataset is acquired in a controlled environment and has the same complexity as the KTH dataset.

In [SG08], the authors concatenate Gabor filters features and OF features in order to perform action classification with SVM classifier on KTH and Weizmann datasets. They show the superiority of their model on both datasets, compared to other methods, and come to the conclusion that one frame is enough to get a satisfactory classification score. Indeed, using only one frame and the posture of the person, actions in both datasets are easily distinguishable.

In 2009, action recognition methods were already reaching very high accuracies on both KTH and Weizmann datasets and the introduction of UCF11 dataset gave more space for improvements. Indeed, this dataset is more challenging since it is recorded "in the wild", that is in natural conditions, under the constraints of camera motion and flickering for example. The UCF101 [SZS12] samples are extracted from the YouTube platform. Along with their dataset, the authors propose a method for classification based on motion features and static features. They use AdaBoost learning method on the histogram-based representation and compare it with k-means clustering method. AdaBoost leads to better results and the hybrid combination resulted in 93.8% of accuracy on KTH dataset against 71.2% on UCF11 showing the higher complexity of the task for such dataset with the same number of classes.

The authors of [CRHV09] introduce at the same moment Histogram of Oriented optical Flow (HOF) features and reach 94.4% of accuracy on Weizmann actions dataset. The method is simple and easy to reproduce and will be used later on in [WKSL11], along with MBH, HOG features to compute dense trajectory features on the basis of dense optical flow field.

In [WS13], the authors improve dense trajectory features by considering camera motion. Camera motion is estimated using dense optical flow and Speeded Up Robust Features (SURF) descriptors. An homography is estimated using Random Sample Consensus (RANSAC) algorithm. The Improved Dense Trajectories (IDT) yield 91.2% of accuracy against 88.6% with regard to the original dense trajectory features. This work is used later in many applications such as action localization. The latter problem consists in defining not only temporal boundaries of actions in the video, but also spatial locus of them.

In [GHS13], the authors redefine actions as "actoms". Actom is a short atomic action with discriminative visual information, such as opening a door. It is therefore useful for action localization but can also be applied to classification-by-localization. The definition of actoms is important in the field of action recognition to decompose an action in individual parts. Actoms thus can be present across different actions such as entering or leaving a room with "opening door" as an actom, for example. Their understanding can lead to better video representation and accordingly to a better classification. However, a too great number of actoms might lead to the situation when they are well presented in the training set, but they would not be present in the test set and be unrelated to the action performed. The number of actoms to consider becomes then a variable to control according to the complexity of the actions to classify.

Another way to perform action recognition is developed by [JvGJ+14] who introduce the concept of Tubelets. It is a sampling method to produce 2D+T sequences of bounding boxes where the action is localized. This method, which tackles the localization and classification problem of actions at the same time, is based on super voxel generation through an iterative process using color, texture and motion to finally create tubelets. Those tubelets are then described by MBH features, and one BoW per class method is used for classification. The classifier with the maximum score assigns the class to the tubelet.

9.2.2 The Move to DNNs in Action Recognition Problem

Deep Neural Networks have very quickly outperformed all handcrafted feature-based methods, due to the strong generalization capacity of these classifiers. The method described in [JXYY13], was one of the first to use Deep Learning via a 3D CNN for action recognition, but they did not obtain better results than the state-of-the-art methods on the KTH dataset. It is only from 2014 and the innovative *two-stream network* approach of [SZ14] that temporal coherence will be exploited in CNN and that Deep Learning approaches will begin to supplant other methods. In the development of Deep Learning Methods for Action Recognition we could observe two trends: Deep Convolutional Neural Networks (CNNs) and recurrent neural networks (RNN) such as LSTM briefly presented in Chap. 2 (Sect. 2.5.2). Nevertheless, according to [VLS18] and the own experience of the authors, these networks are more difficult to train than 3D CNNs integrating spatial information along the time dimension in video. They have also difficulties to handle long term temporal interactions. Hence in this chapter we will not focus on them.

Better performances of Deep Neural Networks (DNNs) with inherent feature extraction from raw video in the end-to-end training and generalization process does not mean that engineered features have to disappear. On the contrary, human understanding of visual scenes influences the choice of the designed features such as e.g. OF. Still recent work may use handcrafted features as a baseline or fuse them with the deep features extracted by a DNN. [BGS+17] confirm that SVM does not perform better than a partially retrained Deep Convolutional Neural Network (DCNN) and that the learned features lead to better results than engineered ones; however the fusion outperforms the single modalities. These works confirm the findings of the community: the fusion of multiple features allows improving recognition scores for complex visual or multi-modal content understanding [IBPQ14]. Action recognition with Deep CNNs was first fulfilled with 2D CNN architectures. Here we will briefly discuss some of them.

9.2.3 2D Convolutional Neural Networks for Action Classification

2D convolution refers to the fact that convolutions are performed on a 2D spatial support of the image. For RGB data, 2D convolution actually uses 3D kernels to weight each color channel differently. However the way that the kernel will move in the image will only be in two dimensions.

In the scope of action recognition, it is what [SZ14] perform. They introduce a Two-Stream Convolutional Network which takes one single RGB frame for one stream, and for the other stream, several frames of the computed OF. Each stream is respectively called "Spatial stream ConvNet" and "Temporal stream ConvNet".

They notice that "Temporal stream ConvNet" reaches much better performances compared to "Spatial stream ConvNet". This could easily be explained, as in addition to the dynamic genre of the input data, the temporal stream uses up to ten frames against only one for the spatial stream branch. Performances are much more alike when the temporal branch uses only one frame. Of course, the fusion of the two streams using a SVM method performs the best.

An action can be considered as a volume, i.e. tube in the video. This concept of tube—which expresses a homogenous content, a singular action in our case, is used in [LWLW16]. A Tube Convolutional Neural Network (T-CNN) is introduced which can be sequentially decomposed into two distinct networks. They first create motion-segmented tubes using a Residual Convolutional Neural Network (R-CNN) [RHGS17]. Then those tubes feed a VGG-like network using 20 motion amplitude frames distributed along the channel dimension.

Until now we were speaking about CNNs. Nevertheless, temporal (recurrent) neural networks also represent an alternative to CNNs with windowing approaches. The focus of our work is on CNNs hence we will just very briefly mention them.

Long-term Recurrent Convolutional Network (LRCN) models are introduced in [DHR+17] for action recognition. The authors extract features using 2D CNN for each image which feed a LSTM from start to end. The decision is based on the average score. This simple model is tested with OF and a single RGB image. The fusion of the two modalities perform obviously the best.

Numerous works use models based on temporal networks such as Recurrent Neural Network (RNN) and Long Short-Term Memory (LSTM) [UAM+18]. However, RNN may be harder to train depending on the application context. Besides, LSTMs are more efficient when they are coupled to the output of a CNN [NHV+15].

Now coming back to the convolutional NNs for action recognition, we can state that 2D CNNs are often used as feature extractors and 3D convolutions are performed on the extracted features. Thus the temporal dimension is taken into account. This leads us to now focus on 3D CNN based methods for action recognition.

9.2.4   From 2D to 3D ConvNets in Action Classification

We can consider videos as 3D data with the third dimension along the time axis and either process them similarly to 2D images, or treat the temporal dimension differently, or extract temporal information such as dynamic data that can feed a DNN.

However, most methods need to consider extra information, which obviously leads to larger networks, greater number of parameters and the need of a greater number of GPUs with stronger capacities. This might not be possible for every research team, and brought some of them to try attaining accurate results with restrictions on the model size or computation time. This aspect thus brings many shades in the performances, and has brought to light many different methods which shall not be compared only in terms of performances, but also by their means to achieve them. In addition to such limitations, the choice of the architecture for a specific task remained open, leading to numerous implementation attempts.

3D convolutional neural networks are a good alternative as well for capturing long-term dependencies [CZ17], and involve 3D convolutions in space and time. When doing 3D convolution with C_{in} channels on a signal of depth D, width W and height H, the output value of the layer with input size (C_{in}, D, H, W) and output $(C_{out}, D_{out}, H_{out}, W_{out})$ can be precisely described as:

$$out(C_{out_j}) = bias(C_{out_j}) + \sum_{k=0}^{C_{in}-1} weight(C_{out_j}, k) \star input(k) \tag{9.1}$$

where C_{out_j} is the $j^t h$ output channel, and $\star$ is the valid 3D cross-correlation operator.

9.2.5   3D Convolutional Neural Networks for Action Classification

The Convolutional 3D (C3D) model [TBF+15] consists of eight consecutive convolutional layers using $3 \times 3 \times 3$ kernels and five max-pooling layers. In [HCS17], they use this model in a two stream T-CNN. Here videos are first divided into clips of equal length and are segmented using 3D R-CNN to create tube proposals. Tubes are then classified and linked together. By using the features extracted from the segmented video tubes with C3D model, they increase the performance compared to a direct application of the C3D model.

The authors of [FPW17] extend what was done in 2D by [SZ14], into 3D to introduce their Spatio-Temporal ResNet (ST-ResNet). They replace simple CNN branches by R-CNN with one connection between the two branches. Their results prove that RGB stream processed alone gets better performances than the OF stream. By processing them together, they reach an accuracy of 93.4% on UCF101 dataset.

A major breakthrough was proposed by the method of [CZ17], with much higher scores obtained on action classification. They present their Two-Stream I3D model as the combination of RGB-I3D and Flow-I3D models trained separately. Each of their models uses inflated inception modules, inspired from the 2D inception modules [SLJ+15]. The major strength of their model is the pretraining on

ImageNet first and then on Kinetics-400 dataset [KCS+17], more complex than UCF101 with 400 classes. By using Kinetics, they boost performances from 93.4% to 98% on UCF101. They reach also 74.2% of accuracy on Kinetics-400 dataset. Inception modules have already proven their efficiency for image classification on ImageNet dataset and thus have been successfully used in the I3D models.

Long-term Temporal Convolutions (LTC) CNN were introduced by [VLS18]. They experiment different temporal sizes for input video clips, in order to improve classification. Better accuracies are obtained when considering a greater number of frames as input, especially on long-lasting actions which have a longer temporal support.

[WGW+18] introduce Spatial-Temporal Pyramid Pooling Layer (STPP) using 3D convolutions in a two-stream like network fed by RGB and OF streams. The output becomes the input of a LSTM network. The use of LSTM allows classification of videos of arbitrary size and length. Each modality performs similarly: 85% and 83.8% of accuracy for RGB and OF stream respectively. When fused together, the method reaches 92.6% of accuracy.

[KKA20] introduce the Bidirectional Encoder Representations from Transformers (BERT) layer to better make use of the temporal information of BERT's attention mechanism firstly used for language understanding [VSP+17]. The BERT layer is based on the use of the Multi-Head Attention layer, which comprises a Scaled Dot Product layer. The Multi-Head attention layer is a part of a bigger network, the Transformer model which is dedicated to translation tasks. The incorporation of the BERT layer in the REsNeXT, R(2+1)D and I3D models, previously described, improve their performances. They reach the state-of-the-art results on UCF101 dataset with 98.7% of accuracy using the R(2+1)D architecture [TWT+18]. It is a ResNet-type architecture with separable temporal and spatial convolutions and a final BERT layer in order to better use the obtained features. One important point to stress is also the use of IG65M dataset [GTM19] for pre-training the model. IG65M dataset is build from the Kinetics-400 [KCS+17] class names. Those class names are then used as hashtags on Instagram and lead to 65M clips from 400 classes. Their dataset is however not publicly available.

9.2.6 Video Understanding for Racket Sports

Our interest is fine-grained action recognition with application in table tennis. We therefore present methods focusing on video classification and/or segmentation in the domain of racket sports.

[EBMM03] propose a motion descriptor based on optical flow in order to classify actions in sports. For this purpose they consider three different datasets: Ballet, Football and Tennis datasets. They track the player (or the person performing the action) and build a 3D volume based on their motion. Their motion descriptor has four channels: the positive and negative values for horizontal and vertical motions.

Then classification is performed following a nearest neighbour approach using similarity metric:

$$S = \sum_k C_1(k)C_2(k) \qquad (9.2)$$

with C_1 and C_2 being two cuboid samples based on the motion descriptors at coordinate k.

Even if Ballet dataset and Tennis dataset are acquired in a controlled environment, performances for the Tennis dataset are more limited. Football dataset comes from broadcast source which explains the limited performances. Moreover, the number of classes for the Tennis dataset is lesser than the two others, however, it is where their method is the less efficient. This underlines the greater complexity of racket sport and their fine-grained aspect.

Another research field in video classification aims at identifying the different parts of tennis broadcasting. To do so HMMs are applied to tennis action recognition by [KGOG06]. Their model is statistic and integrate the structure of tennis match. They combine audio and key frame features to be able to segment, with a good accuracy, the different parts of the tennis broadcasting such as the first serves, rallies, replays and breaks. The sound of the crowd such as applause, the sound of the ball or the commentator speech combined with key frames which capture visual information lead to 86% of segmentation accuracy compared to 65% and 77% with only respectively visual features and audio features. Such applications are interesting for sport coaches who wish to comment and examine only sequences of sports.

[dCBM+11] present a new dataset for tennis actions. This one contains only three types of classes: "hit", "serve" and "non-hit-class". The dataset is build from TV broadcasts of tennis games (matches of females in the Australian Open championships). They are interested in action localization and their classification. To do so, they introduce a local BoW method on the Spatio-Temporal gradients HOG3D features which are an extension of the classical 2D HOG features in three dimensions. They also use STS features. Both features are from the located actor and classification is performed using Fisher discriminant analysis. They obtain an accuracy of 77.6% using STS model based. Their confusion matrix is represented in Fig. 9.1.

One can see that "serve" samples are easier to classify than "hit" or "non-hit" samples. This is certainly due to the time that a service takes and its decomposition in time, which starts by large movement of the player when launching the ball. Hit and non-hit classes are then harder to distinguish because the hit class is very limited in time. Looking only at the player shape, the ball might not be visible, and features might look the same as when the player is simply moving in the field.

Recently, deeply related to our domain, [CPM19] use the OF Singularities with BoW and SVM in order to classify very similar actions. This task is also called fine-grained action classification. They apply their method on the TTStroke-21 dataset which contains 20 different strokes and a negative class. Their method is

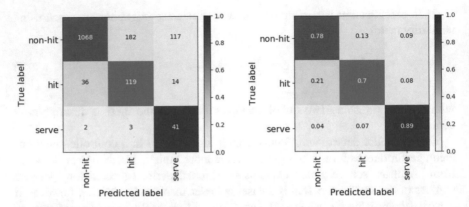

Fig. 9.1 Confusion matrix of tennis game from TV broadcast [dCBM+11]. Left with number of samples and right normalized

inspired from [BPM14] which uses the trajectory of critical points for classification. In this case, the actions to recognize are the different types of strokes preformed during table tennis training session. However, the scores remain low due to the high similarity of the different strokes and the limited amount of video samples. It makes generalization of extracted features harder.

Table Tennis stroke recognition is also performed by [LWS+19]. It is based on their Body Sensor Network (BSN). Their sensors collect acceleration and angular velocity information from the upper arm, lower arm and the back of the player. From the recorded signals, they extract Principal Component Analysis (PCA) features which are then fed to a SVM. They reach an accuracy of 97.4%, however they use only five classes: "forehand drive, "block shot", "forehand chop", "backhand chop" and "smash". Similarly, [XWX+20] recently proposed classification from integrated wearable sensors using K-means and DBScan clustering. Their taxonomy is more limited than in TTStroke-21 since they use only nine classes across badminton and table tennis sports. They reach an accuracy of 86.3% when considering all the classes. This score reaches 92.5% when considering only table tennis but this classification is limited to four classes: "Service", "Stroke", "Spin" and "Picking up". The extent of their taxonomy is thus limited and does not contribute much to the player experience. Furthermore, using such sensors, strongly limits the application possibilities and has a greater cost regarding training equipment adaptation. Also their system does not offer visualisation of the stroke performed since it is based on sensors, and it limits the feedback for the player.

Recently, a method is introduced by [WZD+20] to get the tactics of the players based on their performance in past matches. Their model is based on Hidden Markov Model (HMM) and aims at characterizing and simulating the competition process in table tennis. They use richer taxonomy and terms of stroke techniques than the previously presented methods: 13 different classes and 4 player positions which can be combined. Compared to TTStroke-21, we consider 10 classes with 2 player positions: "Forehand" and "Backhand". Their goal is therefore, not to classify an input, but to simulate matches between two different players. It is not directly linked to action recognition methods, but it does give a tool for players to simulate sport encounters and give credits to the TTStroke-21 dataset which propose much richer taxonomy than previous datasets.

Thus having analysed the two kinds of approaches: with handcrafted features and with Deep Neural Networks we can state the following. Despite the use of temporal information coming from OF and derived features, the methods with handcrafted features allow a good classification on datasets that remain simple: either with a low number of classes or with classes that are easily separable. It does become more complicated when the task focuses on one particular sport with different actions within or in case, when the complexity of "in-the-wild" scenes is higher. Here, already earlier research work have given a direction to follow: the use of Deep Neural Networks. Before we afford our solutions, it is interesting to review existing datasets which are used by the community for action recognition in video.

9.3 Datasets for Action Recognition

The need of datasets for action recognition has grown those last years. These datasets can change in terms of number of videos starting from a few videos up to millions of them. In addition to their size, the number of categories and their complexity also vary from few classes up to few hundreds, or even thousands of them. Each dataset can be labelled with annotations either by enriching the terminology, localising the action in space and time or by adding modalities information such as skeleton.

9.3.1 Annotation Processes

One can distinguish two ways to annotate a dataset: automatic or "by hand", that is by a human operator. Label propagation techniques were an attempt to leverage the gap between fully "manual" annotation and automatic one [HOSS13]. Here the dataset is represented as a big graph with samples-nodes and associated similarity metric between them. The label propagation is fulfilled from manually annotated nodes to unlabelled ones by optimal search on graphs. Nevertheless, the practices in annotation of video datasets are such that, the automatic annotation by a concept recognition method at hand is a first step and then human intervention is required to filter out automatic annotation errors. Also, it is common to have a dataset split into "auto" and "clean" sets. The "auto" being the one annotated using automatic methods and "clean" the one automatically labeled, verified and adjusted by hand. The two annotation processes are first described before presentation of the datasets.

9.3.1.1 Automatic Annotation

Tags from social network platforms can be used to collect rich datasets. It is the case, for example, with the IG65M dataset [GTM19]. To collect it, the authors use Kinetics-400 [KCS+17] dataset class names as hashtags on Instagram and collect 65M clips from 400 classes. Such annotation process requires filtering in order to refine the annotations.

In movies, the script can also help to automatically label the sequence. It is for example what [MLS09] have done for the Hollywood2 dataset. They generate the samples this way and clean them manually for the test set. Similarly, datasets can be constructed from the description of the videos from online platforms hosting them. Then, according to a description, datasets can be generated in an automatic way.

9.3.1.2 Manual Annotation

The most common way to annotate a dataset, especially when it is not large, is to label all the samples by hand. Some tools might be used to help in the process such as a pre-classification if a model already exists, or localization segment candidates of the actions when the video is untrimmed. One can distinguish two ways in the hand-labelling process: if the annotation is done by one person or several. By one person, the risk is that an inattention might lead to errors in the dataset or make it biased according to the point of view of the annotator. To overcome this issue, a crowdsourcing method can be used.

Crowdsourcing is based on the annotation of the same segments by different persons. It relies on the collective intelligence and should give better results than with only one person annotating. Outliers in annotations are not considered in the final decision. Different rules might apply, e.g take the mean of the annotators when possible or consider only the annotator that performed the best today; in order to take an annotation decision. There are also datasets which provide gross crowdsourced annotations and it is the team working on the dataset that decides which decision to make.

A new trend appeared recently: the use of Amazon Mechanical Turk (AMT).[1] AMT, also called "MTurk", is a crowdsourcing marketplace for individuals and businesses to outsource their processes and jobs to a distributed workforce who can perform these *micro-tasks* virtually. Here it is applied to annotation and AMT are paid according to the number of annotations performed. It started to be used with ImageNet dataset dedicated to image classification with 3.2 millions of images over 5247 classes. The strategy is twofolds:

AMT workers verify the presence of the action in the video candidates and they can also temporally annotate them. It is often coupled with a crowdsourced method

---

[1] https://www.mturk.com/.

meaning that each video will be annotated by several AMT workers. This allows the construction of a large dataset in a short amount of time.

9.3.2   Datasets for Action Classification

Datasets of actions classification problem can be categorized in many ways. In this section, the datasets are grouped according to the acquisition process: acquired in a controlled environment, extracted from movies or recorded "in the wild". Obviously, such a categorization is not perfect since some datasets mix different types of videos. Without being exhaustive we will present some of them focusing mainly on sport video datasets.

9.3.2.1   The Acquisition-Controlled Datasets

These datasets are often self made by the authors who decide in what type of environment the actions will be performed. It does not always mean that the dataset is easier than "in the wild" since difficulties can be added on purpose. The databases from broadcasts or recordings not meant for action recognition task are also considered in this subsection because the acquisition environment can be taken into account in the classification process.

KTH and Weizmann Datasets These two datasets were the most popular at the early ages of action recognition research. Despite their simplicity, some researchers are continuing using them as a benchmark.

KTH introduced in [SLC04][2] stands for "Kungliga Tekniska Högskolan" in Swedish which is the Royal Institute of Technology (Stockholm, Sweden), institution of the authors. The dataset is composed of six classes: "Walking", "Jogging", "Running", "Boxing", "Handwaving" and "Hand clapping". The acquisition was done in a controlled environment, with a homogeneous background, static camera at 25 fps, with 25 actors and has 2391 video clips across 600 videos. Videos are recorded outdoors and indoors. Weizmann dataset[3] [GBS+07] is quite similar but is enriched with more actions, e.g. 'jumping- jack", "galloping-sideways"... increasing the number of action classes to 9. It is constituted of 81 video sequences recorded at 25 fps at low resolution (180 × 144).

The necessity of action recognition "in-the-wild" yielded production of much more complex datasets described below.

---

[2]https://www.csc.kth.se/cvap/actions/.

[3]http://www.wisdom.weizmann.ac.il/~vision/SpaceTimeActions.html.

ACASVA [dCBM+11] "Adaptive Cognition for Automated Sports Video Annotation" (ACASVA) introduce a tennis action dataset.[4] Their objectives is to evaluate classical action recognition approaches with regard to player action recognition in tennis games. The data are collected from tennis TV broadcasts. The videos are then spatially segmented on the players and temporally annotated using three classes: "hit", "serve" and "non-hit". The complexity of the dataset remains simple.

FineGym Dataset The FineGym dataset[5] [SZDL20], is a fine-grained action dataset with a special focus on gym sport. The authors use a rich taxonomy to decompose each actom of structured gymnastic figures. They use three-level semantics and analyse four different gymnastic routines: balance-beam, uneven-bars, vault and floor exercise. They have a total of 530 element categories but only 354 have at least one instance. This rich amount of categories is due to all the combinations of possible actoms. The authors offer two settings: Gym288 with 288 classes but of very unbalanced distribution and Gym99, more balanced but with "only" 99 classes. The total number of samples considering all classes reaches 32,697. The 708 h of videos are hosted on YouTube with most of them in high resolution.

TUHAD The TUHAD dataset [LJ20] is also a dataset dedicated to fine-grained action recognition on Taekwondo sport. The dataset was recorded with the help of 10 Taekwondo experts using two Kinect cameras with front and side view. The number of classes is low: with only eight Taekwondo moves. Thousand nine hundred and thirty-six action samples are recorded with depth and IR images along with the RGB data. The classes are very similar in many ways but a foot position, which might be overcome with proper features.

9.3.2.2 Datasets from Movies

Despite these datasets do not generally contain sport actions, they are also interesting as they comprise recordings of natural behaviour of actors in cluttered environments. Thus Drinking and Smoking Dataset [LP07] is composed of sequences from Jim Jarmush Movie "Coffee and Cigarettes". It was designed for joint action detection and classification for the two classes: "smoking" and "drinking" with respectively 141 and 105 samples.

The **Hollywood2 dataset**[6] [MLS09] was designed for action and scene classification. It contains 10 scene classes and 12 actions: "Answer phone", "Drive car", "Eat", "Fight person", "Get out car", "Hand shake", "Hug person", "Kiss", "Run", "Sit down", "Sit up" and "Stand up"; over 7 h of video from 69 movies. They have

---

[4]https://www.cvssp.org/acasva/Downloads.html.

[5]https://sdolivia.github.io/FineGym/.

[6]https://www.di.ens.fr/~laptev/actions/hollywood2/.

a total of 1694 actions samples. The difficulty lies in the fact that different actions can happen in the same sequence.

9.3.2.3   In-the-Wild Datasets

"In-the-wild" means that the videos are from different sources and can be recorded by professionals or amateurs. They thus may contain camera motion, strong blur, occlusions...everything that can make the action recognition task harder. However, videos can also contain much background information, which might be an exploitable source for training of classification models.

The UCF Datasets The UCF datasets [SZS12] have become very popular for developing and benchmarking methods for action recognition in sport video. UCF title comes from the name of the university in which the datasets have been developed: University of Central Florida.

The first UCF dataset was UCF-Sports. It contains various sequences from broadcast TV channels across nine different sports: "diving", "golf swinging", "kicking", "lifting", "horseback riding", "running", "skating", "swinging a baseball bat", and "pole vaulting". Pole vaulting is split in two classes: "Swing-Bench" and "Swing-Side" totaling 10 classes. It first contained 200 sequences (reduced to 150 later) with an image resolution of 740 × 480 at 10 fps.

Later, the UCF YouTube Action also called UCF11 dataset[7] is introduced. It consists of 11 classes from 1160 videos from the online video platform YouTube.

UCF50 is an extension of UCF11 with a total of 50 action classes. This version is then extended to make UCF101 dataset.[8] UCF101 includes a total number of 101 action classes which can be divided into five domains: "Human-Object Interaction", "Body-Motion Only", "Human-Human Interaction", "Playing Musical Instruments" and "Sports". Constructed from 2500 videos "in-the-wild", they extract a total of 13,320 clips in order to have at least 101 clips per class. The dataset is widely used by the scientific community and led to the THUMOS challenge[9] held in 2013, 2014 and 2015. The dataset was cleaned and enriched with temporal annotations in 2015 in order to provide qualitative benchmark for different methods and be used also for spatio-temporal localization and temporal detection only.

The Kinetics Datasets The kinetics datasets:[10] Kinetics-400 [KCS+17], Kinetics-600 and Kinetics-700 [CNHZ19] consider respectively 400, 600 and 700 action classes. They are all financed by DeepMind company, specialized in AI which, from

---

[7]www.crcv.ucf.edu/data/UCF_YouTube_Action.php.

[8]www.crcv.ucf.edu/data/UCF101.php.

[9]www.thumos.info.

[10]https://deepmind.com/research/open-source/kinetics.

2014, belongs to Google. In the taxonomy of actions they contain, we find sport actions as well, e.g. "playing squash or racquetball".

The videos are collected from YouTube video platform, automatically annotated and candidates are refined using AMT. The difference between the versions of the datasets lies in:

- the number of classes: the number of classes has increased over time. New classes were added and pre-existing classes were refined. Some were merged.
- the amount of videos: the number of videos started with 306,245 clips and more than doubled in the last version
- the splits between the different sets: training, validation and test sets have been modified over time. For example, samples belonging to the training set in the first version might belong to the test set in the last version.

AVA and AVA-Kinetics In [GSR+18], the authors introduced the AVA dataset[11] in order to perform joint localization and classification of actions. It contains 437 videos gathered from YouTube, 15 min are extracted from them and annotated every second. They use a vocabulary of 80 atomic actions. The difficulty in this dataset is the overlapping actions in time and their localization. They offer a split of the dataset by extracting 900 video segments of 3 s from all the 15 min videos. By doing so, the 55 h of video are split in 392,000 overlapping segments. The AVA-Kinetics dataset [LTR+20].[12] Is the merger of the two: Kinetics-700 and AVA datasets. Kinetics videos were annotated using AVA protocol. The dataset thus contains over 230,000 video clips spatially and temporally annotated using the 80 AVA action classes.

SAR4 [FVD+19] present the SAR4 dataset which focuses on action in football sport (or soccer). They track and label the players from available videos on YouTube. The actions performed by the tracked players are then annotated using a taxonomy of four classes: "dive", "shoot", "pass received" and "pass given". The total number of sequences is 1292 with actions lasting from 5 up to 59 frames. The discussed datasets are summarized in Table 9.1

9.3.3   The `TTStroke-21` Dataset

The `TT-Stroke21` dataset was recorded for fine-grained recognition of sport actions, in the context of the improvement of sport performance for amateurs or professional athletes. Our case study is table tennis, and our goal is the temporal segmentation and classification of strokes performed. The low inter-class variability makes the task more difficult for this content than for more general action databases such as UCF or Kinetics.

---

[11] https://research.google.com/ava/.

[12] https://deepmind.com/research/open-source/kinetics.

Table 9.1 Presentation of the different action datasets in terms of number of classes, acquisition process, the amount of videos and the number of extracted clips

Datasets	# classes	Acquisition	# videos	# clips
KTH [SLC04]	6	Controlled	600	2391
Weizmann [GBS+07]	9	Controlled	–	81
Coffee and Cigarettes [LP07]	2	Film	1	246
UCFSports	10	Broadcast	–	150
UCF11	11	In the wild	–	1160
Hollywood2 [MLS09]	12	Films	69	1694
UCF101 [SZS12]	101	In the wild	2500	13,320
AVA [GSR+18]	80	In the wild	437	392,000
SAR4 [FVD+19]	4	Broadcast	–	1292
Kinetics-700 [CNHZ19]	700	In the wild	–	650,317
FineGym [SZDL20]	354	Broadcast	–	32,697
AVA-Kinetics [LTR+20]	80	In the wild	–	230,000

Twenty stroke classes and an additional rejection class have been established based on the rules of table tennis. The filmed athletes are students, and their teachers supervise the exercises performed during the recorded sessions. The objective of table tennis stroke recognition is to help the teachers to focus on some of these strokes to help the students in their practice.

Table tennis strokes are most of the time visually similar. Action recognition in this case requires not only a tailored solution, but also a specific expertise to build the ground truth. This is the reason why annotations were carried out by professional athletes. They use a rather rich terminology that allows the fine-grained stroke definition. Moreover, the analysis of the annotations shows that, for the same video and the same stroke, professionals do not always agree. The same holds for defining temporal boundaries of a stroke, which may differ for each annotator. This variability cannot be considered as noise, but shows the ambiguity and complexity of the data and has to be taken into account. We call this new database TTStroke-21, TT standing for Table Tennis and 21 for the number of classes.

9.3.3.1   TTStroke-21 Acquisition

TTStroke-21 is composed of videos of table tennis games with 17 different players. The sequences are recorded indoors without markers using artificial light and light-weight cameras. The recording setting is illustrated in Fig. 9.2a.

The player is filmed in three situations:

- performing repetition of the same stroke. However those repetitions might fail once or several times in the video and the player might do another stroke than the one expected.

a. Video acquisition with   b. Annotation platform
aerial view from the ceiling

Fig. 9.2 Overview of the TTStroke-21 dataset

- simple exchanges between two players: those exchanges are meant to practise the different techniques.
- in match conditions: the players are meant to mark points. The game speed is much faster and strokes are shorter in time.

9.3.3.2 TTStroke-21 Annotation

The annotation process was designed as a crowdsourcing method. The annotation sessions were supervised by professional table tennis players and teachers. A user-friendly web platform has been developed by our team for this purpose (Fig. 9.2b), where the annotator spots and labels strokes in videos: starting frame, end frame and the stroke class. The annotator also indicates if the player is right-handed or left-handed. The taxonomy is built upon a shake-hand grip of the racket leading to forehand and backhand stroke according to the side of the racket used.

The taxonomy comprises 20 table tennis stroke classes. All the strokes can, as well, be divided in two super-classes: **Forehand** and **Backhand**. The linguistic analysis of annotations shows that for the same video and the same stroke, professionals do not employ the same degree of details in their annotations. The same problem occurs with temporal analysis: for instance, a service (first stroke when the player releases the ball) might be considered to start (1) when the player is in position, (2) when the ball is released or (3) when the racket is moving.

Since a video can be annotated by several annotators, temporal annotations needed to be filtered. An overlap between each annotation of 25% of the annotated stroke duration is allowed. Above this percentage, the annotations are considered to be part of the same stroke and are temporally fused.

Another filter is applied by checking if labels of the same stroke are consistent. If not, this portion of video is not considered in our classification task. This filtering, based on multiple annotations for the same recorded video, can still leave some labeling errors since multiple labeling of the same clip by different annotators was not always easy to meet in practice.

9.3.3.3 Negative Samples Extraction

Negative samples are created from videos with more than 10 detected strokes. This was decided after noticing how some videos were poorly annotated and could lead to include actual strokes as negative samples.

The negative samples are video sub-sequences between each detected stroke. We allow the overlap with the previous and the subsequent stroke of 10% of our target time window length: 0.83 s, which allows to capture short strokes without considering another one. This represents 100 frames at 120 fps.

9.3.3.4 Data for Evaluation

Hundred and twenty-nine videos at 120 fps have been considered. This content represents 94 min of table tennis games, totalling 675,000 video frames and 1387 annotations. After filtering, 1074 annotations were retained. The peak statistics of stroke duration are $min = 0.64$ s, $max = 2.27$ s and the average duration is 1.46 s with standard deviation of 0.36. Accordingly, a total of 1048 strokes were extracted with a min duration of 0.83 s, a max duration of 2.31 s and an average duration of 1.47 s with standard deviation of 0.36. Some annotations were merged making the statistical duration a bit longer. After these steps, 681 negative (non-stroke) samples were extracted. They have a mean duration of 2.34 s and standard deviation of 2.66 s. This high standard deviation comes from the non game activity of long period between strokes, which can be due to a ball lost or talks of players between games. However, as presented in Table 9.2 representing the distribution over the split of the dataset, not all negative samples are considered to avoid biases in the training and evaluation processes.

9.4 TSTCNN: A Twin Spatio-Temporal 3D Convolutional Neural Network for Action Recognition

After having reviewed a bunch of methods for action recognition in video and of reference datasets, we introduce here our solution to the problem of fined-grained action recognition in video with a 3D CNN we call TSTCNN—a twin spatio-temporal CNN.

A Two Stream Architecture

The Twin Spatio-Temporal Convolutional Neural Network (TSTCNN), denoted as illustrated in Fig. 9.3, is a two stream 3D Convolutional Network constituted of two

Table 9.2 Datasets taxonomy of TTStroke-21

Table tennis strokes	# Samples				# Frames		
	Train	Val	Test	Sum	Min	Max	Mean[a]
Def. Backhand Backspin	22	6	3	31	121	233	189 ± 25
Def. Backhand Block	19	5	3	27	100	261	131 ± 37
Def. Backhand Push	6	2	1	9	121	229	155 ± 31
Def. Forehand Backspin	29	8	4	41	129	229	177 ± 25
Def. Forehand Block	8	2	2	12	100	137	115 ± 14
Def. Forehand Push	23	7	3	33	105	177	143 ± 19
Off. Backhand Flip	25	7	3	35	100	265	195 ± 49
Off. Backhand Hit	28	8	4	40	100	173	134 ± 21
Off. Backhand Loop	21	6	3	30	100	229	155 ± 32
Off. Forehand Flip	31	9	5	45	113	269	186 ± 44
Off. Forehand Hit	45	13	6	64	100	233	158 ± 34
Off. Forehand Loop	23	7	3	33	101	277	177 ± 43
Serve Backhand Backspin	56	16	8	80	133	261	188 ± 31
Serve Backhand Loop	43	12	6	61	100	265	186 ± 42
Serve Backhand Sidespin	60	17	9	86	129	269	193 ± 33
Serve Backhand Topspin	57	16	8	81	100	273	175 ± 48
Serve Forehand Backspin	58	17	8	83	125	269	182 ± 35
Serve Forehand Loop	56	16	8	80	100	273	171 ± 51
Serve Forehand Sidespin	57	16	9	82	101	273	192 ± 39
Serve Forehand Topspin	67	19	9	95	100	273	184 ± 52
Non strokes samples	74	21	11	106	100	1255	246 ± 154
Total	808	230	116	1154	100	1255	182 ± 65

[a] Mean value $\pm$ standard deviation

branches. Each branch follows the same structure: three blocks constituted of a 3D convolutional layer using kernels of size $(3 \times 3 \times 3)$, with stride and padding 1 in all directions and "ReLU" as activation function, feeding a 3D Max-Pooling layer using kernels of size $(2 \times 2 \times 2)$ and floor function. From input to output, the convolutional layers use 30, 60 and 80 filters. Each branch ends with a fully connected layer of size 500. The two branches are combined using a bilinear transformation with Softmax function to output a classification score of size 21 corresponding to the number of classes considered in our task.

Dynamic Data

The use of dynamic data, such as optical flow, gives extra information to the network and an understanding of the physical world that the network does not have with only the rgb video stream. The OF, which represents the movement through the displacement of the pixels from one image to another, is encoded in the Cartesian coordinate system $V = (v_x, v_y)^T$ with v_x the horizontal displacement and v_y the vertical displacement. The optical flow values are then normalized between -1 and 1.

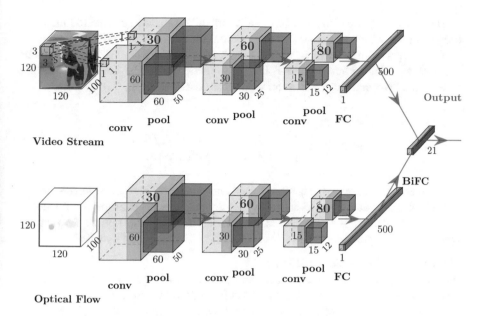

Fig. 9.3 TSTCNN—Twin Spatio-Temporal Convolutional Neural Network architecture. The video stream and its optical flow are processed in parallel through the network composed of successive 3D convolutional layers and 3D pooling layers and the extracted features a merge at the end through a bilinear transformation giving as output a classification score

In [MBPM19], several optical flow estimation methods are compared using Mean Squared Error (MSE) of motion compensation, angular error (AE) and end-point error (EPE) quality metrics on two datasets:

- Sintel Benchmark [MIH+16], dataset of synthetic videos with available reference optical flows (compared with MSE, AE and EPE),
- and TTStroke-21 [MBPM20], recorded in natural conditions with strong flickering due to synthetic light (compared with MSE only since reference flow is not available).

In the same work, different normalization methods are also tested for classification. According to their results, both of these modalities, optical flow estimator and normalization method, are of primary importance for classification. Indeed, the accuracy of the classification on the test sets of TTStroke-21 varies, according to the best performances for each optical flow estimator, from 41.4% for DeepFlow method [WRHS13] to 74.1% for Beyond Pixel method [Liu09] (BP). Even if BP is sensitive to flickering, it is able to capture fine details, such as the motion of the ball, contrarily to DeepFlow method, explaining such gap between performances of those two estimators. On the other hand, the normalization methods allow for boosting performances from 44% with "Max" normalization method to 74.1% with "Normal" normalization method. The "Max" normalization method strongly reduces the magnitude of most of motion vectors and therefore increases also the inter-similarity of the strokes; while the "Normal" normalization method increases the magnitude of most vectors and leaves room for inter-dissimilarity.

In the light of these results, we use the Beyond Pixel method [Liu09], based on iterative re-weighted least square solver, to estimate the optical flow and normalize it using the "Normal" normalization method as described in Eq. 9.3.

$$v' = \frac{v}{\mu + 3 \times \sigma}$$
$$v^N(i, j) = \begin{cases} v'(i, j) & \text{if } |v'(i, j)| < 1 \\ SIGN(v'(i, j)) & \text{otherwise.} \end{cases} \tag{9.3}$$

where v and v^N represent respectively one component of the optical flow $\mathbf{V}$ and its normalization, μ and σ are mean value and standard deviation of the component.

The optical flow is then filtered by considering only the optical flow of the foreground using the method of Zivkovic and Van der Heijden [ZvdH06].

3D Attention: What Could It Bring?

Attention mechanisms, in classification problem from rgb images, are used to determine which part of information is useful and/or needed to classify an image. Such attention can be obtained by recording the gaze fixations of individuals when looking at the image and performing the same classification task [OBGR19] as a CNN classifier, to create saliency map on image and then propagate it through the layers of the CNN. In DNNs, internal attention mechanisms have become popular. We distinguish two of them: (1) global attention which expresses the contribution of feature channels along convolutional layers into decision making for image classification task [HSA+20] and (2) local attention, which focuses on important features in the channels.

When importance of feature channels has to be computed, the processing consists of three steps: (1) *squeeze* (synthesis), (2) *excitation* and (3) *feature scaling*. Thus a small network of neurons learns a weighting coefficient for each channel at each layer and outputs the characteristic channels thus weighted to the next layer.

For the local attention, the authors of [WJQ+17] use the principles of residual neural networks to propose "residual" learning of the attention masks incorporated in the convolution layers in both forward and backward propagation, which leads to better robustness to noise. By minimizing the objective function by gradient descent, the attention mechanisms are implicitly introduced via the derivative calculation where the weighted characteristics are used. "Teacher-student" networks is another way to introduce attention in the layers of a classification network. In [ZK17] the "Teacher" network is the one that learns attention and guides the "student" network for the image classification task. Such works which focus on image classification inspire the design of spatio-temporal attention to tackle the action classification problem.

In [DYL+18], the authors use spatial attention mechanism based on feature pyramids and construct the temporal attention by aggregation of the attention maps obtained spatially, in the temporal domain. Then, based on their work, the authors of [DK19] build 3D convolutional blocks and incorporate them in a 3DResNet

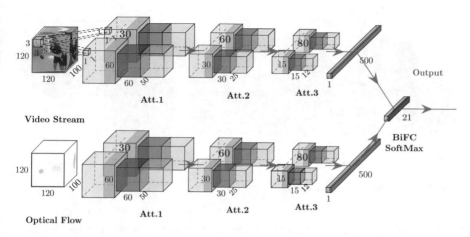

Fig. 9.4 TSTCNN with 3D attention blocks after each Max Pool layer

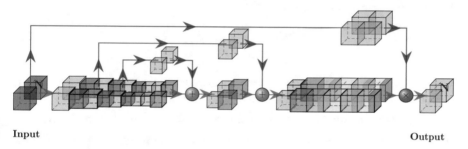

Input   Output

Fig. 9.5 3D attention block architecture

network for 3D gesture recognition. However, as in [DYL+18], the authors do not use the motion information explicitly. In [MBPM21], we introduce the attention blocks through our TSTCNN into the two video streams: the branch containing the spatial information (RGB) and the branch containing the temporal information (optical flow) as depicted in Fig. 9.4. The optical flow plays the discriminating role for classification and does not act only as localization information.

As depicted in Fig. 9.5, the attention mechanism uses several 3D residual blocks, illustrated in Fig. 9.6.

The implemented residual block is inspired from the work carried out in 2D in [HZRS16] and has been extended and adapted in 3D with a 4D data block of size $(N \times W \times H \times T)$ representing respectively the number of channels, the two spatial dimensions and the temporal dimension. Input data are then processed by three successive layers $f_{conv_i}, i = 1, \ldots, 3$, see Eq. 9.10, with respectively $\frac{N}{4}, \frac{N}{4}$ and N filters of size $(1 \times 1 \times 1)$, $(3 \times 3 \times 3)$ and $(1 \times 1 \times 1)$. Their output is then summed with the input data to build the output of the 3D residual block (Fig. 9.6).

$$res(x) = f_{conv_3}(f_{conv_2}(f_{conv_1}(x))) + x \qquad (9.4)$$

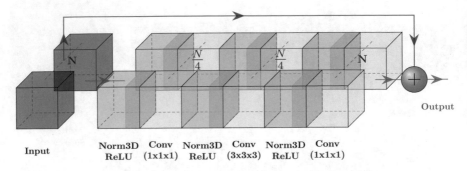

Fig. 9.6 3D residual block architecture

The 3D batch normalization is performed channel by channel over the batch of data. If we have $x = (x_1, x_2, \ldots, x_{N_{channels}})$, then the normalization is $F_n(x) = (f_n(x_1), f_n(x_2), \ldots, f_n(x_{N_{channels}}))$ with:

$$f_n(x_i) = \frac{x_i - \mu_i}{\sqrt{\sigma_i^2 \oplus \epsilon}} * \gamma_i \oplus \beta_i \qquad (9.5)$$

with $i = 1, \ldots N_{channels}$, μ_i and σ_i the mean and standard deviation vectors of x_i computed over the batch,$\oplus$ is addition of a scalar to each vector coordinate, γ_i and β_i are learnable parameters per channel and the division by $\sqrt{\sigma_i^2 \oplus \epsilon}$ is element-wise. Here, $N_{channels} = N$ or $\frac{N}{4}$, depending on the normalization position in the residual block.

The 3D attention blocks illustrated in Fig. 9.5 are inspired from the work carried out in 2D in [WJQ+17].

A 3D attention block takes as input a 4D data block of size ($N \times W \times H \times T$). In this block, all convolutions are performed with the same number of filters, N, to maintain the dimension of the processed data. The input data are processed by the first 3D residual block, denoted as *res*. The network then splits in two branches: the trunk branch consisting of two successive 3D residual blocks, described by Eq. 9.6; and the soft floating mask branch, see lowest position in Fig. 9.5, described through Eqs. 9.7, 9.8, 9.10. Its role is to accentuate the features generated by the trunk branch. Those two branches are merged as described in Eq. 9.11.

$$branch_{trunk}(.) = res(res(.)) \qquad (9.6)$$

The soft mask branch is constituted of several 3D residual blocks followed by Max Pooling layers, denoted as *MaxP*. It increases the reception field of convolutions using a bottom-up architecture, denoted as $f_{bu}(.) = res(MaxP(.))$. The lowest resolution is obtained after three Max Pooling steps.

$$x_1 = f_{bu}(res(Input))$$
$$x_2 = f_{bu}(x_1))$$
$$x_3 = f_{bu}(x_2))$$
(9.7)

The information is then extended by a symmetrical top-down architecture, $f_{td}(.) = Inter(res(.))$, to project the input features of each resolution level. *Inter* denotes the trilinear interpolations used for up-sampling. Two skipped connections are used for collecting information at different scales.

$$y_1 = f_{td}(x_3) + res(x_2)$$
$$y_2 = f_{td}(y_1) + res(x_1)$$
$$y_3 = f_{td}(y_2)$$
(9.8)

The soft mask branch is then composed of two successive layers. Each includes a 3D batch normalization, denoted as $F_n(.)$ as described by Eq. 9.5, followed by a ReLU activation function and a convolution layer with kernel sizes $(1 \times 1 \times 1)$. This is expressed by Eq. 9.9:

$$f_{conv}(.) = conv(ReLU(F_n(.)))$$
(9.9)

It ends with a sigmoid function, denoted as "Sig", to scale values between 0 and 1. These two layers are depicted on the right of the lowest branch in Fig. 9.5 and are expressed by Eq. 9.10.

$$branch_{fmask}(Input) = Sig(f_{conv}(f_{conv}(y_3)))$$
(9.10)

The output of the trunk branch is then multiplied term by term by $(1 \oplus branch_{fmask}(Input))$ where $branch_{fmask}(Input)$ is the output of the mask branch. The result is then processed by the last 3D residual block $res(.)$ which ends the attention block, see Eq. 9.11.

$$y = res(branch_{trunk}(Input) \odot (1 \oplus branch_{fmask}(Input)))$$
(9.11)

Here the $\odot$ is an element-wise multiplication and $\oplus$ is an addition of a scalar to each vector component as defined above.

9.4.1   Results

In Table 9.3, we compare the models in terms of accuracy for the pure classification task. In order to have an overall view, comparison is done with the models using three attention blocks (one after each max pooling layer) with the models and without them.

Table 9.3 Comparison of the classification performances for models using attention mechanism after convergence in terms of accuracy. The best model and test accuracy are reported in bold for each modality

Models	Epochs	Accuracies in %		
		Train	Val	Test
RGB-I3D	778	98.3	72.6	69.8
RGB-STCNN	1665	96.7	88.7	89.8
RGB-STCNN with attention	524	96.5	88.3	**93.2**
Flow-I3D	1112	98.8	74.8	73.3
Flow-STCNN	1449	97.5	79.6	75.9
Flow-STCNN with attention	732	96.4	83.5	**90.7**
Two-Stream I3D	–	99.2	76.2	75.9
LF-STCNN	–	97	88.7	89.8
LF-STCNN with attention	–	97	88.7	94.9
T-STCNN	1784	95.8	87.8	93.2
T-STCNN with attention	591	97.3	87.8	**95.8**

The best model and test accuracy are reported in bold for each modality.

The presented models using spatio-temporal convolutions shallower than the I3D models prove to be more efficient to classify fine-grained actions on a challenging dataset with limited amount of samples. The fine-grained aspect of the task seems to be dealt better if the model is not too deep. Indeed, the deepness of the I3D models is efficient on task were a great number of objects and scenes need to be recognized in order to classify coarse-grained actions, but it seems to be less effective for the fine-grained task.

Furthermore, the amount of samples and the fact that the models are not pre-trained, makes the task even harder. The overfitting problem is more noticeable on the very deep models I3D. However, the STCNN models using attention mechanisms do not suffer from this: the performances are improved and the convergence is faster. Best performances are observed for the Twin model using attention mechanism. The intermediate fusion allows a better combination of each branches in order to perform classification. A late fusion approach seems efficient but is 1% behind.

9.5 Conclusion and Perspectives

In this chapter, we have shown that the features developed to perform for action classification in videos have evolved from handcrafted to deep learnt with the increasing complexity of datasets and their increasing number of video samples. Furthermore, the increasing capacity of CPUs and GPUs to process large amount of data have allowed deep learning methods to move from 2D to 3D convolutions. 3D convolutions for action classification from videos have proven to be efficient tools to capture efficiently spatio-temporal features in order to perform classification.

The incorporation of attention mechanisms, through attention blocks, helps in the process by highlighting the discriminant features, boosting the convergence speed and classification performances. Such behaviour was observed on the fine-grained dataset TTStroke-21. It has also been noted that the deepness of the implemented models needs to be adapted to the classification task and the dataset in order to avoid overfitting.

References

[BGS+17]   Mateusz Budnik, Efrain-Leonardo Gutierrez-Gomez, Bahjat Safadi, Denis Pellerin, and Georges Quénot. Learned features versus engineered features for multimedia indexing. *Multim. Tools Appl.*, 76(9):11941–11958, 2017.

[BPM14]   Cyrille Beaudry, Renaud Péteri, and Laurent Mascarilla. Action recognition in videos using frequency analysis of critical point trajectories. In *2014 IEEE International Conference on Image Processing, ICIP 2014, Paris, France, October 27–30, 2014*, pages 1445–1449, 2014.

[CDF+04]   Gabriella Csurka, Christopher R. Dance, Lixin Fan, Jutta Willamowski, and Cédric Bray. Visual categorization with bags of keypoints. In *In Workshop on Statistical Learning in Computer Vision, ECCV*, pages 1–22, 2004.

[CNHZ19]   João Carreira, Eric Noland, Chloe Hillier, and Andrew Zisserman. A short note on the kinetics-700 human action dataset. *CoRR*, abs/1907.06987, 2019.

[CPM19]   Jordan Calandre, Renaud Péteri, and Laurent Mascarilla. Optical flow singularities for sports video annotation: Detection of strokes in table tennis. In *Working Notes Proceedings of the MediaEval 2019 Workshop, Sophia Antipolis, France, 27–30 October 2019*, 2019.

[CRHV09]   Rizwan Chaudhry, Avinash Ravichandran, Gregory D. Hager, and René Vidal. Histograms of oriented optical flow and binet-cauchy kernels on nonlinear dynamical systems for the recognition of human actions. In *2009 IEEE Computer Society Conference on Computer Vision and Pattern Recognition (CVPR 2009), 20–25 June 2009, Miami, Florida, USA*, pages 1932–1939, 2009.

[CZ17]   João Carreira and Andrew Zisserman. Quo vadis, action recognition? A new model and the kinetics dataset. In *2017 IEEE Conference on Computer Vision and Pattern Recognition, CVPR 2017, Honolulu, HI, USA, July 21–26, 2017*, pages 4724–4733, 2017.

[dCBM+11]   Teofilo de Campos, Mark Barnard, Krystian Mikolajczyk, Josef Kittler, Fei Yan, William J. Christmas, and David Windridge. An evaluation of bags-of-words and spatio-temporal shapes for action recognition. In *IEEE Workshop on Applications of Computer Vision (WACV 2011), 5–7 January 2011, Kona, HI, USA*, pages 344–351, 2011.

[DHR+17]   Jeff Donahue, Lisa Anne Hendricks, Marcus Rohrbach, Subhashini Venugopalan, Sergio Guadarrama, Kate Saenko, and Trevor Darrell. Long-term recurrent convolutional networks for visual recognition and description. *IEEE Trans. Pattern Anal. Mach. Intell.*, 39(4):677–691, 2017.

[DK19]   Naina Dhingra and Andreas M. Kunz. Res3ATN - deep 3D residual attention network for hand gesture recognition in videos. In *2019 International Conference on 3D Vision, 3DV 2019, Québec City, QC, Canada, September 16–19, 2019*, pages 491–501, 2019.

[DYL+18]   Yang Du, Chunfeng Yuan, Bing Li, Lili Zhao, Yangxi Li, and Weiming Hu. Interaction-aware spatio-temporal pyramid attention networks for action classification. In *ECCV (16)*, volume 11220 of *Lecture Notes in Computer Science*, pages 388–404. Springer, 2018.

[EBMM03] Alexei A. Efros, Alexander C. Berg, Greg Mori, and Jitendra Malik. Recognizing action at a distance. In *9th IEEE International Conference on Computer Vision (ICCV 2003), 14–17 October 2003, Nice, France*, pages 726–733, 2003.

[FPW17] Christoph Feichtenhofer, Axel Pinz, and Richard P. Wildes. Spatiotemporal multiplier networks for video action recognition. In *2017 IEEE Conference on Computer Vision and Pattern Recognition, CVPR 2017, Honolulu, HI, USA, July 21–26, 2017*, pages 7445–7454, 2017.

[FVD+19] Mehrnaz Fani, Kanav Vats, Christopher Dulhanty, David A. Clausi, and John S. Zelek. Pose-projected action recognition hourglass network (PARHN) in soccer. In *16th Conference on Computer and Robot Vision, CRV 2019, Kingston, ON, Canada, May 29–31, 2019*, pages 201–208, 2019.

[GBS+07] Lena Gorelick, Moshe Blank, Eli Shechtman, Michal Irani, and Ronen Basri. Actions as space-time shapes. *IEEE Trans. Pattern Anal. Mach. Intell.*, 29(12):2247–2253, 2007.

[GHS13] Adrien Gaidon, Zaïd Harchaoui, and Cordelia Schmid. Temporal localization of actions with actoms. *IEEE Trans. Pattern Anal. Mach. Intell.*, 35(11):2782–2795, 2013.

[GSR+18] Chunhui Gu, Chen Sun, David A. Ross, Carl Vondrick, Caroline Pantofaru, Yeqing Li, Sudheendra Vijayanarasimhan, George Toderici, Susanna Ricco, Rahul Sukthankar, Cordelia Schmid, and Jitendra Malik. AVA: A video dataset of spatio-temporally localized atomic visual actions. In *2018 IEEE Conference on Computer Vision and Pattern Recognition, CVPR 2018, Salt Lake City, UT, USA, June 18–22, 2018*, pages 6047–6056, 2018.

[GTM19] Deepti Ghadiyaram, Du Tran, and Dhruv Mahajan. Large-scale weakly-supervised pre-training for video action recognition. In *IEEE Conference on Computer Vision and Pattern Recognition, CVPR 2019, Long Beach, CA, USA, June 16–20, 2019*, pages 12046–12055, 2019.

[HCS17] Rui Hou, Chen Chen, and Mubarak Shah. Tube convolutional neural network (T-CNN) for action detection in videos. In *IEEE International Conference on Computer Vision, ICCV 2017, Venice, Italy, October 22–29, 2017*, pages 5823–5832, 2017.

[HOSS13] Michael E. Houle, Vincent Oria, Shin'ichi Satoh, and Jichao Sun. Annotation propagation in image databases using similarity graphs. *ACM Trans. Multim. Comput. Commun. Appl.*, 10(1):7:1–7:21, 2013.

[HSA+20] Jie Hu, Li Shen, Samuel Albanie, Gang Sun, and Enhua Wu. Squeeze-and-excitation networks. *IEEE Trans. Pattern Anal. Mach. Intell.*, 42(8):2011–2023, 2020.

[HZRS16] Kaiming He, Xiangyu Zhang, Shaoqing Ren, and Jian Sun. Deep residual learning for image recognition. In *2016 IEEE Conference on Computer Vision and Pattern Recognition, CVPR 2016, Las Vegas, NV, USA, June 27–30, 2016*, pages 770–778, 2016.

[IBPQ14] Bogdan Ionescu, Jenny Benois-Pineau, Tomas Piatrik, and Georges Quénot, editors. *Fusion in Computer Vision - Understanding Complex Visual Content*. Advances in Computer Vision and Pattern Recognition. Springer, 2014.

[Joh73] G. Johansson. Visual perception of biological motion and a model for its analysis. *Perception and Psychophysics*, 14:pp. 201–211, 1973.

[JvGJ+14] Mihir Jain, Jan C. van Gemert, Hervé Jégou, Patrick Bouthemy, and Cees G. M. Snoek. Action localization with tubelets from motion. In *2014 IEEE Conference on Computer Vision and Pattern Recognition, CVPR 2014, Columbus, OH, USA, June 23–28, 2014*, pages 740–747, 2014.

[JXYY13] Shuiwang Ji, Wei Xu, Ming Yang, and Kai Yu. 3d convolutional neural networks for human action recognition. *IEEE Trans. Pattern Anal. Mach. Intell.*, 35(1):221–231, 2013.

[KCS+17] Will Kay, João Carreira, Karen Simonyan, Brian Zhang, Chloe Hillier, Sudheendra Vijayanarasimhan, Fabio Viola, Tim Green, Trevor Back, Paul Natsev, Mustafa Suleyman, and Andrew Zisserman. The kinetics human action video dataset. *CoRR*, abs/1705.06950, 2017.

[KGOG06] Ewa Kijak, Guillaume Gravier, Lionel Oisel, and Patrick Gros.   Audiovisual integration for tennis broadcast structuring. *Multim. Tools Appl.*, 30(3):289–311, 2006.

[KKA20] M. Esat Kalfaoglu, Sinan Kalkan, and A. Aydin Alatan. Late temporal modeling in 3d CNN architectures with BERT for action recognition. *CoRR*, abs/2008.01232, 2020.

[Lap13] Ivan Laptev. *Modeling and visual recognition of human actions and interactions.* Habilitation à diriger des recherches, Ecole Normale Supérieure de Paris - ENS Paris, July 2013.

[Liu09] Ce Liu. *Beyond Pixels: Exploring New Representations and Applications for Motion Analysis.* PhD thesis, Massachusetts Institute of Technology, 5 2009.

[LJ20] Jinkue Lee and Hoeryong Jung. Tuhad: Taekwondo unit technique human action dataset with key frame-based CNN action recognition. *Sensors*, 20(17):4871, 2020.

[LP07] Ivan Laptev and Patrick Pérez.   Retrieving actions in movies.   In *IEEE 11th International Conference on Computer Vision, ICCV 2007, Rio de Janeiro, Brazil, October 14–20, 2007*, pages 1–8, 2007.

[LTR+20] Ang Li, Meghana Thotakuri, David A. Ross, João Carreira, Alexander Vostrikov, and Andrew Zisserman. The ava-kinetics localized human actions video dataset. *CoRR*, abs/2005.00214, 2020.

[LWLW16] Zhihao Li, Wenmin Wang, Nannan Li, and Jinzhuo Wang. Tube convnets: Better exploiting motion for action recognition. In *2016 IEEE International Conference on Image Processing, ICIP 2016, Phoenix, AZ, USA, September 25–28, 2016*, pages 3056–3060, 2016.

[LWS+19] Ruichen Liu, Zhelong Wang, Xin Shi, Hongyu Zhao, Sen Qiu, Jie Li, and Ning Yang. Table tennis stroke recognition based on body sensor network. In *Internet and Distributed Computing Systems - 12th International Conference, IDCS 2019, Naples, Italy, October 10–12, 2019, Proceedings*, pages 1–10, 2019.

[Mac67] J. Macqueen. Some methods for classification and analysis of multivariate observations. In *In 5-th Berkeley Symposium on Mathematical Statistics and Probability*, pages 281–297, 1967.

[MBPM19] Pierre-Etienne Martin, Jenny Benois-Pineau, Renaud Péteri, and Julien Morlier. Optimal choice of motion estimation methods for fine-grained action classification with 3d convolutional networks. In *2019 IEEE International Conference on Image Processing, ICIP 2019, Taipei, Taiwan, September 22–25, 2019*, pages 554–558, 2019.

[MBPM20] Pierre-Etienne Martin, Jenny Benois-Pineau, Renaud Péteri, and Julien Morlier. Fine grained sport action recognition with twin spatio-temporal convolutional neural networks. *Multim. Tools Appl.*, 79(27–28):20429–20447, 2020.

[MBPM21] Pierre-Etienne Martin, Jenny Benois-Pineau, Renaud Péteri, and Julien Morlier. 3d attention mechanisms in twin spatio-temporal convolutional neural networks. application to action classification in videos of table tennis games. In *25th International Conference on Pattern Recognition (ICPR2020) - MiCo Milano Congress Center, Italy, 10–15 January 2021*, 2021.

[MIH+16] Nikolaus Mayer, Eddy Ilg, Philip Häusser, Philipp Fischer, Daniel Cremers, Alexey Dosovitskiy, and Thomas Brox. A large dataset to train convolutional networks for disparity, optical flow, and scene flow estimation. In *2016 IEEE Conference on Computer Vision and Pattern Recognition, CVPR 2016, Las Vegas, NV, USA, June 27–30, 2016*, pages 4040–4048, 2016.

[MLS09] Marcin Marszalek, Ivan Laptev, and Cordelia Schmid. Actions in context. In *2009 IEEE Computer Society Conference on Computer Vision and Pattern Recognition (CVPR 2009), 20–25 June 2009, Miami, Florida, USA*, pages 2929–2936, 2009.

[NHV+15] Joe Yue-Hei Ng, Matthew J. Hausknecht, Sudheendra Vijayanarasimhan, Oriol Vinyals, Rajat Monga, and George Toderici. Beyond short snippets: Deep networks for video classification. In *IEEE Conference on Computer Vision and Pattern Recognition, CVPR 2015, Boston, MA, USA, June 7–12, 2015*, pages 4694–4702, 2015.

[OBGR19] Abraham Montoya Obeso, Jenny Benois-Pineau, Mireya Saraí García-Vázquez, and Alejandro Alvaro Ramírez-Acosta. Forward-backward visual saliency propagation in deep NNS vs internal attentional mechanisms. In *Ninth International Conference on Image Processing Theory, Tools and Applications, IPTA 2019, Istanbul, Turkey, November 6–9, 2019*, pages 1–6, 2019.

[RHGS17] Shaoqing Ren, Kaiming He, Ross B. Girshick, and Jian Sun. Faster R-CNN: towards real-time object detection with region proposal networks. *IEEE Trans. Pattern Anal. Mach. Intell.*, 39(6):1137–1149, 2017.

[RMB+10] A Rokszin, Z Márkus, G Braunitzer, A Berényi, G Benedek, and A Nagy. Visual pathways serving motion detection in the mammalian brain. *Sensors*, 10(4):3218–3242, 2010.

[SFBC16] Andrei Stoian, Marin Ferecatu, Jenny Benois-Pineau, and Michel Crucianu. Fast action localization in large-scale video archives. *IEEE Trans. Circuits Syst. Video Techn.*, 26(10):1917–1930, 2016.

[SG08] Konrad Schindler and Luc Van Gool. Action snippets: How many frames does human action recognition require? In *2008 IEEE Computer Society Conference on Computer Vision and Pattern Recognition (CVPR 2008), 24–26 June 2008, Anchorage, Alaska, USA*, 2008.

[SLC04] Christian Schüldt, Ivan Laptev, and Barbara Caputo. Recognizing human actions: A local SVM approach. In *17th International Conference on Pattern Recognition, ICPR 2004, Cambridge, UK, August 23–26, 2004*, pages 32–36, 2004.

[SLJ+15] Christian Szegedy, Wei Liu, Yangqing Jia, Pierre Sermanet, Scott E. Reed, Dragomir Anguelov, Dumitru Erhan, Vincent Vanhoucke, and Andrew Rabinovich. Going deeper with convolutions. In *IEEE Conference on Computer Vision and Pattern Recognition, CVPR 2015, Boston, MA, USA, June 7–12, 2015*, pages 1–9, 2015.

[SZ14] Karen Simonyan and Andrew Zisserman. Two-stream convolutional networks for action recognition in videos. In *Advances in Neural Information Processing Systems 27: Annual Conference on Neural Information Processing Systems 2014, December 8–13 2014, Montreal, Quebec, Canada*, pages 568–576, 2014.

[SZDL20] Dian Shao, Yue Zhao, Bo Dai, and Dahua Lin. Finegym: A hierarchical video dataset for fine-grained action understanding. In *2020 IEEE/CVF Conference on Computer Vision and Pattern Recognition, CVPR 2020, Seattle, WA, USA, June 13–19, 2020*, pages 2613–2622, 2020.

[SZS12] Khurram Soomro, Amir Roshan Zamir, and Mubarak Shah. UCF101: A dataset of 101 human actions classes from videos in the wild. *CoRR*, abs/1212.0402, 2012.

[TBF+15] Du Tran, Lubomir D. Bourdev, Rob Fergus, Lorenzo Torresani, and Manohar Paluri. Learning spatiotemporal features with 3d convolutional networks. In *2015 IEEE International Conference on Computer Vision, ICCV 2015, Santiago, Chile, December 7–13, 2015*, pages 4489–4497, 2015.

[TWT+18] Du Tran, Heng Wang, Lorenzo Torresani, Jamie Ray, Yann LeCun, and Manohar Paluri. A closer look at spatiotemporal convolutions for action recognition. In *2018 IEEE Conference on Computer Vision and Pattern Recognition, CVPR 2018, Salt Lake City, UT, USA, June 18–22, 2018*, pages 6450–6459, 2018.

[UAM+18] Amin Ullah, Jamil Ahmad, Khan Muhammad, Muhammad Sajjad, and Sung Wook Baik. Action recognition in video sequences using deep bi-directional LSTM with CNN features. *IEEE Access*, 6:1155–1166, 2018.

[VLS18] Gül Varol, Ivan Laptev, and Cordelia Schmid. Long-term temporal convolutions for action recognition. *IEEE Trans. Pattern Anal. Mach. Intell.*, 40(6):1510–1517, 2018.

[VSP+17]   Ashish Vaswani, Noam Shazeer, Niki Parmar, Jakob Uszkoreit, Llion Jones, Aidan N. Gomez, Lukasz Kaiser, and Illia Polosukhin. Attention is all you need. In *Advances in Neural Information Processing Systems 30: Annual Conference on Neural Information Processing Systems 2017, 4–9 December 2017, Long Beach, CA, USA*, pages 5998–6008, 2017.

[WGW+18]   Xuanhan Wang, Lianli Gao, Peng Wang, Xiaoshuai Sun, and Xianglong Liu. Two-stream 3-d convnet fusion for action recognition in videos with arbitrary size and length. *IEEE Trans. Multimedia*, 20(3):634–644, 2018.

[WJQ+17]   Fei Wang, Mengqing Jiang, Chen Qian, Shuo Yang, Cheng Li, Honggang Zhang, Xiaogang Wang, and Xiaoou Tang. Residual attention network for image classification. In *CVPR*, pages 6450–6458. IEEE Computer Society, 2017.

[WKSL11]   Heng Wang, Alexander Kläser, Cordelia Schmid, and Cheng-Lin Liu. Action recognition by dense trajectories. In *The 24th IEEE Conference on Computer Vision and Pattern Recognition, CVPR 2011, Colorado Springs, CO, USA, 20–25 June 2011*, pages 3169–3176, 2011.

[WRHS13]   Philippe Weinzaepfel, Jérôme Revaud, Zaïd Harchaoui, and Cordelia Schmid. Deep-flow: Large displacement optical flow with deep matching. In *IEEE International Conference on Computer Vision, ICCV 2013, Sydney, Australia, December 1–8, 2013*, pages 1385–1392, 2013.

[WS13]   Heng Wang and Cordelia Schmid. Action recognition with improved trajectories. In *IEEE International Conference on Computer Vision, ICCV 2013, Sydney, Australia, December 1–8, 2013*, pages 3551–3558, 2013.

[WZD+20]   Jiachen Wang, Kejian Zhao, Dazhen Deng, Anqi Cao, Xiao Xie, Zheng Zhou, Hui Zhang, and Yingcai Wu. Tac-simur: Tactic-based simulative visual analytics of table tennis. *IEEE Trans. Vis. Comput. Graph.*, 26(1):407–417, 2020.

[XWX+20]   Kun Xia, Hanyu Wang, Menghan Xu, Zheng Li, Sheng He, and Yusong Tang. Racquet sports recognition using a hybrid clustering model learned from integrated wearable sensor. *Sensors*, 20(6):1638, 2020.

[ZK17]   Sergey Zagoruyko and Nikos Komodakis. Paying more attention to attention: Improving the performance of convolutional neural networks via attention transfer. In *5th International Conference on Learning Representations, ICLR 2017, Toulon, France, April 24–26, 2017, Conference Track Proceedings*, 2017.

[ZvdH06]   Zoran Zivkovic and Ferdinand van der Heijden. Efficient adaptive density estimation per image pixel for the task of background subtraction. *Pattern Recognit. Lett.*, 27(7):773–780, 2006.

Chapter 10
Deep Learning for Audio and Music

Geoffroy Peeters and Gaël Richard

10.1 Introduction

As in computer vision (CV) or natural language processing (NLP), deep learning has now become the dominant paradigm to model and process audio signals. While the term "deep learning" can designate any algorithm that performs deep processing, we define it here as a deep stack of non-linear projections, obtained by non-linearly connecting layers of neurons; a process vaguely inspired by biological neural networks. Such algorithms were denoted by Artificial Neural Network (ANN) in the past and by DNN since [HOT06].

Deep Learning encompasses a large set of different architectures and training paradigms, distinguishing the way neurons are connected to each others, how spatial or temporal information is taken into account, which criteria are being optimized, for which task the network is supposed to be used for. This is often considered as a "zoo" of possible architectures. However most of these architectures share the common use of one form of gradient descent and back-propagation algorithm [RHW86] to estimate the best parameters of the non-linear projections. In this chapter, we describe the DNN building blocks used for audio processing. While some of them are audio translations of CV or NLP networks, others are specific to audio processing.

We focus on two types of audio content which are less commonly addressed in the literature: music and environmental sounds. For a recent and good overview of DNN applied to speech processing we refer the reader to [KLW19].

Differences Between Speech, Music and Environmental Sounds While a speech audio signal usually contains a single speaker (single source), both music and

G. Peeters (✉) · G. Richard
Telecom Paris, IP-Paris, Paris, France
e-mail: geoffroy.peeters@telecom-paris.fr; gael.richard@telecom-paris.fr

© Springer Nature Switzerland AG 2021
J. Benois-Pineau, A. Zemmari (eds.), *Multi-faceted Deep Learning*,
https://doi.org/10.1007/978-3-030-74478-6_10

environmental sounds audio signals are made of several simultaneous sources. In the case of music, some sources are polyphonic (as the piano) and can then produce several pitches simultaneously. This makes the analysis of music and environmental sounds particularly challenging. Speech is highly structured over time (or *horizontally* in reference to the commonly used conventions in time-frequency representations of audio signals). This structure arises from the use of a vocabulary and a grammar specific to a language. Music is also highly structured both horizontally (over time) and vertically (various simultaneous sound events). This structure arises from the music composition rules specific to a culture (harmony for Western music, modes/raga for Eastern/Indian music). In the opposite, environmental sounds have no specific temporal structure.

Deep Learning for Music Processing Music processing is associated with an interdisciplinary research field known as Music Information Research (MIR).[1] This field is dedicated to the understanding, processing and generation of music. It combines theories, concepts, and techniques from music theory, computer science, signal processing perception, and cognition. MIR deals with the development of algorithms for

- *describing the content of the music* from the analysis of its audio signal. Examples of this are the estimation of the various pitches, chords, rhythm, the identification of the instruments being used in a music, the assignment of "tags" to a music (such as genres, mood or usage) allowing to recommend music from catalogues, the detection of cover/plagiarism in catalogue or in user-generated contents.
- *processing the content of the music*. Examples of this are enhancement, source separation.
- *generating new audio signals or music pieces*, or transferring properties from one signal to another.

Deep Learning for Environmental Sounds Processing Environmental sound processing is associated with the research field known as Detection and Classification of Acoustic Scenes and Events (DCASE).[2] The latter deals with the development of algorithms for

- *classifying acoustic scenes* (identify where a recording was made—for example in a metro station, in an office or in a street –),
- *detecting sound events* (detect which events occur over time in an audio scene—a dog barking, a car passing, an alarm ringing –),
- *locating these events in space* (in azimuth and elevation angles).

Historical Perspectives Using DNN algorithms to represent the audio signal has been proposed as early as [WHH+90] where Time-Delay Neural Network

---

[1] http://ismir.net.
[2] http://dcase.community.

(TDNN) where proposed to allow the representation of the time-varying natures of phonemes in speech. Later, [BM94] in their "connectionist speech recognition" convincingly demonstrated the use of the discriminative projection capabilities of DNN to extract audio features. This has lead, among others, to the development of the "tandem features" [HES00] which use the posterior probabilities of a trained Multi-Layer-Perceptron (MLP) as audio features or the "bottleneck features" [GKKC07] extracted from the bottleneck part of a MLP. This has lead today to the end-to-end speech recognition systems which inputs are directly the raw audio waveforms and the output the transcribed text [SWS+15, SVSS15]. As 2012 is considered a landmark year for CV (with the AlexNet [KSH12] network wining the ImageNet Large Scale Visual Recognition Challenge), it is also one for speech recognition with the publication of the seminal paper [HDY+12], jointly written by the research groups of the University of Toronto, Microsoft-Research, Google, and IBM-Research demonstrating the benefits of DNN architectures for speech processing.

The same year [HBL12] published a manifesto promoting the use of DNN for non-speech audio processing (MIR and DCASE). In this paper, the authors demonstrated that any hand-crafted feature (such as MFCC or Chroma) or algorithms (such as pitch, chord or tempo estimation) used so far are just layers of non-linear projections and pooling operations and can therefore be profitably replaced by the trainable non-linear projections of DNN. DNN has now become the dominant paradigm in MIR and DCASE.

Chapter Organization In Sect. 10.2, we first review the commonly used DNN architectures, meta-architectures and training paradigms used for audio processing. In Sect. 10.3, we review the various types of audio representations used as input to DNN and the proposals made to adapt the first layers of the DNN to take into account the audio specificities. In Sect. 10.4, we present a set of common MIR and DCASE applications for content description, processing and generation. We also discuss how Semi-Supervised Learning and Self-Supervised Learning are currently developed in these fields to face the lack of large annotated datasets. Finally, in Sect. 10.5, we discuss future directions for deep learning applied to audio processing.

10.2 DNN Architectures for Audio Processing

A DNN architecture defines a function f with parameters θ which output $\hat{y} = f_\theta(x)$ approximates a ground-truth value y according to some measurements. The parameters θ are (usually) trained in a supervised way using a set of of N inputs/outputs pairs $(x^{(i)}, y^{(i)})$ $i \in \{1, \ldots, N\}$. The parameters θ are then estimated using one variant of the Steepest Gradient Descent algorithm, using the well-known back-propagation algorithm to compute the gradient of a Loss function w.r.t. to the parameters. The function f defines the architecture of the network. We first review the most popular architectures.

10.2.1 DNN Architectures

Multi-Layer-Perceptron (MLP) An MLP is an extension of the Perceptron [Ros57] in which many perceptrons[3] are organized into layers in a Fully-Connected (FC) way. FC denotes the fact that each neuron $a_j^{[l]}$ of a layer $[l]$ is connected to all neurons $a_i^{[l-1]}$ of the previous layer $[l-1]$. The connection is done through multiplication by weights $w_{ij}^{[l]}$, addition of a bias $b_j^{[l]}$ and passing through a non-linearity activation g (the common sigmoid, tanh or ReLu functions): $a_j^{[l]} = g(\vec{a}^{[l-1]}\vec{w}_j^{[l]} + b_j^{[l]})$. Each $\vec{w}_j^{[l]}$ therefore defines a specific projection j of the neurons of the previous layers.

Convolutional Neural Network (CNN) The FC architecture does not assume any specific organisation between the neurons of a given layer $[l]$. This is in contract with Vision where neurons representing nearby pixels are usually correlated (the adjacent pixels that form a "cat's ear") and far away ones uncorrelated. It would therefore be beneficial to consider a *local connectivity* of the neurons i. Also in the FC architecture the weights are specific to each connection and never re-used. This is in contrast with Vision where the neurons representing the two "cat's ears" would benefit from having the same projection, hence from *sharing their weights*. These two properties led to the development of the CNN architecture [FM82, LBBH98]. In this, the projections are defined by J small filters/kernels $\vec{W}_j$ (which size (h, w) is usually $(3,3)$ or $(5,5)$) which are convolved along the two spatial dimensions (height and width) of the input images $\vec{X}$ (or previous layer output $\vec{A}^{[l-1]}$). These filters are the trainable parameters of the network. In classic CV, such filters would allow to detect edges or corners. The output of the convolution is a new set of J images $\vec{A}_j^{[l]}$ considered as a 3D tensor $\vec{A}^{[l]}$ of depth J. This tensor then serves as input to the following layers: $\vec{A}_{j'}^{[l+1]} = g(\vec{A}^{[l]} \circledast \vec{W}_{j'}^{[l+1]} + b_{j'}^{[l+1]})$ (where g denotes a non-linear activation, $\circledast$ the convolution operator, $\vec{W}_{j'}^{[l+1]}$ is a tensor of dimensions (h, w, J)). Spatial invariance (such as detecting the presence of a "cat's ear" independently of its position in the image) is achieved by applying pooling operators. The most popular pooling operator is the max-pooling which only keeps the maximum value over a spatial region. CNN is the most popular architecture in CV.

Temporal Convolutional Networks (TCN) While attempts have been made to apply CNN to a 2D representation of the audio signal (such as its spectrogram), recent approaches [DS14] use **1D-Convolution** directly applied on the raw audio

---

[3]While the Perceptron uses a Heaviside step function, MLP uses non-linear derivable functions.

waveform $x(n)$. The filters $\vec{W}_{j'}$ have then only one dimension (the time) and are convolved only over the time axis of the input waveform. The motivation of using such convolution is to learn better filters than the ones of usual spectral transforms (for example the sinus and cosinus of the Fourier transform). However, compared to images, audio waveforms are of much higher dimensional. To understand this we consider their respective Receptive field (RC). The RC is defined as the portion of the input data to which a given neuron responds. Because images are usually low dimensional (256×256 pixels), only a few layers is necessary in CV to make the RC of a neuron cover the whole input image. In contrast, because input audio waveform are very high dimensional (1 s of audio leads to 44,100 samples), the number of layers to make the RC covers the whole signal becomes very large (as the number of parameters to be trained). To solve this issue, [vdODZ+16] have proposed in their WaveNet model the use of **1D-Dilated-Convolutions** (also named convolution-with-holes or atrous-convolution). For a 1D-filter w of size l and a sequence $x(n)$, the usual convolution is written $(x \circledast w)(n) = \sum_{i=0}^{l-1} w(i)x(n - i)$; the dilated convolution with a dilatation factor d is written $(x \circledast_d w)(n) = \sum_{i=0}^{l-1} w(i)x(n - (d \cdot i))$, i.e. the filter is convolved with the signal only considering one over d values. This allows to largely extend the RC and then allows the model to capture the correlations over longer time ranges of audio samples. The 1D-Dilated-Convolutions is at the heart of the **Temporal Convolutional Networks (TCN)** [BKK18] which is very popular in audio today. The TCN adds a causality constraint (only data from the past are used in the convolution) and, similar to the ResNet cells, stacks two dilated-convolutions on top of each other (each followed by a weight normalization, ReLu and DropOut) with a parallel residual path.

Recurrent Neural Network (RNN)   While CNN allows representing the spatial correlations of the data, they do not allow to represent the sequential aspect of the data (such as the succession of words in a text, or of images in a video). RNN [RHW86] is a type of architecture, close to the Hopfield networks, in which the internal/hidden representation of the data at time t, $\vec{a}^{<t>}$, does not only depend on the input data $\vec{x}^{<t>}$ but also on the internal/hidden representation at the previous time $\vec{a}^{<t-1>}$: $\vec{a}^{<t>} = g(\vec{x}^{<t>}\vec{W}_{xa} + \vec{a}^{<t-1>}\vec{W}_{aa} + \vec{b}_a)$. Because of this, RNN architectures have become the standard for processing sequences of words in NLP tasks.[4] While RNN can theoretically represent long-term dependencies, because of a problem known as the vanishing gradient through time, they cannot in practice. For this reason, they have been replaced by the more sophisticated cells Long Short Term Memory (LSTM)[HS97] or Gated Recurrent Units (GRU)[CVMG+14] in which a set of gates (sigmoids) allow the storage and delivery of information from a memory over time.

---

[4]They are then often combined with representation of the vocabulary using word-embedding techniques.

10.2.2 DNN Meta-architectures

The above MLP, CNN and RNN architectures can then be combined in "meta-architectures" which we describe here.

Auto-Encoder (AE) AE is a type of network made of two sub-networks. The encoding network ϕ_e projects the input data $\vec{x} \in \mathbb{R}^M$ in a latent space $\vec{z} \in \mathbb{R}^d$ of smaller dimensionality $(d << M)$: $\vec{z} = \phi_e(\vec{x})$. The decoder network then attempts to reconstruct the input data from the latent dimension $\hat{\vec{y}} = \phi_d(\vec{z})$. The encoding and decoding networks can be any of the architectures described above (MLP, CNN, RNN). The training is considered unsupervised since it does not necessitate ground-truth labels. We train the network such that $\hat{\vec{y}}$ is a good reconstruction (usually according to a Mean Square Error (MSE) loss) of the input $\vec{x}$: $\arg\min_{\phi_e,\phi_d} ||\vec{x} - (\phi_d \circ \phi_e(\vec{x}))||^2$. AEs are often used for feature learning (learning a representation, a latent space, of the input data). Many variations of this vanilla AE have been proposed which allow improving the properties of the latent space, such as Denoising AE, Sparse AE or Contractive AE.

Variational Auto-Encoder (VAE) For generation, the most popular form of AE is probably today the VAE [KW14a]. In contrast to the vanilla AE, the VAE is a generative model, i.e. a model in which one can sample points $\vec{z}$ in the latent space to generate new data $\hat{y}$. In a VAE, the encoder models the posterior $p_\theta(\vec{z}|\vec{x})$ while the decoder (the generative network) models the likelihood $p_\theta(\vec{x}|\vec{z})$. However because $p_\theta(\vec{z}|\vec{x})$ is untractable, it is approximated by $q_\phi(z|x)$ (variational Bayesian approach) which is set (for mathematical simplicity) to a Gaussian distribution which parameters $\vec{\mu}$ and $\vec{\Sigma}$ are the outputs of the encoder. Minimizing the Kullback-Leibler divergence between $q_\phi(z|x)$ and $p_\theta(\vec{z}|\vec{x})$ is mathematically equivalent to maximizing an ELBO (Evidence Lower BOund) criteria. For the later, a prior $p_\theta(\vec{z})$ needs to be set. It is set (again for mathematical simplicity) to $\mathcal{N}(0, 1)$. The goal is then to maximize $\mathbb{E}_q[\log p(x|z)]$. This can be estimated using a Monte-Carlo method, i.e. maximizing $\log p(x|z)$ (the reconstruction error) over samples $z \sim q_\phi(z|x)$ given to the decoder. Given the smoothness of the latent space $\vec{z}$ obtained (in contrast to the one of vanilla AE) it is adequate for sampling and generation.

Generative Adversarial Network (GAN) Another popular type of network for generation is the GAN [GPAM+14]. GAN only contains the decoder part of an AE here named "Generator" G. Contrary to the VAE, z is here explicitly sampled from a chosen distribution $p(z)$. Since z does not arise from any existing real data, the Generator $G(z)$ must learn to generate data that look real, i.e. the distribution of the generated data p_G should look similar to the ones of real data p_{data}. Rather than imposing a distribution (as in VAE), this is achieved here by defining a second network, the "Discriminator" D, which goal is to discriminate between real and fake (the generated ones) data. D and G are trained in turn using a minmax optimisation.

For G fixed, D is trained to recognize real data from fake ones (the ones generated by G).[5] For D fixed, G is then trained to fool D.[6]

Encoder/Decoder (ED) While the goal of AE is to encode the data into a latent space $\vec{z}$ such that it allows reconstructing the input, ED [CVMG+14] or Sequence-to-Sequence [SVL14] architectures aim at encoding an input sequence $\{\vec{x}^{<1>} \dots \vec{x}^{<t>} \dots \vec{x}^{<T_x>}\}$ into $\vec{z}$ which then serves as initialization for decoding a sequence $\{\vec{y}^{<1>} \dots \vec{y}^{<\tau>} \dots \vec{y}^{<\tau_y>}\}$ into another domain. Such architectures are for example used for machine translation where an input English sentence is translated into an output French sentence. Both sequences have usually different length $T_x \neq \tau_y$. In machine translation both encoder and decoder are RNNs (or their LSTM or GRU versions). In image captioning [VTBE15], a deep CNN is used to encode an input image into $\vec{z}$; $\vec{z}$ then serves as initialization of a RNN decoder trained to generate the text of image captions.

Attention Mechanism In the original ED for machine translation [CVMG+14], $\vec{z}$ is defined as the internal states of the RNN after processing the whole input sequences, i.e. at the last encoding time step $\vec{a}^{<T_x>}$. It quickly appeared that doing so prevents from correctly translating long sentences. [BCB14] therefore proposed to add to the ED architecture, an attention mechanism. The latter provides a mechanism to let the decoder chose at each decoding time τ the most informative times t of the encoding internal states $\vec{a}^{<t>}$. This mechanism is a small network trained to align encoding and decoding internal states.

Transformer Recently it has been shown [VSP+17] that only the attention mechanism was necessary to perform machine translation. The transformer still has an encoder and a decoder part but those are now simple stacks of so-called self-attention mechanisms coupled with a FC. At each layer, the self-attention mechanisms encode each element of the sequence taking into account its relationship with the other elements of the sequence. This is done using a simple query, key and value mechanism. The transformer has become very popular for sequence processing.

10.2.3 DNN Training Paradigms and Losses

The most popular training paradigms for DNN are classification, reconstruction and metric learning.

Classification The simplest case of classification, is the *binary classification*. In this, the network has a single output neuron (with sigmoid activation) which predicts the likelihood of the positive class $\hat{y} = p(y = 1|x)$. The training of the network is achieved by minimizing the Binary-Cross-Entropy (BCE) between y and $\hat{y}$ over

---

[5] $D(x \sim p_{data})$ should output "real" while $D(G(z))$ should output "fake".
[6] $D(G(z))$ should output "real".

the N training examples: $\mathcal{L} = -\sum_{i=1}^{N}[y^{(i)}\log(\hat{y}^{(i)}) + (1 - y^{(i)})\log(1 - \hat{y}^{(i)})]$. The goal of *multi-class classification* is to predict a given class c among C mutually exclusive classes. Each class c is represented by an output neuron y_c (with a softmax activation) which predicts $\hat{y}_c = p(y = c|x)$ The training of the network is then achieved by minimizing the general cross-entropy between the y_c and the $\hat{y}_c$. The goal of *multi-label classification* is to predict a set of class $\{c_i\}$ among C non-mutually exclusive classes. The most usual solution to this problem is to consider each class c as an independent binary classifier (with sigmoid activation) and then train the network by minimizing the sum of the BCE of each class c.

Reconstruction When the goal of the network is to reconstruct the input data (such as with AE), the simple MSE between the output and input data is used: $MSE = \sum_{i=1}^{N}||\vec{x}^{(i)} - \hat{\vec{y}}^{(i)}||^2$.

Metric Learning Metric learning aims at automatically constructing distance metrics from data, in a machine-learning way. DNN provides a nice framework for this. In this, the parameters θ of a network are learnt such that a distance function $g(f_\theta(x), f_\theta(y))$ is minimized for similar training samples x and y and maximized for dissimilar samples. Methods proposed for that, mainly differ on the way these two constrains are represented: they are represented in turns in Siamese networks [BGL+94] and contrastive loss [HCL06], they are represented simultaneously in the triplet loss [SKP15]). In the later, three data are simultaneously considered: an anchor a, a positive p (similar to a) and a negative n (dissimilar to a). The goal is to train the network such that $P = f_\theta(p)$ will be closer to $A = f_\theta(a)$ than $N = f_\theta(n)$ is to A. For safety a margin α is added leading to the definition of the triplet loss to be minimized $\mathcal{L} = \max(0, g(A, P) + \alpha - g(A, N)$. g can be a simple Euclidean distance.

10.3 DNN Inputs for Audio Processing

A wide variety of audio representations are used as input for DNN. These representations can be broadly classified in (1) time and frequency representations; (2) waveform representations (3) knowledge-driven representations and (4) perceptual-driven representation. The latter is not discussed in details in this chapter but the interested readers are referred to [RSN13] for an overview of popular perceptually-based representations for audio classification tasks.

10.3.1 Using Time and Frequency Representations as Input

A recorded audio signal $x(t)$ represents the evolution of the sound pressure x over time t. In its discrete version, the time dimension is discretized in samples m resulting in a discrete sequence $x(m)$. The number of samples sampled from x during one second is named "sampling rate". A common value for it is 44,100 Hz. One second of audio signal is then represented by the sequence $\{x(1), \ldots x(44,100)\}$. To represent a piece of music of 4 min duration, this would lead to a very high number of values.

For a discrete non-periodic signal, the Discrete-Fourier-Transform (DFT) is used to represent $x(m)$ over discrete frequencies $k \in [0, N - 1]$:

$$X(k) = \sum_{m=0}^{N-1} x(m)e^{-j2\pi \frac{k}{N} m}$$

Since the content of the audio signal varies over time (for example it is assumed that the phoneme rate in speech is around 4Hz), DFTs are computed over successive time frames of the signal (obtained through multiplication with an analysis window $h(m)$) leading to the well-known Short-Time Fourier Transform (STFT):

$$X(k, n) = \sum_{m=0}^{N-1} x(m)h(n - m)e^{-j2\pi \frac{k}{N} m}$$

$X(k, n)$ represents the content of the audio signal at frequency k and around time n.

The complex value STFT matrix $X(k, n)$, can be represented by its real and imaginary parts or by its amplitude (which represents the amount of periodicity at a given frequency) and phase (which represents the location at a given frequency). Most approaches that use the STFT to represent the audio, only consider its amplitude. It is then often denoted as the **spectrogram**. Since the later can be displayed as an "image", the first audio-DNN used standard computer vision CNNs applied to this spectrogram-image.

Recently, it has been proposed to use directly the **complex STFT** as input to DNN with the goal of benefiting from the location information contained in the phase. For this, either a (real,imaginary) or a (amplitude, instantaneous frequency) representation have been tested.

Before the rise of deep learning for audio, the most popular audio representation for speech tasks (recognition/identification/diarization), MIR or DCASE tasks was the **Mel-Frequency-Cepstral-Coefficients (MFCC)s**. Those are obtained by computing the real cepstrum representation (Discrete Cosine Transform (DCT) applied to the logarithm-amplitude of the DFT) on a Mel-scale representation.[7] It can be shown that in the case of a source-filter sound production model (see Sect. 10.3.3), the cepstrum allows to separate the contribution of the filter (the lowest coefficients of the cepstrum) from the source (highest coefficients). These lowest coefficients are therefore usually used to obtain a compact representation

---

[7]The Mel scale is a perceptual scale of pitch height perception. A mel-filter bank is then a set of filters whose bandwidth center frequencies are equally spaced on the Mel scale (or logarithmically spaced in Hertz).

of the spectrum envelope (or formants of the various vowels in vocal signals or the timbre of musical instruments) independently of their pitch. In the MFCCs computation, the DCT is used to make the various dimensions of the MFCC somehow decorrelated. This is needed since those are often represented in speech acoustical models using Gaussian mixture distributions with diagonal covariance matrices. Because this de-correlation of the input is not required in the case of DNN, the **Log-Mel-Spectrogram (LMS)** (hence without the DCT de-correlation) has been widely adopted. This leads to a time versus mel-band-frequency matrix representation.

In the DFT, the time and frequency resolution (we mean by resolution the possibility provided by the representation to distinguish two adjacent time or frequency components) remains constant over time and frequency. This limitation led to the development of the wavelet analysis [Mal89] which allows for a finer spectral resolution at low-frequencies and finer temporal resolution at high-frequency. The **Constant-Q-Transform (CQT)** [Bro91] has been proposed as a form of wavelet analysis adapted to musical signals, i.e. which allows distinguishing the various possible pitches of the musical scale. As for the wavelet representation, this is achieved by using analysis windows $h(m)$ which durations are inversely proportional to the various musical pitch frequencies. The CQT follows a logarithmic frequency scale (as the musical pitches). It is therefore said to be shift-invariance in pitch, i.e. transposing a note (changing its pitch) simply results in a shift of its harmonic pattern (the sequence of its harmonics) along the log-frequency axis. This is however not entirely true as we will discuss later considering the source/filter decomposition.

10.3.1.1 Spectrogram Images Versus Natural Images

While spectrograms are often processed using CNN and hence considered as images, there is a large difference between this image and a natural image, such as a cat picture.

In **natural images** (see Fig. 10.1 left), the two axis x and y represent the same concept (spatial position). The elements of an image (such as a cat' ear) have the same meaning independently of their positions over x and y. Also neighboring pixels of an image are usually highly correlated and often belong to the same object (such as the cat's ear). The use of CNN, and its inherent properties (hidden neurons are only locally connected to the input image, parameters are shared between the various hidden neurons of a same feature map and max pooling allows spatial invariance) are therefore highly appropriate to process such data.

In **time-frequency audio representations** (such as the spectrogram, the LMS or the CQT) (see Fig. 10.1 right), the two axis x and y represent profoundly different concepts (time and frequency). The elements of a spectrogram (such as a time-frequency area representing a sound source) has the same meaning independently of its position over time but not over frequency. There is therefore no invariance over y, even in the case of log-frequencies. Neighboring pixels of a spectrogram are not

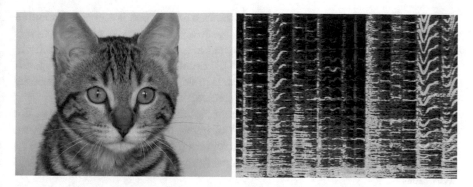

Fig. 10.1 [Left part] Natural image of cats, [Right part] image of a spectrogram

necessarily correlated since a given sound source (such has an harmonic sound) can be distributed over the whole frequency in a sparse way (the harmonics of a given sound can be spread over the whole frequency axis). It is therefore difficult to find a local structure using a CNN.

10.3.1.2   DNN Models for Time and Frequency Representations as Inputs

It is generally considered that DNNs learn a hierarchical feature representation of the input. However one can easily consider that the first layers are more associated with this feature learning while the last layers to the task at hand, e.g. a classification task. We review here which choices have been made so far to allow these first layers to deal with audio inputs as time and frequency representations.

In **speech**, one of the first attempt to apply DNN to the audio signal is using a so-called Time-Delay Neural Network (TDNN) [WHH+90]. This architecture is similar to a 1-D convolution operating only over time. In [WHH+90], this convolution is applied to a Mel-gram (16 normalized Mel-scale spectral coefficients). No convolution are performed over the frequency axis. In the works following the "connectionist speech recognition" approaches [BM94] ("tandem features" [HES00] or "bottleneck features" [GKKC07]), a context window of several successive frames of a feature vector (such as MFCC) is used as input to an MLP. Here the convolutions over time is replaced by a context-window. No convolution are performed over the frequency axis. In [LPLN09], a Convolutional Deep Belief Networks (CDBN)[8] is used to process the audio input. The audio input is a 160 dimensional spectrogram which is then PCA-whitened to 80 dimensions.[9]

---

[8] A CDBN is a stack of Restricted Boltzman Machine (RBM) with convolutions operations, hence trained in an unsupervised way.

[9] Each of the whitened dimension is therefore a combination of the initial 160 dimensions of the spectrogram.

The filters (named bases in [LPLN09]) of the first and second layers are of length 6 and are convolved over the PCA-whitened spectrogram. By visual comparison, it is shown that the learned filters (bases) are related to the different phonemes of speech. Following this, the seminal paper [HDY+12] defines the new baseline for speech recognition system as the DNN-HMM model. In this, the acoustic model part of the system is defined as a DNN model (more precisely as stacked RBMs).

In **music**, [Die14] also consider a 1D-convolution operating only over time. For a task of predicting latent representation[10] of music tracks (a regression problem), they use as input of a 1D-CNN a Mel-Spectrogram (MS) of 128 frequency bins. The filters of the first layer are of shape (time=4,frequency=128) and only convolved over time.

In the opposite [CFS16] consider time/frequency representation as natural images and apply a computer vision CNN to it. The network is a VGG-Net [SZ15], i.e. a deep stack of convolution layers with small (3,3) filters convolved over the time and frequency axis. With this architecture, they show that using MS as input performs better than STFT or MFCC.

However, as described in Sect. 10.3.1.1, time/frequency representations cannot be considered as a natural image. When using CNN architectures, one should carefully choose the shape of the filters and the axis along which the convolution is performed.

One of the first work to deal with this necessary adaptation is [SB13]. For a task of onset detection (detecting the start of a musical events) they carefully design the filters to allow highlighting mid-duration variations over small-frequency ranges. For this, they specify filters of shape (time=7,frequency=3). An LMS representation is then convolved over time and frequency with these filters. Another specificity of their approach is to allow the representation of multi-scale analysis, i.e. STFT computed using various window durations (23 ms, 46 ms and 93 ms) to better face the time/frequency resolution trade-off. They use the depth of the input layer[11] to represent the various scales. The resulting onset detection algorithm has remain the state-of-the-art for a long time.

The work presented in [PLS16] is entirely devoted to this musically-motivated filter design. In their work, the shapes of the CNN filters are carefully chosen to allow representing the timbre (vertical filters extending over the frequency axis) or the rhythm (horizontal filters extending over the time axis) content of a music track. They show that carefully choosing the shape of the filters allows to obtain equivalent performances than the CV-based approach of [CFS16] (here renamed "black-box") but with much less parameters.

---

[10]The latent representation resulting from a collaborative filtering model.

[11]In CV the depth is used to represent the RGB channels of an image.

10.3.2   Using Waveform Representations as Input

While musically-motivated CNN filter shape is a promising path, one still has to manually design this shape for a given application. Also one has to decide what is the most appropriate 2D representation (STFT, LMS or CQT) and its parameters (window size, hop size, number of bands) for a given application.

For these reasons, the so-called "end-to-end" approaches have been developed. Those consider directly the raw audio waveform as input.

In **speech**, one of the first end-to-end approaches is the one of [JH11] where a RBM is used to model the raw speech signals.

In **music**, one of the first end-to-end approaches is the one of [DS14] who proposed, for a music auto-tagging task, to use 1D-convolution (a convolution over time with 1D-filters) on the waveform as a replacement to the spectrogram input. To compare both, [DS14] actually reproduce the computation of the spectrogram using 1D-convolution. While a spectrogram is computed using a succession of DFTs each computed on an audio frame of length N and each separated by a hop size S, the 1D-convolution is computed using 1D-filters of length N[12] and a stride of S.[13] However, their "end-to-end" approach under-performed the traditional spectrogram-based one. This may be due to the lack of Time Translation Invariance (TTI) of their representation.

Time Translation Invariance (TTI) is a property of a transform that makes it insensitive to time translation (or phase shift) of the input. The amplitude of the DFT (as used in the spectrogram) is TTI. This is because the DFT projects the waveform on orthogonal cosinus and sinus basis, and the modulus of the resulting vectors remain invariant to time translation (the phase of the vectors are however shifted according to the time translation). Mimicking this property with 1D-convolution would require (a) reducing the stride to S=1 (and using a very high sampling rate) or (b) having a different 1D-filter for each possible time translation. One will still needs to perform a max-pooling over time-steps for (a) or over filters for (b). Both are however computationally prohibitive.

Sample-CNN One way to improve the TTI is to reduce the size of the 1D-convolution filters (hence also the stride). If the filters are smaller, then the number of time translation to be learned is also reduced. This is the idea developed in the Sample-CNN [LPKN17] [KLN18] network. The later can be considered as an equivalent to the VGG-Net for 1D-convolution applied to waveforms. It is a deep stack of 1D-convolution of small (3,1) filters applied to the waveform. Sample-CNN was shown to slightly outperforms the 2D-CNN on the spectrogram.

Multi-Scale When computing a spectrogram, the choice of the window size fixes the trade-off between time and frequency resolution. One can think of the same for

---

[12]In his experiments $N \in 256, 512, 1024$ for a sampling rate of 16 kHz.

[13]$S \in 256, 512, 1024$.

the choice of the filter size N of 1D-convolution. To get around this choice, [ZEH16] propose a multi-scale approach where the waveform is simultaneously convolved in parallel with filters of different sizes (1ms, 5ms and 10ms). The resulting outputs are then concatenated. This idea follows the one of the Inception network [SLJ+15] in computer vision.

10.3.3 Using Knowledge-Driven Representations as Input

When one has some knowledge of the sound production process it is possible to use this knowledge to better shape the input and/or the first layer of the network. Such commonly used sound production processes are the source/filter and the harmonic models. The *source/filter model* considers that the sound $x(t)$ results from the convolution of a periodic (excitation) source signal $e(t)$ (such as the glottal pulses in the case of voice) with a filter $v(t)$ (such as the vocal track in the case of voice): $x(t) = (v \circledast e)(t)$. The *harmonic model* considers that a sound with a pitch f_0 can be represented in the spectral domain as the sum of harmonically related components at frequencies hf_0, $h \in \mathbb{N}^+$ with amplitudes a_h.

[LC16] were among the first to use such models for a task of musical instrument recognition. Below a cut-off frequency, they consider the harmonic model: the spectrum of harmonic sounds is sparse and co-variant with pitch. It is therefore processed using convolution filters which only have values at octave intervals (mimicking Shepard pitch spiral array). Above this cut-off frequency, they consider the source/filter model: the spectrum is dense and independent of pitch (according to the source/filter model, transposed sounds have similar spectra). It is therefore processed with filters that extent over the whole upper-part of the spectrum.

Harmonic CQT [BMS+17] also use the harmonic assumption; this for a task of dominant melody and multi-pitch estimation. Contrary to natural images (where neighboring pixels usually belong to the same source), the harmonics of a given sound source are spread over the whole spectrum and can moreover be interleaved with the harmonics of other sound sources. [BMS+17] propose to bring back this vicinity of the harmonics by projecting each frequency f into a third dimension (the depth of the input) which represents the values of the spectrum at the harmonics hf. Convolving this representation with a small time and frequency filter but which extends over the whole depth allow then to model easily the specific harmonic series of pitched sounds hence detecting the dominant melody. This representation is named Harmonic CQT and led to excellent results in this context. This approach has been extended with success by [FP19] for a task case of tempo estimation. In this, pitch frequencies are replaced by tempo frequencies and the CQT of the audio signal by the one of onset-energy-functions.

Source/Filter Still for a task of dominant melody estimation, [BEP18] use the source/filter assumption. Rather than considering the audio as input to the network, they consider the output of a Non Negative Matrix Factorization (NMF) model.

They use the NMF source/filter model of [DRDF10] and use the source activation matrix as input to the network. They show that including the knowledge of the production model allows to drastically reduce the size of the training set.

SincNet In the end-to-end approaches mentioned above, the filters of the 1D-convolution are often difficult to interpret since their shape are not constrained. While being learned in the temporal domain, authors often display them in the frequency domain to demonstrate that meaningful filters have been learned (such as Gamma-tone filters in [Sai15]). With this in mind, the SincNet model [RB18] proposes to define the 1D-filters as parametric functions g which theoretical frequency responses are parameterizable band pass filters. To do so g is defined in the temporal domain as the difference between two sinc functions which learnable parameters define the low and high cutoff frequencies of the band-pass filters. They show that not only the obtained filters are much more interpretable but also the performances for a task of speaker recognition is much improved. This idea has been extended recently to the complex domain in the Complex Gabor CNN [NPM20].

HarmonicCNN Combining the idea of SincNet with the harmonic model lead to the HarmonicCNN of [WCNS20]. In this the 1D-convolution is performed with filters constrained as for SincNet but extended to the harmonic dimensions (stacking band-pass filters at harmonic frequencies hf_c).

Neural Autoregressive Models A source/filter model $x(n) = (v \circledast e)(n)$ can be associated to an autoregressive model, i.e. the value $x(n)$ can be predicted as a linear combination of its P preceding values: $x(n) = \sum_{p=1}^{P} a(p)x(n-p)$. Neural Auto-regressive models are a non-linear form of auto-regressive models in which the linear combination is replaced by a DNN. The two most popular models are probably the Wavenet [vdODZ+16] and the SampleRNN [MKG+17] architectures. In WaveNet, the conditional probability distribution $p(x_n|x_1, \ldots, x_{n-1})$ is modeled by a stack of dilated 1D convolutions. To facilitate the training, the problem is considered as a classification problem. For this $x(n)$ is discretized into 256 possible values (8 bits using μ-law) considered as classes to be predicted by a softmax. The model has been developed for speech generation and can be conditioned on side information **h** such as speaker identity or text: $p(x_n|x_1, \ldots, x_n - 1, \mathbf{h})$. While WaveNet relies on dilated convolutions to allow both short term and long term dependencies in $p(x_n|x_1, \ldots, x_{n-1})$, SampleRNN, uses a stack of RNNs each operating at a different temporal scale.[14]

DDSP The recently proposed Differentiable Digital Signal Processing (DDSP) [EHGR20] is probably the DNN models which relies the most on the prior knowledge of the sound production process. Just as SincNet defines the 1D-filters as parametric functions g and the training consists in finding the parameters of g, DDSP defines the sound production model and the training consists in finding its

---

[14]RNN layers operate at different temporal resolutions and are followed by up-sampling for the next scale.

parameters. The model considered here is the Spectral Modeling Synthesis (SMS) model [SS90]. It combines harmonic additive synthesis (adding together many harmonic sinusoidal components) with subtractive synthesis (filtering white noise); it also adds room acoustics to the produced sound through reverberation. In DDSP, the input audio signal x is first encoded into its pitch f_0 and a latent representation $\vec{z}$. Time-varying loudness $l(t)$, $f_0(t)$ and $\vec{z}(t)$ are then fed to a decoder which estimates the control parameters of the additive and filtered noise synthesizers.

10.4 Applications

10.4.1 Music Content Description

As described in Sect. 10.1, Music Information Research (MIR) encompasses a large set of tasks related to the description of the music from the analysis of its audio signal. Since almost all possible audio front-ends and DNN algorithm have been tested for each task, it is useless to describe them all. We rather focus here on some iconic MIR tasks and an iconic DNN algorithm proposed to solve each.

Beat-Tracking "Beat" or "pulse" is the basic unit of time in music. It is often defined as the rhythm listeners would tap their foot to when listening to a piece of music. Beat-tracking is the task of estimating the temporal position of the beats within a music track. As far as 2011, i.e. before the rise of deep learning for audio, [BS11] already proposed a fully DNN system to estimate the beat positions. The input to the network is made of three Log-Mel-Spectrogram (LMS) computed with window sizes of 23.2 ms, 46.4 ms, 92.8 ms and their corresponding positive first order median difference. Since "beat" is a temporal phenomenon, [BS11] proposes to use an RNN architecture to estimate it. The network is made of three layers of bi-directional LSTM units. The last layer has a softmax activation that predicts at each time if the input time is a beat (1) or not (0). A peak-picking algorithm is then applied on the softmax output to detect the beats. This algorithm led to excellent results in the MIREX benchmark.[15]

For the estimation of more high-level rhythm concepts such as the downbeat, which is considered to be the first beat of each bar, it is often necessary to rely on multiple representations (or features). For example in [DBDR17], four musical attributes contributing to the grouping of beats into a bar, namely harmony, rhythmic pattern, bass content, and melody are estimated by well designed representations which are in turn fed to parallel specific CNN.

Onset Detection An "onset" denotes the starting time of a musical event (pitched or non-pitched). Onset detection is the task of estimating the temporal positions of

---

[15]MIREX (Music Information Retrieval Evaluation eXchange) is an annual evaluation campaign for MIR algorithms.

all onsets within a music track. The system proposed by [SB13] is a typical MIR DNN system. It uses a stack of convolution/max-pooling layers to progressively reduce the time and frequency dimensions and transfer those to the depth. It is then flattened and fed to a stack of FC layers with a sigmoid or a softmax output which perform the prediction. The novel idea proposed by [SB13] is to feed the network with chunks of spectrogram (each chunk represents 15 successive time frames of the spectrogram) and associate to it a single output y which represents the ground-truth for the middle frame of the chunk ($y = 1$ means that the middle frame of the chunk is an onset). These chunks can be considered as the "context windows" of [HES00] but benefit for the convolutional process. Contrary to the use of RNN, a music track is here processed as a bag of chunks which can be independently processed in parallel. The input to the network is made of the same three LMS computed with window sizes of 23.2 ms, 46.4 ms and 92.8 ms. This algorithm led to excellent results in the MIREX benchmark.

Music Structure "Music Structure" denotes the global temporal organization of a music track into parts (such as intro, verse, chorus, bridge for popular music or movements for classical music). Music boundary detection is the task of estimating the transition times between these parts. To solve this, [SUG14] actually follow the same idea as for the onset detection [SB13]: a large temporal chunk is taken as input to a deep CNN which output predicts if the center frame of the chunk is a "music boundary" or not. However, here the input of the network is different: beside the LMS input a so-called Lag-Similarity-Matrix [Got03] is also used to better highlight the large-scale structure of the track. This path have been followed by [CHP17] leading to excellent results.

Dominant Melody and Multi-Pitch Estimation Dominant melody refers to the temporal sequence of notes played by the dominant instrument in a music track (such as the singer in pop-music or the saxophone/trumpet in jazz music). Multi-pitch estimation refers to the estimation of the whole musical score (the temporal sequences of notes of each instrument). This is one of the most studied tasks in MIR. The DNN estimation methods proposed by [BMS+17] can be considered as a breakthrough. This method uses a Harmonic-CQT (already mentioned in Sect. 10.3.3) as input to a deep CNN architecture which output is an image representing the pitch saliency of each time and frequency bins. The network is therefore trained to construct a "pitch saliency map" given an audio signal. A simple peak-picking or thresholding method can then be used to estimate the dominant melody or the multiple-pitches from this map.

Chord Estimation A chord is a set of multiple pitches that are heard as if sounding simultaneously. It is a convenient reduction of the harmonic content of a music track at a given time. Chords give rise to guitar-tabs which are largely used by guitarist or to real-book scores used by jazz players. Their estimation is both a segmentation task (finding the start and end time of each chord) and a labeling task (finding the correct chord label, such a C-Major, C7 or Cm7). Given its close relationship to speech recognition, the first chord estimation systems [SE03] relied on an acoustic model (usually a Gaussian Mixture Model (GMM) representation of Chroma features [Wak99]) connected to a language model (a hidden Markov model representing the chord transition rules specific to Western music.[16]) [MB17] has proposed to solve the problem using a single DNN system. The specificity of this approach is to exploit the structural relationships between chord classes, i.e. the fact that while the label C-Major and Cm7 are different, their underlying chord construction share a large amount of notes. To do so, a CQT input is first encoded (using a convolutional-recurrent network architecture, i.e. a CNN followed by a bi-GRU) into the triplet of {root, pitches and bass} labels corresponding to the chord to be estimated. The outputs of those are then combined with the one of the encoder to estimate the final chord label. The authors show that constraining the training to learn the underlying structure of chords, allows increasing the chord recognition accuracy especially for the under-represented chord labels.

Auto-tagging Auto-tagging is probably the most popular MIR task. It consists on estimating a set of tags to be applied to describe a music track. Such tags can relate to the track's music-genre (such as rock, pop, classical), mood (such as happy, sad, romantic), instrumentation (such as piano, trumpet, electric guitar) or in other descriptive information. Some tags can be mutually exclusive (such as singing/instrumental) some other not (such as piano and drum which may occur together). One of the most cited DNN system for auto-tagging is the one of [CFS16]. The system is a Fully Convolutional Network (no FC layer are used) inspired by the VGG-Net architecture [SZ15]: it is a stack of 2D convolution layers with small (3,3) kernels followed by max-pooling layers with small (2,4) kernels. This progressively transfers the time dimension of the input to the depth which is finally connected to 50 sigmoid outputs (multi-label classification task). Among the various input representations tested, the Mel-Spectrogram provides the best results. On the Magna-Tag-A-Tune dataset [LWM+09], their approach outperforms any pre-existing systems. While being the most cited auto-tagging paper, this model has also been criticized by [Pon19] for its lack of consideration of the audio specificities. It is basically a computer vision network applied to an audio representation. Unexplainedly, it works very well.

Music Recommendation by Audio Similarity Music recommendation by audio similarity aims at recommending a ranked list of music tracks to a user. THe ranking is based on their audio similarity with a target music track. This kind of recommendation allows to get around the "cold start" problem.[17] To compute such an audio similarity, past approaches modelled the content of a track using generative models (such as GMM) of hand-crafted features (through MFCC). The audio similarity of two tracks was then computed as the Earth mover's

---

[16]For example, the "II-V-I" (two-five-one) cadential chord progression is very common and particularly popular in jazz music.

[17]when no meta-data (as used for tag-based recommendation) or usage data (as used in collaborative filtering recommendation) are available.

distance—Kullback-Leibler divergence between their respective GMMs [APS05]. This approach was computationally expensive and did not allowed to reproduce a ground-truth ranked list. Recently [PRP20] have proposed to apply DNN metric learning to this problem. Starting from ground-truth ranked lists, they first define a set of ranked triplets *Tr={anchor, positive and negative}* using their relative positions in the ranked lists. Using those, a triplet loss [SKP15] is then used to train a CNN similar to [CFS16] (VGG-Net). It is fed with chunks of 512 CQT frames. The network learns to project each track in a 128-dimensions "audio-similarity embedding" space. In this, the similarity between two tracks is obtained as their Euclidean distance.

Cover Detection "Covers" denotes the various recorded interpretations of a musical composition (for example "Let It Be" performed by The Beatles or performed by Aretha Franklin). The problem has received a lot of attention recently due to the large amount of User-Generated Content which necessitates scalable copyright monitoring systems. While it is hard to define exactly why two tracks can be considered "covers" of each other, it is easy to provide examples and counter-examples of those. This is the approach proposed by [DP19, DP20, DYS+20]. They propose to represent the content of a music track using jointly the CQT, the estimated dominant pitch and estimated multi-pitch representations. Those are fed to deep CNN networks. The networks are then trained using also a triplet loss paradigm [SKP15] using sets of anchor tracks, positive examples (covers of the anchors) and negative examples (non-covers of the anchors). The output of the networks are considered as track embeddings and it is shown that, once trained, the distance between the embedding of two tracks indicate their cover-ness. This algorithms has provide a large increase in cover-detection performances.

10.4.2 *Environmental Sounds Description*

The research field associated to the *Detection and classification of Acoustic Scene and Events (DCASE)* has received a steep growing interest with high industrial expectations. Similarly to other fields, recent progress in environmental sounds recognition has been largely fuelled by the emergence of Deep Neural Networks (DNN) frameworks [Abe20],[VPE17],[MHB+18]. Nearly all the concepts and architectures described above have been used on specific DCASE problems such as *Urban scene analysis* (traffic events recognition, scene recognition, etc.), *bioacoustic sounds recognition* (bird songs recognition, sea mammals identification, etc.) or *biological sounds* (deglutition, digestion, etc.).

However, the extreme diversity of potential sounds in natural soundscapes has favoured the development of specific methods which can more easily adapt to this variability. An interesting strategy is to rely on **feature learning approaches** which are proven to be more efficient than traditional time or time-frequency audio representations [SBER18]. Sparse representations, matrix factorizations and dictionary learning are some of the emblematic examples of this strategy. For example, some methods aim to decompose the audio scene recordings into a combination of basis components which can be obtained using **non-negative matrix factorization** (NMF)[BSER17] or shift-invariant probabilistic latent component analysis (SIPLCA) [BLD12]. In [BSER17], it was in particular shown that such a strategy when associated to DNN in a problem of acoustic scene classification allows to opt for simpler neural architectures and to use smaller amount of training data.

In terms of network structure and architectures, **Resnets** and shallow inception models have been shown to be particularly efficient on Acoustic source classification [SSL20] [MG20]. Resnets are specific networks in which each layer consists of a residual module and a skip connection bypassing this module [HZRS16]. It was recently shown that they can be interpreted as an ensemble of smaller networks which may be an explanation for their efficiency [VWB16].

For applications of predictive maintenance (anomalous sound detection), architectures based on auto-encoders are getting particularly popular due to their capacity to be learned in an unsupervised way. This is particularly interesting for this problem since there is usually a very low number of observations, if any, of the anomalous sounds to be detected [KSU+19].

Another interesting avenue for environmental sound recognition is around approaches that are suitable for few-shot learning or transfer learning such as relation networks (prototypical networks [SSZ17] or Matching networks [VBL+16]). Matching networks use an attention mechanism over a learned latent space to predict classes for the unlabelled points and can be interpreted as a weighted nearest-neighbour classifier applied within an embedding space. In prototypical networks, the core idea is that there exists a latent space (e.g. embedding) described by a single prototype representation for each class. More precisely, a non-linear mapping of the input into an embedding space is learned using a neural network and takes a class's prototype to the mean of its support set in the embedding space. Classification can be performed for an embedded query point by simply finding the nearest class prototype. The capacity of prototypical networks to go beyond more straightforward transfer learning approaches and their efficacity for sound event recognition are shown in [PSS19].

10.4.3   Content Processing: Source Separation

Blind Audio Source Separation (BASS) is the field of research dealing with the development of algorithms allowing the recovery of one or several source signals $s_j(t)$ from a given mixture signal $x(t) = \sum_j s_j(t)$ without any additional information (the separation is blind). It has close relationships with speech enhancement/denoising.

For a long time, BASS algorithms relied on the application of Computational Auditory Scene Analysis (CASA) principles [BC94] or matrix decomposition methods. Among the latter, Independent Component Analysis (ICA) assumes that the various sources are non-Gaussian and statistically independent; NMF factorizes the mixture's spectrogram as the product of a non-negative source activation matrix with a non-negative source basis matrix (see [PLDR18] for an overview on music source separation).

In recent years DNN methods for BASS has allowed to largely improved the separation quality. Most of the DNN methods consider the BASS problem as a supervised task: a DNN model is trained to transform an input mixed signal $x(t)$ to an output separated source $s_j(t)$ or to an output separation mask $m_j(t)$ to be applied to the input to get the separated source $s_j(t) = x(t) \odot m_j(t)$.

U-Net Such a DNN model often takes the form of a Denoising Auto-Encoder (DAE) where a model is trained to reconstruct the clean signal from its noisy version. Because of their (theoretically) infinite memory, the first models used RNNs (or their LSTM and GRU variations) for both the encoder and decoder [MLO+12, WHLRS14, EHWLR15]. Since then, it has been demonstrated that non-recurrent architectures, such as CNN, can also be applied successfully at a much lower cost. However, convolutional DAE while successful for image denoising have been found limited for audio reconstruction (the bottleneck layer does not allow to capture the fine details necessary to reconstruct an harmonic spectrogram). To allow the reconstruction of these fine details, the U-Net architecture has been proposed. This architecture was first proposed for the segmentation of biomedical images [RFB15]. It is an AE with added skip connections between the encoder and the encoder to allow the reconstruction of the fine details. In [JHM+17], this architecture has been applied to a spectrogram representation to isolate the singing voice from real polyphonic music largely improving previously obtained results. Precisely, the network is trained to output a Time/Frequency mask $M_j(t, f)$ such that applied to the amplitude STFT of the mixture $|X(t, f)|$, it allows to separate the amplitude STFT of the isolated source $|S_j(t, f)| = |X(t, f)| \odot M_j(t, f)$. The signal $s_j(t)$ is then reconstructed by inverting $|S_j(t, f)|$ using the phase of the initial mixture spectrogram $\phi_X(t, f)$. However, using the phase of the original signal limits the performances of the system.

Complex-U-Net To deal with this limitation, [CKH+19] have proposed in the case of speech enhancement to use the complex-spectrogram as input, and to modify the network, the masks and the loss to deal with complex values. In this case the complex-mask does not only modify the amplitudes $|X(t, f)|$ but also apply changes to the phases $\phi_X(t, f)$ so as to estimate the complex-spectrogram of the isolated source $S_j(t, f)$

Wave-U-Net Another way to deal with the problem of the phase is to by-pass the STFT and process the audio waveform directly. Along this, [SED18] have proposed a Wave-U-Net which applies the U-Net directly to the waveform. In this, the encoder is made of a cascade of 1D-convolution/Decimation to progressively reduce the time-dimension of $x(t)$ to the bottleneck representation z. A cascade of

Up-Sampling/1D-convolution is then used to decode z in the separated signals $s_j(t)$ (no masking filters are used here).

End-to-End [LPS19] also propose to use directly the waveform but without the U-Net architecture. The architecture is here inspired by WaveNet [vdODZ+16] and uses a stack of dilated convolutions with skip connections but while WaveNet aims at predicting the next sample value, it is used here in a non-causal way to predict the set of isolated sources of the center frame.

SEGAN SEGAN (Speech Enhancement Generative Adversarial Network) [PBS17] is an architecture proposed for speech enhancement which also uses the WaveNet blocks to represent the waveform. Moreover it also uses a DAE architecture but here considered as the generator G in a GAN set-up. The generator is trained to generate enhanced signals that look like real signals.

AE as NMF [SV17] reconcile the DNN and the NMF source separation research community by expressing an AE as a non-linear NMF. In NMF a positive observed matrix X is reconstructed as the product of a positive basis-matrix W with a positive activation-matrix H: $\hat{X} = W \cdot H$. Similarly in an AE, X is reconstructed by passing z in the decoder function ϕ_d: $\hat{X} = \phi_d(z)$. Considering only one linear layer for ϕ_d would therefore make ϕ_d play the same role as W and z the same role as H. The encoder part $z = \phi_e(X)$ would then be $H = W^{\ddagger} \cdot X$.[18] They then propose a Non-Negative AE as a stack of non-linear encoding layers $Y_0 = X$, $Y_1 = g(W_1 \cdot Y_0)$, $Y_2 = g(W_2 \cdot Y_1) \ldots H = Y_L$ followed by a stack of non-linear decoding layers $Y_{L+1} = g(W_{L+1} \cdot Y_L) \ldots \hat{X} = Y_{2L}$. g can be chosen to be a positive non-linear functions. The latent representation H can then be considered as an activation matrix which activate the "basis" of the decoder ϕ_d. Based on this, the authors propose various source separation algorithms.

TasNet, ConvTasNet With this in mind, the seminal networks TasNet [LM18] and ConvTasNet [LM19] can also be considered as examples of an encoder which provides the activation's and a decoder which reconstruct the signal. However, both TasNet and ConvTasNet directly process the waveform using 1D-Convolution. The decoder ϕ_d reconstructs the mixture waveform as a non-negative weighted sum of basis signals $\vec{V}$: $\vec{x} = \vec{w}\vec{V}$. The weights $\vec{w}$ are the outputs of a simple encoder ϕ_e of the form $\vec{w} = \mathcal{H}(\vec{x}\vec{U})$ where $\mathcal{H}$ is an optional nonlinear function.[19] The separation is done by masking the weights $\vec{w}$ and keeping only the ones necessary to reconstruct $\vec{s}_j$ from $\vec{x}$: $\hat{s}_j = (\vec{w} \odot \vec{m}_j)$. The masks $\vec{m}_j$ are the outputs of a "separation network" ϕ_s: $\vec{m}_j = \phi_s(\vec{w}) \in [0, 1]$. The latter is a Deep-LSTM in TasNet or stacks of 1D-Conv for ConvTasNet. As opposed to the U-Net approaches described above [JHM+17, CKH+19, SED18] which apply the masks on the original mixture, the masks are here applied on the weights.

---

[18] $\ddagger$ denotes the pseudo-inverse.

[19] for example a ReLU, to make the weights positive.

Deep Clustering [HCLRW16] propose a very different paradigm to train a DNN architecture for source separation. Deep Clustering uses a metric learning approach. For this, a DNN is trained to non-linearly project each time and frequency points (t, f) of a spectrogram in a space such that points that belong to the same source (to different sources) are projected in close neighboring (far away respectively). A simple K-means clustering algorithm of the projected points can then be used to perform the separation.

10.4.4   Content Generation

In statistical classification or machine learning, we often distinguish between discriminative or generative approaches [Jeb04]. Generative approaches are particularly attractive for their capacity to generate new data samples from their model. Some of the most popular models include different forms of autoencoders (including Variational Auto-Encoders (VAEs) [KW14b, CWBv19], Auto-Regressive models [vdODZ+16, PVC19, VSP+17] and Generative Adversarial Networks (GANs) [DMP18, GBC16]. These general models have sparked great interest since their introduction, mainly due to their incredible capabilities to generate new and high quality images [RMC16a, SGZ+16] but have also more recently shown their capacity for audio content generation.

Auto-regressive and Attention-Based Models As already discussed in Sect. 10.3.3, *WaveNet* is clearly one of the most popular neural autoregressive generative models for audio waveform synthesis [vdODZ+16]. It is capable of high quality speech and music synthesis but remains a complex model with a demanding sample-level auto-regressive principle. Nevertheless it is used in many other frameworks and in particular in encoder-decoder architectures such as Nsynth [ERR+17] or Variational Auto-Encoders (VAEs) as further discussed below. Another trend in synthesis, initially introduced for Text-To-Speech (TTS), aims for fully end-to-end generative models, where the signal is directly synthesized from characters. For example, the original *Tacotron* relies on a sequence-to-sequence architecture with attention mechanism to generate a linear-scale spectrogram from which the audio signal can be estimated using Griffin and Lim algorithm [GJ84]. Its extension, *Tacotron2* [SPW+18], combines the advantage of both previous models in using a sequence-to-sequence Tacotron-style model to generate mel-scale spectrograms followed by a modified WaveNet synthesizer.

Variational Auto-encoders VAEs were used in speech synthesis as extensions of wavenet autoencoders where the quantized latent space is conditioned on the speaker identity [vdOVK17]. For music synthesis, a generalisation of the previous concept was proposed in [MWPT19] under the form of an universal music translation network. The main idea is to have a so-called universal encoder that forces the embeddings of all musical domains to lie in the same space but separate reconstructing decoders for each domain exploiting an auxiliary conditioning network.

Several experiments of music domain conversion were described including for example early attempts for orchestral music to piano translation. The regularisation principle at the heart of VAEs can also be extended as in [ECRSB18] to enforce that the latent space exhibits the same topology as perceptual spaces such as musical timbre. One of the main advantages of such approaches is that the latent spaces can be directly used to synthesize sounds with continuous timbre evolution. Such capabilities can also be achieved with Generative Adversarial Networks (GANs) as discussed below with the example of drum synthesis [NLR20b]. Another extension of VAEs is known as the Vector-Quantized VAE (VQ-VAE)[vdOVK17] which aims at learning a discrete latent representation or *codebook*. The VQ-VAE can achieve sharper reconstructions than classic VAEs and can extract high-level interpretable audio features that strongly correlate with audio semantic information such as phonemes, with applications for voice conversion [CWBv19] or such as musical timbre for sound transformation. Another interesting approach in that framework is the Jukebox method presented in [DJP+20]. It is built on a multiscale VQ-VAEs (e.g. operating at different temporal resolutions) and on simplified autoregressive *transformers* with sparse attention. This model was in particular used to synthesize entire songs with vocals.

Adversarial Audio Synthesis Generative Adversarial Networks (GANs) have been initially used with success in speech synthesis [STS18] but their use was rapidly extended to music synthesis. For exemple, WaveGan [DMP18] performs unsupervised synthesis of raw-waveform audio. WavGan is based upon the two-dimensional deep convolutional GAN (DCGAN) architecture initially developed for image synthesis [RMC16b] and adapted to audio in considering intrinsic differences between audio and images (which resulted in the use of larger receptive fields and higher upsampling factors between layers). As discussed above in Sect. 10.3, a number of audio representations have been used in neural audio processing. For example in GANsynth [EAC+19], several audio representations are evaluated including Short-Term Fourier Transform (STFT) representations (log Magnitude, wrapped and unwrapped Phase) and Instantaneous frequency (IF). Some other representations, including the raw audio waveform and a variety of time-frequency representations (such as complex spectrogram, CQT or MFCC), were also compared for the task of adversarial audio synthesis in [NLR20a].

Numerous extensions or adaptations of the concepts of GANs were proposed including Style-GAN [KLA19], Cycle-GAN [ZPIE17] or Progressive Growing GANs [AHPG18, KALL18]. In audio synthesis, for example, [NLR20b] proposed a specific Progressive Growing GAN architecture for drum sound synthesis with a conditional generation scheme using continuous perceptual features describing timbre (e.g., boominess, brightness, depth).

Music Style Transformations Besides audio content generation, changing the style or instrumentation of a given piece of music is receiving a growing interest from the research community. Some research work target a direct style transformation of an input audio signal, as for example in [GDOP18] using Convolutive NN or as in the universal music translation network discussed above [MWPT19].

However, most studies operate on symbolic music such as MIDI and can focus on one or several music attributes such as melody [NSNY19], instrumentation or timbre [HCCY19, HLA+19], accompaniment [CcR19, HSP16] or general arrangement style [BKWW18, LS18]. An interesting work at the crossroads of accompaniment generation and style transfer is the so-called Groove2Groove model [CSR20]. It is a one-shot style transfer encoder-decoder neural network method for symbolic music trained in a supervised fashion using synthetic parallel data. In this model, the input to the style translation model is a full accompaniment but the output is entirely regenerated and does not contain any of the original accompaniment tracks.

10.4.5   Semi-Supervised Learning and Self-Supervised Learning

Supervised learning assumes that labeled data, i.e. data x with associated ground-truth label y, are available to train the parameters θ of a prediction model $\hat{y} = f_\theta(x)$. To train a DNN model, the amount of such labeled data can be very large. While such large labeled datasets exist for image or speech, this is not the case today for audio content such as music or environmental sounds. We review here two popular techniques to deal with this lack of annotated data: semi-supervised learning (teacher-student paradigm) and self-supervised learning.

10.4.5.1   Semi-Supervised Learning

Semi-Supervised Learning (Semi-SL) combines training with a small amount of labeled data and training with a large amount of unlabeled data. One popular form of Semi-SL used the so-called teacher-student paradigm. It is a supervised learning technique in which the knowledge of a teacher (a model trained on clean labeled data) is used to label a large set of unlabeled data which is used in turn to train student models.

SoundNet [AVT16] is one of the first models developed in audio (for a task of environmental sounds recognition) that use the teacher-student technique. The idea is to transfer the knowledge of computer vision (CV) networks to an audio network. For this, a large set of Audio-Video clips are considered. Each clip has a video track and an audio-track. The CV networks are applied to the video-tracks to annotate the corresponding audio-tracks which are then used to train the audio network. The teachers are CV networks previously trained for objects and scenes recognition (an ImageNet CNN and a Places CNN) The audio network is a deep stack of 1-D convolutions. The transfer is done by minimizing the Kullback-Leibler divergence between the output probabilities of the audio and image networks. The training is done using two-millions unlabeled videos. It is shown that using such a trained audio network as feature extractor for typical Acoustic Scene Classification tasks largely outperform previous methods.

In music description, [WL17] were the first to use the teacher-student paradigm. For a task of drum transcription, a teacher (a Partially-Fixed-NMF model) previously trained on clean labeled data is applied on a large unlabeled dataset which is then used to train a DNN student model. The author show that the student largely outperforms the teacher. For a task of singing voice segmentation, [MBCHP18] also propose to use the teacher-student technique but in a different way. A teacher (a deep CNN network) previously trained on a clean but small labeled dataset, is applied to a large set of data grabbed from the web with labels obtained by crowd-sourcing (hence very noisy). The outputs of the teacher are then used to filter out the noise from the data. These cleaned data serve as the training label for the student. The author also report larger performances for the student.

10.4.6   Self-Supervised Learning

Self-Supervised Learning (Self-SL) is a supervised learning technique in which the training data are automatically labeled.

To automatically create labels, one can use the natural temporal synchronization between the various modalities of multi-media data. This is denoted by **Audio-Visual Correspondence (AVC)**. One of the first approach that use the AVC is the "Look, Listen and Learn" L^3 network [AZ17] where videos are decomposed into their image and audio modalities. A vision and an audio sub-networks are then fed to a fusion network with a softmax output which aims at predicting if the input image and audio correspond. Corresponding pairs are the ones taken at the same time from the same video, while mismatched pairs are extracted from different videos It is showed that the two image and audio sub-networks trained in such a way can be used afterward for solving sound classification or visual classification (ImageNet) tasks with very large performances.

The **AVE-Net**[AZ18] is an extension of the L^3 network in which the fusion network is replaced by a simple Euclidean distance. The sub-networks are therefore forced to learn to (non-linearly) project the data in a space where the image content (e.g. a guitar player) and its corresponding sound (e.g. a guitar sound) are projected nearby. Since both audio and video are projected in the same space, cross-modal applications are possible (such as querying an image giving a sound or the opposite) as well as visually locating the "object that sounds" (the sub-part of the image which projection is the closest to the projection of the sound).

In the same spirit, [ZGR+18] propose to train a two branches (image and audio) network for a task of source separation: to provide the "sound of the pixels" which are selected on the image. The audio branch (a U-Net) is trained to separate the audio into a set of isolated components. A Self-SL approach is then used to learn the mapping between each of these components and the various parts of the images.

Another type of Self-SL relies on applying transformations to an audio signal x for which we can predict the effect on the ground-truth labels y. The **SPICE** (Self-supervised Pitch Estimation) [GFR+20] network uses such an approach. In this, a Siamese AE is used. The encoder is first applied to the original audio to obtain a latent variable z_1. The signal is then pitch-transposed by a factor p and encoded to obtain z_2. The network is then trained to allow predicting p from the difference between z_1 and z_2. It is showed that, while trained to predict pitch-transposition, the network can be used to perform pitch-estimation with results very close to networks trained in a fully supervised way.

10.5   Conclusion and Future Directions

The advances in deep learning has strongly impacted the domain of audio analysis and synthesis. For many applications, the current state of the art is exploiting to at least some extent some form of deep neural processing. The emergence of deep neural networks as pure data-driven approaches was facilitated by the access to ever-increasing super-computing facilities, combined with the availability of huge data repositories (although largely unannotated). Nevertheless, this poses a number of challenges especially in terms of complexity, explainability, fairness and needs for data. We would like to sketch below some of our view for future directions in Deep learning for audio and music.

- **Increased explainability using Audio models**. For decades, many audio models have been developed. Such models include perceptual models (only audible information is modelled), Signal-based models (parametric models capturing the nature or structure of the signal) or physics-based (exploiting the knowledge of the sound production mechanisms or sound-propagation characteristics). Besides complexity reduction objectives, relying on appropriate audio source models within the deep architecture allows to constrain or "guide" the network to converge to an appropriate solution or to obtain more interpretable or explainable networks. Some recent works have already exploited some aspects of this view: using non-negative factorization models with CNNs for audio scene classification [BSER17], or for speech separation [LM19] or coupling signal processing modules with deep learning for audio synthesis [EHGR20, WTY20].
- **Increased performance and explainability using Multimodality.** In many situations, the audio signal can be associated with other modalities ranging from videos (e.g; in audiovisual scenes), text (such as lyrics or music scores), body movements or EEG (for example of subjects listening music). Video has proven to be useful for many audio tasks including for example audio-visual music performances analysis [DEL+19] and audio-visual scene/object recognition but there are still important challenges especially when the modalities are not observed synchronously [PEO+20]. As other examples, many Informed source separation approaches [OLBR13, LDDR13] do exploit an additional modality

for separation such as lyrics for singing voice [SDRB19, LOD13, MBP20] score for music remixing [EM12], sketches on spectrogram representations for selective source separation [SM09], or EEG for attention-based music source separation [cER20]. There are clear interest to further exploit concurrent cues, when available, to build better and more explainable models.

- **Increased fairness and ethics.** If this is an obvious problem for the applications of Deep learning in health or justice, it is also of utmost importance in audio. In speech recognition, we certainly do not want systems that are more efficient on male voices than female voices. Similarly in music, since most of the studies are in western music, a clear bias towards this type of music exist. For music recommendation systems, fairness should also be a central goal to avoid bias in terms of gender, ethnicity or commercial inequity. In terms of content, to comply with ethics rules it becomes necessary to be able to filter unappropriate or explicit content [VHM+20].

References

[Abe20] Jakob Abeßer. A review of deep learning based methods for acoustic scene classification. *Applied Sciences*, 10, 03 2020.

[AHPG18] Dinesh Acharya, Zhiwu Huang, Danda Pani Paudel, and Luc Van Gool. Towards high resolution video generation with progressive growing of sliced wasserstein gans, 2018.

[APS05] Jean-Julien Aucouturier, François Pachet, and Mark Sandler. The way it sounds: Timbre models for analysis and retrieval of polyphonic music signals. *IEEE Transactions of Multimedia*, 7(6):1028–1035, 2005.

[AVT16] Yusuf Aytar, Carl Vondrick, and Antonio Torralba. Soundnet: Learning sound representations from unlabeled video. In *NIPS (Conference on Neural Information Processing Systems)*, 2016.

[AZ17] Relja Arandjelovic and Andrew Zisserman. Look, listen and learn. In *Proc. of IEEE ICCV (International Conference on Computer Vision)*, 2017.

[AZ18] Relja Arandjelović and Andrew Zisserman. Objects that sound. In *Proc. of ECCV (European Conference on Computer Vision)*, 2018.

[BC94] Guy J Brown and Martin Cooke. Computational auditory scene analysis. *Computer speech and language*, 8(4):297–336, 1994.

[BCB14] Dzmitry Bahdanau, Kyunghyun Cho, and Yoshua Bengio. Neural machine translation by jointly learning to align and translate. *arXiv preprint arXiv:1409.0473*, 2014.

[BEP18] Dogac Basaran, Slim Essid, and Geoffroy Peeters. Main melody extraction with source-filter nmf and c-rnn. In *Proc. of ISMIR (International Society for Music Information Retrieval)*, Paris, France, September 23–27, 2018.

[BGL+94] Jane Bromley, Isabelle Guyon, Yann LeCun, Eduard Säckinger, and Roopak Shah. Signature verification using a "siamese" time delay neural network. In *Advances in neural information processing systems*, pages 737–744, 1994.

[BKK18] Shaojie Bai, J Zico Kolter, and Vladlen Koltun. An empirical evaluation of generic convolutional and recurrent networks for sequence modeling. *arXiv preprint arXiv:1803.01271*, 2018.

[BKWW18] Gino Brunner, Andres Konrad, Yuyi Wang, and Roger Wattenhofer. MIDI-VAE: Modeling dynamics and instrumentation of music with applications to style transfer. In *ISMIR*, 2018.

[BLD12] Emmanouil Benetos, Mathieu Lagrange, and Simon Dixon.   Characterisation of acoustic scenes using a temporally constrained shit-invariant model.   *15th International Conference on Digital Audio Effects, DAFx 2012 Proceedings*, 09 2012.

[BM94] Hervé A. Bourlard and Nelson Morgan.   *Connectionist Speech Recognition A Hybrid Approach*, volume 247. Springer US, 1994.

[BMS+17] Rachel Bittner, Brian McFee, Justin Salamon, Peter Li, and Juan Pablo Bello. Deep salience representations for f0 estimation in polyphonic music. In *Proc. of ISMIR (International Society for Music Information Retrieval)*, Suzhou, China, October, 23–27 2017.

[Bro91] J. Brown. Calculation of a constant q spectral transform. *JASA (Journal of the Acoustical Society of America)*, 89(1):425–434, 1991.

[BS11] Sebastian Böck and Markus Schedl. Enhanced beat tracking with context-aware neural networks. In *Proc. of DAFx (International Conference on Digital Audio Effects)*, Paris, France, 2011.

[BSER17] V. Bisot, R. Serizel, S. Essid, and G. Richard.   Feature learning with matrix factorization applied to acoustic scene classification. *IEEE/ACM Transactions on Audio, Speech, and Language Processing*, 25(6):1216–1229, 2017.

[CcR19] Ondřej Cífka, Umut Şimşekli, and Gaël Richard. Supervised symbolic music style translation using synthetic data. In *ISMIR*, 2019.

[cER20] Giorgia Cantisani, Slim Essid, and Gael Richard. Neuro-steered music source separation with EEG-based auditory attention decoding and contrastive-NMF. working paper or preprint, October 2020.

[CFS16] Keunwoo Choi, György Fazekas, and Mark Sandler. Automatic tagging using deep convolutional neural networks. In *Proc. of ISMIR (International Society for Music Information Retrieval)*, New York, USA, 2016.

[CHP17] Alice Cohen-Hadria and Geoffroy Peeters. Music structure boundaries estimation using multiple self-similarity matrices as input depth of convolutional neural networks. In *AES Conference on Semantic Audio*, Erlangen, Germany, June, 22–24, 2017.

[CKH+19] Hyeong-Seok Choi, Jang-Hyun Kim, Jaesung Huh, Adrian Kim, Jung-Woo Ha, and Kyogu Lee. Phase-aware speech enhancement with deep complex u-net. *Proc. of ICLR (International Conference on Learning Representations)*, 2019.

[CSR20] O. Cífka, U. Simsekli, and G. Richard. Groove2groove: One-shot music style transfer with supervision from synthetic data. *IEEE/ACM Transactions on Audio, Speech, and Language Processing*, 28:2638–2650, 2020.

[CVMG+14] Kyunghyun Cho, Bart Van Merriënboer, Caglar Gulcehre, Dzmitry Bahdanau, Fethi Bougares, Holger Schwenk, and Yoshua Bengio. Learning phrase representations using RNN encoder-decoder for statistical machine translation. *arXiv preprint arXiv:1406.1078*, 2014.

[CWBv19] J. Chorowski, R. J. Weiss, S. Bengio, and A. van den Oord. Unsupervised speech representation learning using wavenet autoencoders. *IEEE/ACM Transactions on Audio, Speech, and Language Processing*, 27(12):2041–2053, 2019.

[DBDR17] S. Durand, J. P. Bello, B. David, and G. Richard. Robust downbeat tracking using an ensemble of convolutional networks. *IEEE/ACM Transactions on Audio, Speech, and Language Processing*, 25(1):76–89, 2017.

[DEL+19] Z. Duan, S. Essid, C. C. S. Liem, G. Richard, and G. Sharma. Audiovisual analysis of music performances: Overview of an emerging field. *IEEE Signal Processing Magazine*, 36(1):63–73, 2019.

[Die14] Sander Dieleman. Recommending music on spotify with deep learning. Technical report, http://benanne.github.io/2014/08/05/spotify-cnns.html, 2014.

[DJP+20] Prafulla Dhariwal, Heewoo Jun, Christine Payne, Jong Wook Kim, Alec Radford, and Ilya Sutskever. Jukebox: A generative model for music, 2020.

[DMP18] Chris Donahue, Julian McAuley, and Miller Puckette. Adversarial audio synthesis. *arXiv preprint arXiv:1802.04208*, 2018.

[DP19] Guillaume Doras and Geoffroy Peeters. Cover detection using dominant melody embeddings. In *Proc. of ISMIR (International Society for Music Information Retrieval)*, Delft, The Netherlands, November 4–8 2019.

[DP20] Guillaume Doras and Geoffroy Peeters. A prototypical triplet loss for cover detection. In *Proc. of IEEE ICASSP (International Conference on Acoustics, Speech, and Signal Processing)*, Barcelona, Spain, May, 4–8 2020.

[DRDF10] Jean-Louis Durrieu, Gaël Richard, Bertrand David, and Cédric Févotte. Source/filter model for unsupervised main melody extraction from polyphonic audio signals. *IEEE transactions on audio, speech, and language processing*, 18(3):564–575, 2010.

[DS14] Sander Dieleman and Benjamin Schrauwen. End-to-end learning for music audio. In *2014 IEEE International Conference on Acoustics, Speech and Signal Processing (ICASSP)*, pages 6964–6968. IEEE, 2014.

[DYS+20] Guillaume Doras, Furkan Yesiler, Joan Serra, Emilia Gomez, and Geoffroy Peeters. Combining musical features for cover detection. In *Proc. of ISMIR (International Society for Music Information Retrieval)*, Montreal, Canada, October, 11–15 2020.

[EAC+19] Jesse Engel, Kumar Krishna Agrawal, Shuo Chen, Ishaan Gulrajani, Chris Donahue, and Adam Roberts. Gansynth: Adversarial neural audio synthesis. In *Proc. of ICLR (International Conference on Learning Representations)*, 2019.

[ECRSB18] Philippe Esling, Axel Chemla-Romeu-Santos, and Adrien Bitton. Bridging audio analysis, perception and synthesis with perceptually-regularized variational timbre spaces. In *Proc. of ISMIR (International Society for Music Information Retrieval)*, 2018.

[EHGR20] Jesse Engel, Lamtharn Hantrakul, Chenjie Gu, and Adam Roberts. Ddsp: Differentiable digital signal processing. In *Proc. of ICLR (International Conference on Learning Representations)*, 2020.

[EHWLR15] Hakan Erdogan, John R Hershey, Shinji Watanabe, and Jonathan Le Roux. Phase-sensitive and recognition-boosted speech separation using deep recurrent neural networks. In *2015 IEEE International Conference on Acoustics, Speech and Signal Processing (ICASSP)*, pages 708–712. IEEE, 2015.

[EM12] Sebastian Ewert and Meinard Müller. Score-Informed Source Separation for Music Signals. In Meinard Müller, Masataka Goto, and Markus Schedl, editors, *Multimodal Music Processing*, volume 3 of *Dagstuhl Follow-Ups*, pages 73–94. Schloss Dagstuhl–Leibniz-Zentrum fuer Informatik, Dagstuhl, Germany, 2012.

[ERR+17] Jesse Engel, Cinjon Resnick, Adam Roberts, Sander Dieleman, Mohammad Norouzi, Douglas Eck, and Karen Simonyan. Neural audio synthesis of musical notes with wavenet autoencoders. In *Proc. of ICML (International Conference on Machine Learning)*, pages 1068–1077, 2017.

[FM82] Kunihiko Fukushima and Sei Miyake. Neocognitron: A self-organizing neural network model for a mechanism of visual pattern recognition. In *Competition and cooperation in neural nets*, pages 267–285. Springer, 1982.

[FP19] Hadrien Foroughmand and Geoffroy Peeters. Deep-rhythm for global tempo estimation in music. In *Proc. of ISMIR (International Society for Music Information Retrieval)*, Delft, The Netherlands, November 4–8 2019.

[GBC16] Ian Goodfellow, Yoshua Bengio, and Aaron Courville. *Deep Learning*. MIT Press, 2016. http://www.deeplearningbook.org.

[GDOP18] Eric Grinstein, Ngoc Q. K. Duong, Alexey Ozerov, and Patrick Pérez. Audio style transfer. In *ICASSP*, 2018.

[GFR+20] Beat Gfeller, Christian Frank, Dominik Roblek, Matt Sharifi, Marco Tagliasacchi, and Mihajlo Velimirović. Spice: Self-supervised pitch estimation. *IEEE/ACM Transactions on Audio, Speech, and Language Processing*, 28:1118–1128, 2020.

[GJ84] D. Griffin and Jae Lim.   Signal estimation from modified short-time fourier transform.   *IEEE Transactions on Acoustics, Speech, and Signal Processing*, 32(2):236–243, 1984.

[GKKC07] Frantisek Grézl, Martin Karafiát, Stanislav Kontár, and Jan Cernocky. Probabilistic and bottle-neck features for lvcsr of meetings.   In *2007 IEEE International Conference on Acoustics, Speech and Signal Processing-ICASSP'07*, volume 4, pages IV–757. IEEE, 2007.

[Got03] Masataka Goto. A chorus-section detecting method for musical audio signals. In *Proc. of IEEE ICASSP (International Conference on Acoustics, Speech, and Signal Processing)*, pages 437–440, Hong Kong, China, 2003.

[GPAM+14] Ian Goodfellow, Jean Pouget-Abadie, Mehdi Mirza, Bing Xu, David Warde-Farley, Sherjil Ozair, Aaron Courville, and Yoshua Bengio. Generative adversarial nets. In *Advances in neural information processing systems*, pages 2672–2680, 2014.

[HBL12] Eric J. Humphrey, Juan Pablo Bello, and Yann LeCun.   Moving beyond feature design: Deep architectures and automatic feature learning in music informatics. In *Proc. of ISMIR (International Society for Music Information Retrieval)*, Porto, Portugal, 2012.

[HCCY19] Yun-Ning Hung, I Ping Chiang, Yi-An Chen, and Yi-Hsuan Yang.   Musical composition style transfer via disentangled timbre representations. In *IJCAI*, 2019.

[HCL06] Raia Hadsell, Sumit Chopra, and Yann LeCun.   Dimensionality reduction by learning an invariant mapping. In *2006 IEEE Computer Society Conference on Computer Vision and Pattern Recognition (CVPR'06)*, volume 2, pages 1735–1742. IEEE, 2006.

[HCLRW16] John R Hershey, Zhuo Chen, Jonathan Le Roux, and Shinji Watanabe. Deep clustering: Discriminative embeddings for segmentation and separation. In *2016 IEEE International Conference on Acoustics, Speech and Signal Processing (ICASSP)*, pages 31–35. IEEE, 2016.

[HDY+12] Geoffrey Hinton, Li Deng, Dong Yu, George E Dahl, Abdel-rahman Mohamed, Navdeep Jaitly, Andrew Senior, Vincent Vanhoucke, Patrick Nguyen, Tara N Sainath, et al. Deep neural networks for acoustic modeling in speech recognition: The shared views of four research groups. *IEEE Signal processing magazine*, 29(6):82–97, 2012.

[HES00] Hynek Hermansky, Daniel PW Ellis, and Sangita Sharma. Tandem connectionist feature extraction for conventional HMM systems. In *2000 IEEE International Conference on Acoustics, Speech, and Signal Processing. Proceedings (Cat. No. 00CH37100)*, volume 3, pages 1635–1638. IEEE, 2000.

[HLA+19] Sicong Huang, Qiyang Li, Cem Anil, Xuchan Bao, Sageev Oore, and Roger B. Grosse. TimbreTron: A WaveNet(CycleGAN(CQT(Audio))) pipeline for musical timbre transfer. In *ICLR*, 2019.

[HOT06] Geoffrey E. Hinton, Simon Osindero, and Yee-Whye Teh. A fast learning algorithm for deep belief nets. *Neural Computation*, 18(7):1527–1554, 2006.

[HS97] Sepp Hochreiter and Jürgen Schmidhuber.   Long short-term memory.   *Neural computation*, 9(8):1735–1780, 1997.

[HSP16] Gaëtan Hadjeres, Jason Sakellariou, and François Pachet. Style imitation and chord invention in polyphonic music with exponential families. *ArXiv*, abs/1609.05152, 2016.

[HZRS16] K. He, X. Zhang, S. Ren, and J. Sun. Deep residual learning for image recognition. In *2016 IEEE Conference on Computer Vision and Pattern Recognition (CVPR)*, pages 770–778, 2016.

[Jeb04] T. Jebara. *Machine Learning: Discriminative and Generative*. 2004.

[JH11] Navdeep Jaitly and Geoffrey Hinton.   Learning a better representation of speech soundwaves using restricted boltzmann machines.   In *2011 IEEE International Conference on Acoustics, Speech and Signal Processing (ICASSP)*, pages 5884–5887. IEEE, 2011.

[JHM+17] Andreas Jansson, Eric J. Humphrey, Nicola Montecchio, Rachel Bittner, Aparna Kumar, and Tillman Weyde. Singing voice separation with deep u-net convolutional networks. In *Proc. of ISMIR (International Society for Music Information Retrieval)*, Suzhou, China, October, 23–27 2017.

[KALL18] Tero Karras, Timo Aila, Samuli Laine, and Jaakko Lehtinen. Progressive growing of gans for improved quality, stability, and variation, 2018.

[KLA19] Tero Karras, Samuli Laine, and Timo Aila. A style-based generator architecture for generative adversarial networks, 2019.

[KLN18] Taejun Kim, Jongpil Lee, and Juhan Nam. Sample-level CNN architectures for music auto-tagging using raw waveforms. 2018.

[KLW19] Uday Kamath, John Liu, and James Whitaker. *Deep learning for NLP and speech recognition*, volume 84. Springer, 2019.

[KSH12] Alex Krizhevsky, Ilya Sutskever, and Geoffrey E Hinton. Imagenet classification with deep convolutional neural networks. In *Advances in neural information processing systems*, pages 1097–1105, 2012.

[KSU+19] Y. Koizumi, S. Saito, H. Uematsu, Y. Kawachi, and N. Harada. Unsupervised detection of anomalous sound based on deep learning and the Neyman–Pearson lemma. *IEEE/ACM Transactions on Audio, Speech, and Language Processing*, 27(1):212–224, 2019.

[KW14a] Diederik P Kingma and Max Welling. Auto-encoding variational bayes. In *Proc. of ICLR (International Conference on Learning Representations)*, 2014.

[KW14b] Diederik P Kingma and Max Welling. Auto-encoding variational bayes. In *Proc. of ICLR (International Conference on Learning Representations)*, 2014.

[LBBH98] Yann LeCun, Léon Bottou, Yoshua Bengio, and Patrick Haffner. Gradient-based learning applied to document recognition. *Proceedings of the IEEE*, 86(11):2278–2324, 1998.

[LC16] Vincent Lostanlen and Carmine-Emanuele Cella. Deep convolutional networks on the pitch spiral for music instrument recognition. *arXiv preprint arXiv:1605.06644*, 2016.

[LDDR13] A. Liutkus, J. Durrieu, L. Daudet, and G. Richard. An overview of informed audio source separation. In *2013 14th International Workshop on Image Analysis for Multimedia Interactive Services (WIAMIS)*, pages 1–4, 2013.

[LM18] Yi Luo and Nima Mesgarani. Tasnet: time-domain audio separation network for real-time, single-channel speech separation. In *2018 IEEE International Conference on Acoustics, Speech and Signal Processing (ICASSP)*, pages 696–700. IEEE, 2018.

[LM19] Yi Luo and Nima Mesgarani. Conv-tasnet: Surpassing ideal time–frequency magnitude masking for speech separation. *IEEE/ACM transactions on audio, speech, and language processing*, 27(8):1256–1266, 2019.

[LOD13] L. Le Magoarou, A. Ozerov, and N. Q. K. Duong. Text-informed audio source separation using nonnegative matrix partial co-factorization. In *2013 IEEE International Workshop on Machine Learning for Signal Processing (MLSP)*, pages 1–6, 2013.

[LPKN17] Jongpil Lee, Jiyoung Park, Keunhyoung Luke Kim, and Juhan Nam. Sample-level deep convolutional neural networks for music auto-tagging using raw waveforms. *arXiv preprint arXiv:1703.01789*, 2017.

[LPLN09] Honglak Lee, Peter Pham, Yan Largman, and Andrew Y Ng. Unsupervised feature learning for audio classification using convolutional deep belief networks. In *Advances in neural information processing systems*, pages 1096–1104, 2009.

[LPS19] Francesc Lluís, Jordi Pons, and Xavier Serra. End-to-end music source separation: is it possible in the waveform domain? In *Proc. of Interspeech*, Graz, Austria, September 15–19 2019.

[LS18] Wei-Tsung Lu and Li Su. Transferring the style of homophonic music using recurrent neural networks and autoregressive models. In *ISMIR*, 2018.

[LWM+09] Edith Law, Kris West, Michael I Mandel, Mert Bay, and J Stephen Downie. Evaluation of algorithms using games: The case of music tagging. In *ISMIR*, pages 387–392, 2009.

[Mal89] Stephane Mallat. A theory for multiresolution signal decomposition: The wavelet representation. *IEEE transactions on pattern analysis and machine intelligence*, 11(7):674–693, 1989.

[MB17] Brian McFee and Juan Pablo Bello. Structured training for large-vocabulary chord recognition. In *Proc. of ISMIR (International Society for Music Information Retrieval)*, Suzhou, China, October, 23–27 2017.

[MBCHP18] Gabriel Meseguer Brocal, Alice Cohen-Hadria, and Geoffroy Peeters. Dali: A large dataset of synchronized audio, lyrics and pitch, automatically created using teacher-student. In *Proc. of ISMIR (International Society for Music Information Retrieval)*, Paris, France, September, 23–27 2018.

[MBP20] Gabriel Meseguer Brocal and Geoffroy Peeters. Content based singing voice source separation via strong conditioning using aligned phonemes. In *Proc. of ISMIR (International Society for Music Information Retrieval)*, Montreal, Canada, October, 11–15 2020.

[MG20] M. D. McDonnell and W. Gao. Acoustic scene classification using deep residual networks with late fusion of separated high and low frequency paths. In *ICASSP 2020 - 2020 IEEE International Conference on Acoustics, Speech and Signal Processing (ICASSP)*, pages 141–145, 2020.

[MHB+18] A. Mesaros, T. Heittola, E. Benetos, P. Foster, M. Lagrange, T. Virtanen, and M. D. Plumbley. Detection and classification of acoustic scenes and events: Outcome of the DCASE 2016 challenge. *IEEE/ACM Transactions on Audio, Speech, and Language Processing*, 26(2):379–393, 2018.

[MKG+17] Soroush Mehri, Kundan Kumar, Ishaan Gulrajani, Rithesh Kumar, Shubham Jain, Jose Sotelo, Aaron Courville, and Yoshua Bengio. Samplernn: An unconditional end-to-end neural audio generation model. In *Proc. of ICLR (International Conference on Learning Representations)*, 2017.

[MLO+12] Andrew Maas, Quoc V Le, Tyler M O'neil, Oriol Vinyals, Patrick Nguyen, and Andrew Y Ng. Recurrent neural networks for noise reduction in robust ASR. In *Proc. of Interspeech*, 2012.

[MWPT19] Noam Mor, Lior Wolf, Adam Polyak, and Yaniv Taigman. A universal music translation network. In *Proc. of ICLR (International Conference on Learning Representations)*, 2019.

[NLR20a] Javier Nistal, Stefan Lattner, and Gaël Richard. Comparing representations for audio synthesis using generative adversarial networks, 06 2020.

[NLR20b] Javier Nistal, Stephan Lattner, and Gaël Richard. Drumgan: Synthesis of drum sounds with timbral feature conditioning using generative adversarial networks. In *Proc. of ISMIR (International Society for Music Information Retrieval)*, Montreal, Canada, October 2020.

[NPM20] Paul-Gauthier Noé, Titouan Parcollet, and Mohamed Morchid. Cgcnn: Complex gabor convolutional neural network on raw speech. In *Proc. of IEEE ICASSP (International Conference on Acoustics, Speech, and Signal Processing)*, Barcelona, Spain, May, 4–8 2020.

[NSNY19] Eita Nakamura, Kentaro Shibata, Ryo Nishikimi, and Kazuyoshi Yoshii. Unsupervised melody style conversion. In *ICASSP*, 2019.

[OLBR13] A. Ozerov, A. Liutkus, R. Badeau, and G. Richard. Coding-based informed source separation: Nonnegative tensor factorization approach. *IEEE Transactions on Audio, Speech, and Language Processing*, 21(8):1699–1712, 2013.

[PBS17] Santiago Pascual, Antonio Bonafonte, and Joan Serra. Segan: Speech enhancement generative adversarial network. *arXiv preprint arXiv:1703.09452*, 2017.

[PEO+20] S. Parekh, S. Essid, A. Ozerov, N. Q. K. Duong, P. Pérez, and G. Richard. Weakly supervised representation learning for audio-visual scene analysis. *IEEE/ACM Transactions on Audio, Speech, and Language Processing*, 28:416–428, 2020.

[PLDR18] Bryan Pardo, Antoine Liutkus, Zhiyao Duan, and Gaël Richard. *Applying Source Separation to Music*, chapter 16, pages 345–376. John Wiley & Sons, Ltd, 2018.

[PLS16] Jordi Pons, Thomas Lidy, and Xavier Serra. Experimenting with musically motivated convolutional neural networks. In *Proc. of IEEE CBMI (International Workshop on Content-Based Multimedia Indexing)*, 2016.

[Pon19] Jordi Pons. *Deep neural networks for music and audio tagging*. PhD thesis, Music Technology Group (MTG), Universitat Pompeu Fabra, Barcelona, 2019.

[PRP20] Laure Pretet, Gaël Richard, and Geoffroy Peeters. Learning to rank music tracks using triplet loss. In *Proc. of IEEE ICASSP (International Conference on Acoustics, Speech, and Signal Processing)*, Barcelona, Spain, May, 4–8 2020.

[PSS19] J. Pons, J. Serrà, and X. Serra. Training neural audio classifiers with few data. In *ICASSP 2019 - 2019 IEEE International Conference on Acoustics, Speech and Signal Processing (ICASSP)*, pages 16–20, 2019.

[PVC19] R. Prenger, R. Valle, and B. Catanzaro. Waveglow: A flow-based generative network for speech synthesis. In *ICASSP 2019 - 2019 IEEE International Conference on Acoustics, Speech and Signal Processing (ICASSP)*, pages 3617–3621, 2019.

[RB18] Mirco Ravanelli and Yoshua Bengio. Speaker recognition from raw waveform with sincnet. In *2018 IEEE Spoken Language Technology Workshop (SLT)*, pages 1021–1028. IEEE, 2018.

[RFB15] Olaf Ronneberger, Philipp Fischer, and Thomas Brox. U-net: Convolutional networks for biomedical image segmentation. In *International Conference on Medical image computing and computer-assisted intervention*, pages 234–241. Springer, 2015.

[RHW86] David E Rumelhart, Geoffrey E Hinton, and Ronald J Williams. Learning representations by back-propagating errors. *nature*, 323(6088):533–536, 1986.

[RMC16a] A. Radford, Luke Metz, and Soumith Chintala. Unsupervised representation learning with deep convolutional generative adversarial networks. *CoRR*, abs/1511.06434, 2016.

[RMC16b] Alec Radford, Luke Metz, and Soumith Chintala. Unsupervised representation learning with deep convolutional generative adversarial networks, 2016.

[Ros57] Frank Rosenblatt. *The perceptron, a perceiving and recognizing automaton Project Para*. Cornell Aeronautical Laboratory, 1957.

[RSN13] G. Richard, S. Sundaram, and S. Narayanan. An overview on perceptually motivated audio indexing and classification. *Proceedings of the IEEE*, 101(9):1939–1954, 2013.

[Sai15] Tara N. Sainath. Towards end-to-end speech recognition using deep neural networks. In *Proc. of ICML (International Conference on Machine Learning)*, 2015.

[SB13] Jan Schlüter and Sebastian Böck. Musical onset detection with convolutional neural networks. In *6th International Workshop on Machine Learning and Music (MML) in conjunction with the European Conference on Machine Learning and Principles and Practice of Knowledge Discovery in Databases (ECML/PKDD)*, Prague, Czech Republic, 2013.

[SBER18] Romain Serizel, Victor Bisot, Slim Essid, and Gaël Richard. *Acoustic Features for Environmental Sound Analysis*, pages 71–101. 01 2018.

[SDRB19] K. Schulze-Forster, C. Doire, G. Richard, and R. Badeau. Weakly informed audio source separation. In *2019 IEEE Workshop on Applications of Signal Processing to Audio and Acoustics (WASPAA)*, pages 273–277, 2019.

[SE03] A. Sheh and Daniel P. W. Ellis. Chord segmentation and recognition using em-trained hidden markov models. In *Proc. of ISMIR (International Society for Music Information Retrieval)*, pages 183–189, Baltimore, Maryland, USA, 2003.

[SED18] Daniel Stoller, Sebastian Ewert, and Simon Dixon. Wave-u-net: A multi-scale neural network for end-to-end audio source separation. In *Proc. of ISMIR (International Society for Music Information Retrieval)*, Paris, France, September, 23–27 2018.

[SGZ+16] Tim Salimans, Ian Goodfellow, Wojciech Zaremba, Vicki Cheung, Alec Radford, and Xi Chen. Improved techniques for training gans. In *Proceedings of the 30th International Conference on Neural Information Processing Systems*, NIPS'16, page 2234–2242, Red Hook, NY, USA, 2016. Curran Associates Inc.

[SKP15] Florian Schroff, Dmitry Kalenichenko, and James Philbin. Facenet: A unified embedding for face recognition and clustering. In *Proc. of IEEE CVPR (Conference on Computer Vision and Pattern Recognition)*, pages 815–823, 2015.

[SLJ+15] Christian Szegedy, Wei Liu, Yangqing Jia, Pierre Sermanet, Scott Reed, Dragomir Anguelov, Dumitru Erhan, Vincent Vanhoucke, and Andrew Rabinovich. Going deeper with convolutions. In *Proceedings of the IEEE conference on computer vision and pattern recognition*, pages 1–9, 2015.

[SM09] P. Smaragdis and G. J. Mysore. Separation by "humming": User-guided sound extraction from monophonic mixtures. In *2009 IEEE Workshop on Applications of Signal Processing to Audio and Acoustics*, pages 69–72, 2009.

[SPW+18] Jonathan Shen, Ruoming Pang, Ron J Weiss, Mike Schuster, Navdeep Jaitly, Zongheng Yang, Zhifeng Chen, Yu Zhang, Yuxuan Wang, Rj Skerrv-Ryan, et al. Natural TTS synthesis by conditioning wavenet on MEL spectrogram predictions. In *Proc. of IEEE ICASSP (International Conference on Acoustics, Speech, and Signal Processing)*, pages 4779–4783. IEEE, 2018.

[SS90] Xavier Serra and Julius Smith. Spectral modeling synthesis: A sound analysis/synthesis system based on a deterministic plus stochastic decomposition. *Computer Music Journal*, 14(4):12–24, 1990.

[SSL20] Youngho Jeong Sangwon Suh, Sooyoung Park and Taejin Lee. Designing acoustic scene classification models with CNN variants. In *DCASE challenge, technical report*, 2020.

[SSZ17] Jake Snell, Kevin Swersky, and Richard Zemel. Prototypical networks for few-shot learning. 03 2017.

[STS18] Y. Saito, S. Takamichi, and H. Saruwatari. Statistical parametric speech synthesis incorporating generative adversarial networks. *IEEE/ACM Transactions on Audio, Speech, and Language Processing*, 26(1):84–96, 2018.

[SUG14] Jan Schlüter, Karen Ullrich, and Thomas Grill. Structural segmentation with convolutional neural networks MIREX submission. In *MIREX (Extended Abstract)*, Taipei, Taiwan, 2014.

[SV17] Paris Smaragdis and Shrikant Venkataramani. A neural network alternative to non-negative audio models. In *Proc. of IEEE ICASSP (International Conference on Acoustics, Speech, and Signal Processing)*, pages 86–90. IEEE, 2017.

[SVL14] Ilya Sutskever, Oriol Vinyals, and Quoc V Le. Sequence to sequence learning with neural networks. In *Advances in neural information processing systems*, pages 3104–3112, 2014.

[SVSS15] Tara N Sainath, Oriol Vinyals, Andrew Senior, and Haşim Sak. Convolutional, long short-term memory, fully connected deep neural networks. In *2015 IEEE International Conference on Acoustics, Speech and Signal Processing (ICASSP)*, pages 4580–4584. IEEE, 2015.

[SWS+15] Tara N Sainath, Ron J Weiss, Andrew Senior, Kevin W Wilson, and Oriol Vinyals. Learning the speech front-end with raw waveform CLDNNs. In *Sixteenth Annual Conference of the International Speech Communication Association*, 2015.

[SZ15] Karen Simonyan and Andrew Zisserman. Very deep convolutional networks for large-scale image recognition. In *Proc. of ICLR (International Conference on Learning Representations)*, 2015.

[VBL+16] Oriol Vinyals, Charles Blundell, Timothy Lillicrap, Koray Kavukcuoglu, and Daan Wierstra. Matching networks for one shot learning. 06 2016.

[vdODZ+16] Aaron van den Oord, Sander Dieleman, Heiga Zen, Karen Simonyan, Oriol Vinyals, Alex Graves, Nal Kalchbrenner, Andrew Senior, and Koray Kavukcuoglu. Wavenet: A generative model for raw audio. *arXiv preprint arXiv:1609.03499*, 2016.

[vdOVK17] Aaron van den Oord, Oriol Vinyals, and Koray Kavukcuoglu. Neural discrete representation learning. In *Proceedings of the 31st International Conference on Neural Information Processing Systems*, NIPS'17, page 6309–6318, Red Hook, NY, USA, 2017. Curran Associates Inc.

[VHM+20] Andrea Vaglio, Romain Hennequin, Manuel Moussallam, Gael Richard, and Florence d'Alché Buc. Audio-Based Detection of Explicit Content in Music. In *ICASSP 2020 - 2020 IEEE International Conference on Acoustics, Speech and Signal Processing (ICASSP)*, pages 526–530, Barcelona, France, May 2020. IEEE.

[VPE17] Tuomas Virtanen, Mark Plumbley, and Dan Ellis. *Computational Analysis of Sound Scenes and Events*. 09 2017.

[VSP+17] Ashish Vaswani, Noam Shazeer, Niki Parmar, Jakob Uszkoreit, Llion Jones, Aidan N Gomez, Łukasz Kaiser, and Illia Polosukhin. Attention is all you need. In *Advances in neural information processing systems*, pages 5998–6008, 2017.

[VTBE15] Oriol Vinyals, Alexander Toshev, Samy Bengio, and Dumitru Erhan. Show and tell: A neural image caption generator. In *Proceedings of the IEEE conference on computer vision and pattern recognition*, pages 3156–3164, 2015.

[VWB16] Andreas Veit, Michael J. Wilber, and Serge J. Belongie. Residual networks behave like ensembles of relatively shallow networks. In *NIPS*, 2016.

[Wak99] Gregory H. Wakefield. Mathematical representation of joint time-chroma distributions. In *Proc. of SPIE conference on Advanced Signal Processing Algorithms, Architecture and Implementations*, pages 637–645, Denver, Colorado, USA, 1999.

[WCNS20] M Won, S Chun, O Nieto, and X Serra. Data-driven harmonic filters for audio representation learning. In *Proc. of IEEE ICASSP (International Conference on Acoustics, Speech, and Signal Processing)*, Barcelona, Spain, May, 4–8 2020.

[WHH+90] Alexander Waibel, Toshiyuki Hanazawa, Geoffrey Hinton, Kiyohiro Shikano, and Kevin J Lang. Phoneme recognition using time-delay neural networks. In *Readings in speech recognition*, pages 393–404. Elsevier, 1990.

[WHLRS14] Felix Weninger, John R Hershey, Jonathan Le Roux, and Björn Schuller. Discriminatively trained recurrent neural networks for single-channel speech separation. In *2014 IEEE Global Conference on Signal and Information Processing (GlobalSIP)*, pages 577–581. IEEE, 2014.

[WL17] Chih-Wei Wu and Alexander Lerch. Automatic drum transcription using the student-teacher learning paradigm with unlabeled music data. In *Proc. of ISMIR (International Society for Music Information Retrieval)*, Suzhou, China, October, 23–27 2017.

[WTY20] X. Wang, S. Takaki, and J. Yamagishi. Neural source-filter waveform models for statistical parametric speech synthesis. *IEEE/ACM Transactions on Audio, Speech, and Language Processing*, 28:402–415, 2020.

[ZEH16] Zhenyao Zhu, Jesse H Engel, and Awni Hannun. Learning multiscale features directly from waveforms. *arXiv preprint arXiv:1603.09509*, 2016.

[ZGR+18] Hang Zhao, Chuang Gan, Andrew Rouditchenko, Carl Vondrick, Josh McDermott, and Antonio Torralba. The sound of pixels. In *Proceedings of the European conference on computer vision (ECCV)*, pages 570–586, 2018.

[ZPIE17] J. Zhu, T. Park, P. Isola, and A. A. Efros. Unpaired image-to-image translation using cycle-consistent adversarial networks. In *2017 IEEE International Conference on Computer Vision (ICCV)*, pages 2242–2251, 2017.

Chapter 11
Explainable AI for Medical Imaging: Knowledge Matters

Pascal Bourdon, Olfa Ben Ahmed, Thierry Urruty, Khalifa Djemal, and Christine Fernandez-Maloigne

11.1 Introduction

11.1.1 A Matter of Trust

Every decision-making process produces a final choice: the selection of one single belief or action among either just a few or many other options. Whether this final choice results fully from human cognition, fully from algorithmic computations (*artificial cognition*) or from a collaboration between both spheres is a matter of choice too. Indeed, the last decades have witnessed major advances in computer-aided decision-making methods, enabling *Artificial Intelligence* (AI) to find its way into an ever-increasing number of industries and businesses [BMKV18, BM17, Dir15, Mos86]. The benefits of AI adoption are barely questionable when it comes to supposedly low-stakes businesses such as social media advertising: it does not really matter if one has no interest in a given personalized product recommendation, people do not expect them to be 100% reliable. As long as one suggestion hits the mark from time to time and business happens, it is a win, no-loss situation for everyone. Yet there are many other fields where trust is not an option, but a necessity. Healthcare is one of them.

Machine Learning (ML) is often confused with Artificial Intelligence (AI). The aim of ML algorithms is to make machines learn from data (*training*) and solve

P. Bourdon (✉) · O. Ben Ahmed · T. Urruty · C. Fernandez-Maloigne
XLIM Laboratory, UMR CNRS 7252, University of Poitiers, Poitiers, France

Common Laboratory CNRS-Siemens I3M , Poitiers, France
e-mail: Pascal.Bourdon@univ-poitiers.fr

K. Djemal
IBISC Laboratory, University Evry Val d'Essonnes, Évry-Courcouronnes, France
e-mail: Khalifa.Djemal@univ-evry.fr

© Springer Nature Switzerland AG 2021
J. Benois-Pineau, A. Zemmari (eds.), *Multi-faceted Deep Learning*,
https://doi.org/10.1007/978-3-030-74478-6_11

problems of a practical nature, thus automating prediction tasks which partly rely on human cognition. In supervised machine learning, a computational model is built using sample data and an evaluation (*objective*) function designed to assert the quality of any computed solution (*prediction*). The term *supervised* implies the fact that every training sample not only contains the input data (*features*) for the problem to solve, but also the solution as well (*ground truth* data). Therefore, one could argue that in its most basic form, the only purpose of ML is to reproduce known solutions through mathematical rules, which is of little help for new input data with unknown solutions. This assertion is wrong of course, for the very same reasons that companies hire new employees based upon references from academic grades or previous job performances. Once again, it is all a matter of trust, a situation where one party is willing to rely on the actions of another party for the future.

In psychology, trust refers to the belief that the trustee will do what is expected. But what is expected in computer-aided problem-solving? Is it just the user-end solution, or also an acceptable cognitive and/or computational process that leads to the solution i.e. an acceptable *explanation*? Computer-aided decision-making methods have found their way into an ever-increasing number of industries and businesses because of their ability to solve problems that are ever-increasing in terms of complexity. But in order to do so, ML algorithms had to rely on mathematical models with equivalent complexity. The latest Deep Neural Networks (DNN) architectures surely pave the way for previously intractable, science-fiction domain problems such as the autonomous car [LBH15, GBC16, Sch15, DY14, STE13, HDWF+17], yet they rely on highly complex computations performed by a vast network of functions (*nodes*) and parameters (*hyperparameters*). Trust is a heuristic decision rule, allowing the human to deal with complexities that would require tremendous efforts in rational reasoning. Besides, psychologists have suggested that the primary purpose of consciousness and rational thinking was to justify irrational, instinctive drives. Humans have put their trust in other humans for thousands of years, unknowingly relying on unconscious strategies such as facial resemblance or body language. Accordingly, the chances for a new-coming, non-human, faceless, body language-less party such as an AI agent to gain the trust and confidence of humans are very little, especially:

- if the human party (*trustor*) had negative experiences with completely out-of-line computer-generated recommendations in social media websites;
- if the human party cannot relate to a party with supernatural powers such as the ability to deal with very complex computations that have become unexplainable by design;
- if the human party has a tendency to think in counterfactual terms, being very aware of popular AI failures and scientific communications about *deepfooling*, *one-pixel attacks* or *adversarial examples*;
- if the human party, like all humans, is prone to cognitive biases, such as having a tendency to discard any information that does not support his/her prior beliefs or ethics (*confirmation bias*);

- if the decision to be taken involves possibly dangerous, life-threatening tasks such as car driving or medical diagnosis.

11.1.2 The Emergence of XAI

With success comes trouble. Public concern over personal data usage and AI-based decision support systems has become a hot political topic regularly featured in the media [FH17] and massively investigated in social science research [DVKB+17, WS18, Ren19, Smu19, SA19, O'n16]. Fears of an unfair, unsafe, Huxley-like brave new algorithmic society have prompted many governments to take action. This is notably illustrated by recent official communications such as the United States' White House Office of Science and Technology Policy report *Preparing for the Future of Artificial Intelligence* [Fel16], which states that AI systems should be open, transparent and understandable. Another example is the European Commission's *Artificial Intelligence for Europe* and *Ethics guidelines for trustworthy AI* reports [Eur18, DT18, Hig19], which identifies the need for humans to understand the actions of AI systems. Even world-renown AI experts went public about their concerns of using algorithmic predictions for high-stakes decisions. One famous charge came from Ali Rahimi at the Neural Information Processing Systems (NIPS) 2017 conference, when he described the current state of Deep Learning as *alchemy* [RR17, Rah17]:

> Machine learning has become alchemy. Now alchemy is okay (. . .) Alchemy "worked" (. . .) But we're beyond that now. We're building systems that govern healthcare and mediate our civic dialogue.

In the early days of the First Industrial Revolution, factory workers opposed to the introduction of machinery engaged in machine-wrecking as a remedy to fight the inevitable new system [Hob52, N+02]. One could argue that the suggestive terminology behind the concepts of AI transparency is just a peaceful facade for machine-wrecking in the Information Age (e.g. *opening the black box*). But there's more to it. DNN systems were never designed to be black boxes with zillions of hyperparameters, something just happened: for very complex problems to solve, human-induced hypotheses became too restrictive and hypothesis-driven prediction models started to perform poorly. Data-driven models, on the other hand, started to perform very well once given enough training data, computational power, and pragmatic guidance rules (e.g. cross-validation). As pointed out by O'Neil [O'n16], *mathematical models should be our tools, not our masters*. While data-driven models are not initially designed for explanation, they are designed to solve multi-objective tasks through the use of constraints. Therefore they can be updated for the additional task of explanation [Rud19]. Moreover, human hypotheses are drawn from observation: can data-driven prediction models be *observed*?

All parties debating the social acceptance of AI or ML stress the importance of *Explainable AI* as a means to monitor algorithmic decisions [SSWR18]. Explainable

AI, or XAI, is a term coined by Van Lent et al. to describe their system's ability to explain AI-based predictions [VLFM04]. In 2020, XAI may already be considered an umbrella term to suggest many different things. Despite the need for shedding light on black-box decision systems is clear for anyone, it remains either a vague or a heavily multifaceted concept that many authors have already tried to assess under various angles [AB18, Hon18, Lip18, LCG12, LL17, Mil19, Mol20, MSM18, NYC16, RSG16, Gun17, SWM17]. For data analysts who put their interest solely in hypothesis-driven models and discard machine learning algorithms as black-box systems, such a concept may be difficult to cope with, as their own conception of interpretation reduces to hypothesis-checking. Some researchers with too much focus on DNN, on the other hand, will describe interpretability only in terms of mathematical rules or algorithm transparency, assuming actions such as the divulgation of source code under open source public licenses is enough to solve the issue of social acceptance. Of course, such assumptions do not help public debate or the need for citizens to understand AI systems. The definition of explanation is discussed in details in writings such as [CPC19], which explore the many axioms, properties, and most of all the subjectivity related to its very concept.

11.1.3   The Case of Medical Imaging

Medical imaging is a very interesting case when it comes to discussing the outcomes of computer-aided decision systems versus decision-taking abilities and skills inherited from sources such as strictly human-to-human interaction or genetic memory. Indeed, medical imaging as we know it today is de facto advanced technology: there is no ancestral training in microscopic, Magnetic Resonance (MR) or ultrasound image analysis. Most images used for healthcare support are not natural images; they are intelligible visual representations generated from physical or chemical properties of organic tissues in the hope to produce intelligible information. Moreover, the current trend of using an exponentially increasing number of visual data sources for diagnosis poses great challenges, turning the need for AI decision assistance tools into something inevitable [MD13, SR13, SZ16, CHH+17, LY17, EKAK17, WYB+10, RWD+16, Wan16].

Extracting or predicting useful information from big medical data sets is a very tedious task, even for high-level specialists. So far we have assessed the need for XAI mostly under social acceptance terms, considering the end-user as a party with limited expertise. When it comes to medical decisions, there are many ways in which a physician or physician-scientist can benefit from AI-assisted data analytics without having to endorse everything that comes with the Digital Age or being accused of using opaque systems. In this particular case, we will consider the medical expert and not the patient as the end-user. Our stand is that deep learning algorithms are not programmed to ask questions; they are programmed to provide answers once fed with a set of informative clues (*features*) data analysts expect to be exhaustive enough for the problem to solve. Therefore there are *our tools, not our masters*. French philosopher René Descartes' view of rational thinking was that of a computing machine progressing through discrete, logical steps in order to recognize patterns. Given proper scientific treatment, we consider interpretable AI as a means to keep recognizing patterns within complex data sets such as medical images in a similar fashion, to confirm or discover new causal relationships, and generate new hypotheses about the real world before testing them experimentally.

For the rest of this chapter, we will focus on the positive outcomes of deep learning solutions for medical image analysis, using examples that demonstrate how collaboration between medical and data science experts is able to achieve great performance with little opacity, many of which are or could be categorized under the banner of XAI. In other words, we will illustrate how, when it comes to AI for medical imaging, *knowledge matters*.

In the next section, we explore the integration of deep learning within the healthcare science and medical imaging domains, providing examples of how DNN-driven prediction tasks can be monitored either by design through human intelligence-inspired concepts, or in a post-hoc fashion using visualization and robustness improvement strategies. A detailed example of AI-powered analysis and interpretation is given in Sect. 11.3, where a DNN was trained on Magnetic Resonance Spectroscopy data to investigate brain-metabolite changes in Alzheimer's disease (AD). Finally, Sect. 11.4 puts focus on current and upcoming DNN-based breast cancer diagnosis techniques.

11.2 The Augmented Pathologist

Augmented pathologist is a term coined by the authors of [HMK+17] to define the perfect balance between human and artificial intelligence for diagnostic medicine. Deep learning algorithms have many advantages compared to humans, such as the ability to learn and keep memory of the thousands of examples they are fed with. Humans cannot, at least not explicitly, yet they are able to infer complex patterns from just a few selected examples, which is the basic definition of experience or expertise. In that prospect, the opportunity to leverage such capacities with computer-aided, augmented memory is a very tempting one. In this section we provide a selection of some of the most promising perspectives for deep learning adoption in medical imaging.

11.2.1 Explainable Human Intelligence?

Understanding human intelligence, brain mechanisms and decision-making processes has always been a fascinating and challenging problem to scientists and physicians. It is naturally a major key to the conception of ML systems, especially by today's standards when bio-inspired deep artificial neural networks have set new records in accuracy for many important problems in pattern recognition and computer vision. Until very recently, the relationship between artificial intelligence and neuroscience research could be defined as one-way communication, with the latter acting as a source of inspiration for the former. But lately the success of deep learning has exposed new paradigms for neuroscience.

The authors of [GvG15] demonstrate an interesting take on deep learning research, by using a DNN to probe human brain activity. In their study, volunteer human subjects are presented with natural images while their neural activity is recorded as Functional Magnetic Resonance Imaging (fMRI) scans using a 4 T scanner. All scans (anatomical and functional volumes) are co-registered together and assigned to visual areas using retinotopic mapping data. For each presented image, a per-voxel fMRI response amplitude is estimated and fitted with predictions computed from the image's feature representations. While the predictive model is linear, the incoming feature representations are multi-layer, nonlinear responses extracted from a pretrained Convolutional Neural Network (CNN) architecture inspired by AlexNet [KSH12, CSVZ14]. Model performance is carefully assessed through correlation and Signal-to-Noise Ratio (SNR) analysis as well as model refitting and comparison with baseline models.

By examining which DNN layer and/or individual feature is the most predictive for each given voxel in the visual cortex, the authors are able to draw many conclusions about the natural encoding and identification process of the human brain. They demonstrate that:

- not only low-level visual features (i.e. simple patterns in the Gestalt sense such as blobs or edges), but also mid- and high-level ones are important for identification;
- image decoding in the human brain is mostly driven by discriminative information (i.e. the use of unique characteristics), with categorical information (i.e. semantic content) providing additional clues only for upstream areas (layers) of the visual ventral stream;
- the visual ventral stream is hierarchically organized, with downstream areas processing increasingly complex features, and with a degree of overlap between successive internal representations;
- the selectivity of individual voxels (nodes) is a distributed over many individual features and vice-versa.

While such conclusions have been suggested and theorized before, using an artificial network for their analysis helps providing quantitative evidence in *virtual probe* fashion. This is of great help for DNN architecture design, as it exposes details about strategies that have proven their effectiveness (adaptive kernel size, nonlinear activation, dropout regularization, fully-connected layers and skip connections to name just a few) as well as hints regarding new strategies to consider. Except for [FSR+19], neuroscience or computational neuroscience is rarely debated in XAI literature, yet this work and similar studies such as the ones found in [CKP+16, EGVT17, TNM+19] could give more inspiration regarding un-black-boxing efforts. As an example, the distributed selectivity of neurons mentioned earlier is also discussed in details in [SZS+13], where the authors draw similar conclusions without any reference whatsoever to the mechanics of human vision.

11.2.2 Data and Model Visualization

At odds with the relatively low coverage of AI-inspired neuroscience, data and model visualization appears to have a significant value for model behaviour interpretability and explainability in XAI literature. For medical imaging, data visualization is a means to make high-dimensional spaces accessible: multimodal anatomic, functional and metabolic MRI sequences, MR spectra, Computed Tomography (CT) or Positron Emission Tomography (PET) scans, etc. Computer image analysis integrated medical imaging before there was deep learning, helping medical experts with low-level vision tasks such as automatic contrast/brightness enhancement, denoising, segmentation, or partial object detection. Algorithmic tools were used to leverage productivity for human-made decisions by revealing visual information that was already accessible to the decision-maker, just hidden by acquisition artifacts (blur, noise, underexposure...) or masking effects. By relocating focus onto decision rules and not just accuracy, XAI research opens new perspectives for *augmented* visualization, where a new, semantics-based visual information could be generated and fused with traditional image sources.

It has become a common approach to render graphical representations in the hope of discovering semantics behind the implicit decision rules of a DNN models. An illustration of visual CNN analysis is provided on Fig. 11.1, where the CNNVis software tool [LSL+16] is used to show (and hopefully evaluate) the interactions between input samples and neurons on recognition tasks. In [TG19], the authors introduce the global concept of *perceptive interpretability*. Visualization techniques can be described as such, and be divided into two sub-categories defined within the article as *saliency* and *signal methods*.

Saliency focuses on the influence of a given input over model predictions. It consists in generating visual feedback under the form of heatmaps which highlight the importance of each individual pixel or superpixel component regarding decision boundaries. Many saliency models were designed for DNN explanation, such as deep Taylor decomposition [MLB+17], a successor to Layer-Wise Relevance Propagation (LRP) [BBM+15, LWB+19], guided back-propagation [SDBR14], Class Activation Maps (CAM) [ZKL+16], Grad-CAM [SCD+17] or Concept Activation Vectors (CAV) [KWG+17]. An example in medical imaging using CAM will be presented and discussed in details later in this chapter. Another example is provided by [TCD+19], which introduces a CNN-based classification model for Alzheimer's disease pathologies. While their prediction model achieves a high classification accuracy, the authors also investigate its inner decision rules through saliency using guided Grad-CAM. The results obtained show consistency with human expertise, as the artificial networks learn patterns agreeing with accepted pathologic features. Showing evidence that data-driven predictions are able to learn

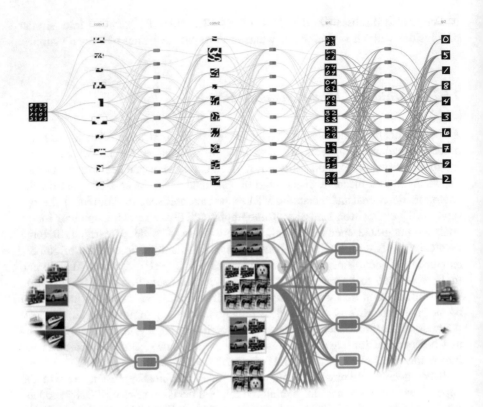

Fig. 11.1 An example of machine learning performance assessment under *why* terms: exploring a convolutional neural network with CNNVis [LSL+16]

previously established hypotheses without prior knowledge is a golden path to establish trust. As discussed earlier, the algorithmic tools developed for interpretable AI are an opportunity for medical research to generate new hypotheses about the real world before testing them experimentally.

Signal methods is a term originally used in [KHA+19] for this context, although many prior network dissection or investigation techniques fall into this category. Such methods do not strictly focus on input samples from the training data set. They consist in observing what causes the stimulation of individual or collective neuronal activities. This can be achieved by generating 2D or 3D reconstructions showing the influence of one convolution filter when applied to a full layer (*feature maps*), or by going as far as addressing the inverse problem of image reconstruction from a given set of output activation values, at least within a limited number of layers [ZF14]. Another take on the inverse problem of image reconstruction over the full network from upstream layer activations is *activation maximization*, an optimization process where noise is iteratively adjusted in order to reach a chosen activation configuration for a set of neurons [EBCV09]. To our knowledge, while artificial recognition models trained and tested on natural images exhibit intriguing properties such as showing neural units dedicated to white flower or round green object discrimination [SZS+13] when investigated with signal methods, the outcomes of similar investigations on DNN models dedicated to medical imaging are still very limited. This may be due to the structural complexity of medical image patterns, that makes them difficult to describe in explicit, semantic terms for medical experts (even less for ML experts). An alternative hypothesis could be that despite the unique character of such patterns, expertise on medical image analysis still partly relies on prior visual recognition knowledge learned from natural images. This assumption could correlate with the findings in [TSG+16], which demonstrated that CNN models pre-trained on natural images and adequate fine-tuned for medical image analysis outperformed CNN models trained from scratch for the same task.

Choosing one method to explain model predictions over the other can be motivated by many factors; sometimes even different methods within a same category or sub-category may complement one another in the same way multiple data sources may be needed for medical evaluation. We refer to Fig. 11.2 for an illustration. In this example we trained a binary classification CNN based on the ResNet-50 architecture [HZRS16] with multi-sequence MR brain images as

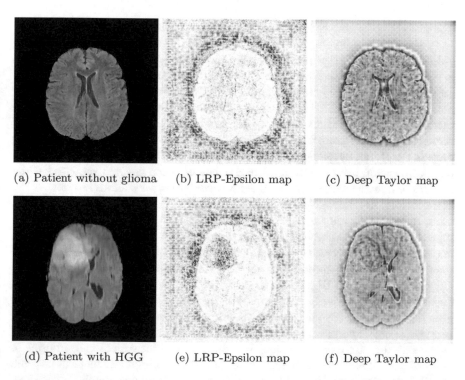

(a) Patient without glioma   (b) LRP-Epsilon map   (c) Deep Taylor map

(d) Patient with HGG   (e) LRP-Epsilon map   (f) Deep Taylor map

Fig. 11.2 Exploring model predictions with saliency maps on MR brain images. Upper images: a patient without glioma. Lower images: a patient with HGG. From left to right: false color RGB recompositions from T2 FLAIR, T2 and T1ce sequences, LRP-Epsilon maps, Deep Taylor maps

inputs. Given an input sample, the model has to predict whether its corresponding patient is considered to have a High-Grade Glioma (HGG) or not. Gliomas are characterized by a number of radiographic features accepted by clinicians, such as FLAIR or T2-weighted hyperintensities or significant mass effect. Alas they are usually just hints. As an example, mass effect has demonstrated prognostic significance for glioblastoma, yet it is poorly quantified [STB+18]. When using LRP-Epsilon [BBM+15] and Deep Taylor [MLB+17] saliency maps to expose prediction rules, one can notice that while it was trained for full-image classification and not segmentation, the model is able to localize white matter hyperintensities on its own (LRP-Epsilon map on Fig. 11.2e). Moreover, it seems to be partially isolating mass effect artifacts (Deep Taylor map on Fig. 11.2f). It is our assumption that while further work is needed to get there, such post-hoc prediction maps could provide the much-needed radiographic feature quantification system mentioned above.

11.2.3 Safety and Robustness Improvement

The discussion in this chapter started with the question of trust as a mandatory condition for AI adoption in diagnostic medicine. In order to endorse machine-aided decisions, the augmented pathologist must first ensure that a prediction model is *right*, in the sense that it achieves good performance in terms of accuracy and generalization. Secondly the model must not just be right but also *right for the right reasons*, to quote [RHDV17], which is usually the most intended meaning in XAI literature as of today. Hints as how to assess whether a model is *right for the right reasons* or not were given in the previous Sect. 11.2.2. But knowing what to do once there is proof that despite good performance in terms of accuracy, a model relies on poor decision rules or unwanted biases, has not been covered yet. Nor has the concern of how to ensure that the model is *never wrong for obvious reasons*.

According to many, given the high predictive power of current DNN architectures, being both *right for the right reasons* or *never wrong for obvious reasons* might be a matter of enforcing *generalization*, i.e. the power to keep good prediction performance for observations outside or beyond the given training data set, with more sample data or heuristic data augmentation schemes. Computer scientist and Google Inc. Research Director Peter Norvig once defended the power of data by claiming "we don't have better algorithms. We just have more data" [HNP09]. But even the most exhaustive datasets and performance metrics need additional, global knowledge and expertise. In [RHDV17], the authors took the works in [CLG+15] as a basis for discussion: a model is trained to predict patient care priority in cases of pneumonia. While patients with asthma are at a greater medical risk, the model defines them as being at a lower risk of dying from pneumonia. A post-hoc, human-given explanation points out that *because* patients with asthma history are automatically given more aggressive treatment, they are indeed at a lower medical risk compared to other patients. This example reflects the need for domain-tuned training capacity, where a model's explanations can be constrained to match domain knowledge. In [RHDV17] this is achieved by regularization through penalty. A common trend is to use simpler models like linear regression or decision trees as surrogate models, sometimes not just for the sake of interpretability but also to induce robustness and ease domain knowledge integration.

As mentioned earlier, humans have a tendency to think in counterfactual terms. Trust involves feelings of safety and security which can not be achieved if a prediction model tends to make obvious mistakes. In Sect. 11.2.2 we described *signal methods* as a category of artificial network exploration techniques which consist in observing what causes neuronal activities, using alterations or full image reconstructions. *Activation maximization* is a way to depict what kind of input singularity favours a prediction over the other. Another angle could be taken in measuring what is needed to favour any prediction over the right one. Methods such as *DeepFool* [MDFF16] or *one-pixel attacks* [SVS19] show how a great number of well-known DNN architectures are unstable to small perturbations on input images. Deep adversarial attack and defense techniques have attracted increasing attention and raised huge concerns about the deployment of DNN systems in clinical settings [FBI+19, MNG+20]. A recent overview is provided in [YHZL19], where the authors highlight the question of knowing if adversarial examples an inherent property of DNN, which brings us back to the question balance between human and artificial intelligence for diagnostic medicine.

11.3 Investigating Alzheimer's Disease with CAM

The first question that healthcare practitioners would ask before using an AI-based software is: "can we trust decisions made by AI models?". Indeed, trained models interpretability is crucial for trusting AI. As discussed earlier, there is a wide variety of techniques to help clinicians better understand the decision rules of AI algorithms. At the end, information provided by those techniques should be in concordance with domain knowledge and findings from clinical studies. Class Activation Map (CAM) [ZKL+16] is a standard model interpreting technique that allows highlighting discriminative regions used by a model to identify a category. CAM have been widely used to explain deep models decision in the medical domain [FYLA18, KAH19, KRA19, KAH19]. Recently, CAM have been investigated for Alzheimer's Disease (AD) diagnosis [YRR18]. Roughly, CAM can be interpreted as "where the deep CNN is looking at" or, more precisely, which areas in spatially organized data are the most important for distinguishing between AD and Normal Control (NC) subjects. Several CAM-based approaches for AD diagnosis have been proposed for different medical imaging modalities such as PET [YPBI20, YPBtADNI20] and structural MRI [FYL+18, LLWS19]. In [YPBtADNI20], the authors propose a 3D CNN with residual connections for AD detection. The proposed CNN generates Class Activation Maps for FDG-PET images interpretation. Figure 11.3 presents the activation maps of AD subjects

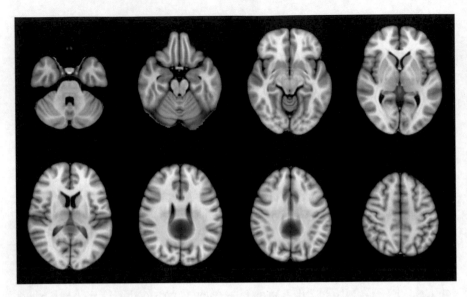

Fig. 11.3 CAMs highlighting the Posterior Cingulate Cortex (PCC) as the region most important for predicting AD [YPBtADNI20]

predicted by the proposed 3D deep CNN [YPBtADNI20] trained on PET-FDG data from the Alzheimer's Disease Neuroimaging Initiative (ADNI)[1] public datasets. From Fig. 11.3, we can see that the generated CAMs clearly highlight the regions of Posterior Cingulate Cortex (PCC), which is in accordance with previous clinical studies that have validated the discriminative capacity of these brain regions for AD diagnosis.

Another promising technique for brain metabolic study in the case of AD is the Magnetic Resonance Spectroscopy (MRS) [FGSO+05]. MRS provides a "virtual biopsy" by detecting small metabolic changes in the brain in a safe and noninvasive manner. Recently 1D CNN has been used for MRS metabolites quantification [DCSM17] and for ghost removal [KDK18]. In the case of AD, this metabolic information correspond to the neuro-degenerative pathologies that affect brain tissues. While patients at risk for AD cannot be always detected with structural MRI or FDG-PET images [WTW+15], MRS-based detection can be successful even when clinical syndromes such as cognitive impairment remain unnoticeable. To the best of our knowledge, few works have investigated the use of CAM for AD analysis based on MRS data.

Recently in [AFG+20] we have proposed the DeepMRS model for Alzheimer disease detection using MRS. Data of 135 subjects, collected in the Poitiers University Hospital, are used to learn the proposed DeepMRS network. The classification of patients with early AD versus NC subjects achieves an Area-Under-the-Curve

---

[1]http://adni.loni.usc.edu.

(AUC) score of 94, 74%, a sensitivity of 100% and a specificity of 89, 47% demonstrating a promising early dementia detection performance compared to the use of structural MRI. The CAMs generated from DeepMRS' hyperparameters provide key insights into the regions of interest found the most significant for information extraction and prediction making by the model. In this particular case, the activation maps highlight which metabolites in the signal were found relevant for the detection of a given class between AD and NC.

Given an output class c, its corresponding CAM is computed as a linear combination of the feature maps from the last convolutional layer, weighted by the class weights learned by the classifier. The CAM computation for class c is written as follows:

$$CAM_{MRS} = \sum_{k=1}^{K} Conv_k.w_{k,c}, \tag{11.1}$$

where $Conv_k$ is the kth feature map of the convolutional layer before the global average pooling layer, and where $w_{k,c}$ is the weight associated with the kth feature map and class c.

Figure 11.4 shows the MRS of an arbitrary chosen patient and the relevance of every part of the signal (in blue). It is visualized with the blue line over a spectrum of a random patient for a given class (Alzheimer's Disease) highlighting the metabolites contributing the most to the DeepMRS model decision. We can see that we have very low (≤ 0) relevance values in some regions that are known to be irrelevant for diagnosis. The most important features that have contributed

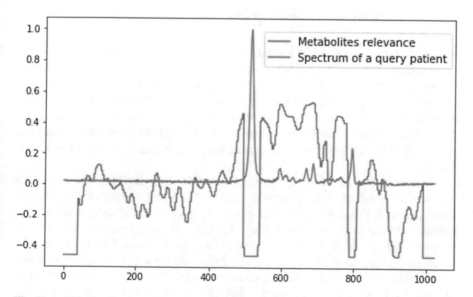

Fig. 11.4 CAM for discriminating metabolite areas [AFG+20]

to the discrimination between the AD and NC groups are in the range between 550 and 800 ppm. Based on the domain knowledge, these regions correspond to metabolites such as Choline-containing compounds (Cho), N-acetyl-aspartate (NAA) and Creatine (Cr). The fact that our model relies on those three metabolites to discriminate between the AD and NC groups is in accordance with the state of the art on clinical criteria [GB14, SZD+12]. The obtained results demonstrate that the model is automatically learning these metabolites for AD detection from MRS raw data without the need to a quantification or data preprocessing steps. We can also see that the model considers signal noise artifacts such as the peak of water signal (high magnitude peak around 500 ppm) as irrelevant features for the discrimination, putting by assigning negative values over the corresponding *ppm* range.

11.4 Breast Cancer Identification Using Deep Learning Approaches

This section presents the method of breast cancer identification and most machine learning approaches. To this end, we discuss the breast cancer descriptors and specific deep learning approaches used in recent years. Indeed, despite the promising performance of deep learning, advanced state of the art analysis identifies several important challenges facing deep learning, such as the size of learning databases in this domain.

11.4.1 Breast Cancer Description

In breast cancer, there are two main types of lesions; on the one hand there are microcalcifications and on the other hand there are masses. In this work, the breast masses are concerned by the description. By definition, a mass is an important opacity occupying a space and seen on two different incidences. In the case where the mass is observed on only one incidence, it is called asymmetry until its three-dimensional nature is confirmed. Different features allow specialists to describe masses in order to determine their nature, these are generally shape, contour and density.

To detect and classify benign and malignant tumors, [BLB90] first segmented the masses contained in the images using a multi-resolution approach based on a fuzzy pyramid. From the segmented masses, they extracted several descriptors of shapes and textures. These descriptors are then used in a four-level hierarchical classification model to categorize breast mass. They resulted in an 85% good classification rate on a 25-images dataset. In the same way, [REFDA97] proposed moment-based shape descriptors to characterize the transition between region of interest (ROI) and mass based on the density of each pixel of the contour of the latter. The results obtained for a benign/malignant classification show a satisfactory performance of 94.9%. On their side, [GRCS07] have developed a method for describing the mass based on polygonal models of the mass contour that preserve the spicules as well as important diagnostic details. Once the modeling is done, they proceed to the extraction of several spiculation descriptors. The evaluation of the relevance of their technique on a dataset composed of malignant and benign masses gives a score of 94%.

In their work, [NR12] preferred to use the Krawtchouk moments extracted on the contours of benign and malignant masses, which they then used as input for a kNN classifier. Then, a comparative study was conducted with the Zernike moments in order to evaluate the performance of the Krawtchouk moments. They then noticed that the Krawtchouk moments were better than the Zernike moments with a rate of 93% against 85%.

[KDM12] proposed three morphology descriptors, namely skeletal end points (SEP), protuberance selection (PS), and spiculate mass descriptor (SMD). After extraction on a database of benign and malignant images from DDSM, they obtained scores of 92%, 93% and 97% respectively for SEP, SP and SMD. They also noted that these performances were superior to those displayed by standard descriptors such as compactness. We compare the results of these last three descriptors with those obtained by our architecture based on Convolutional Neural Networks (AlexNet).

11.4.2 Deep Learning and Breast Cancer Databases

Convolutional Neural Networks (CNNs) are one of the most remarkable approaches to deep learning, in which multiple layers of neurons are formed in a robust manner. They have shown that they are able to demonstrate high generalizability over large datasets of millions of images [KSH12],[SZ14], [RDS+15]. These results come mainly from the particular architecture of the CNNs, which takes into account the specific topology of tasks related to the field of computer vision that exploit two-dimensional images. Other dimensions can also be taken into account when it comes to multi-channel color images. The CNNs make use of the very strong correlation within a local two-dimensional structure by restricting the receiving field of the hidden units to focus on local variations [LBH15]. As a result, a local connectivity scheme is learned on the first hidden layers to describe simple structures such as edges or corners. Stacking convolution layers, i.e. increasing the depth, forces the network to learn more abstract, discriminating and high-level representations, combining within the deeper layers the local descriptors learned in the first layers.

The CNN must learn to distinguish a particular object from other objects in the image. Within the first layer, the convolution kernels detect simple shapes such as borders and this according to several rotations. These convolution kernels are very similar to the Gabor filter that is commonly used in computer vision. The deeper the hidden layer, the more complex and abstract the learned descriptors tend to be. In the second hidden layer, convolution kernels are the result of convolving the previous ones to form much more complex patterns that correspond to specific parts of the object of interest, such as eyes, nose or mouth in face recognition. Finally, the last hidden layer shows several abstract representations of the object. The CNN has the ability to learn high-level descriptors in an unsupervised way and with much better performance than what has been achieved so far in the state of the art. This is proof that the use of CNNs can help us to improve the description and classification of breast cancer.

11.4.2.1 Deep Learning Architecture

Due to the small size of the data sets that exist in mammography, we felt that it would be more appropriate for deep learning to use a pre-trained CNN that will allow unsupervised extraction of descriptors across the training database. There are several models of CNN, including the AlexNet [KSH12]. This choice is justified by the fact that it is the first modern model of CNN which is at the origin of the renewed interest of the scientific community for CNNs [RDS+15].

The composition and functioning of this model can be summarized as follows: first of all, and in order to reduce overlearning, the authors proceeded to an image pre-processing which consists in a translation and a horizontal reflection of the image, by random extraction of patches of size 224×224 on the starting images; these patches then served as a learning database. Then, in order to ensure invariance to illumination and color, they altered the intensities of the RGB color channels by adding noise to the main components of the images in the learning database. As far as layers are concerned, the authors used eight layers, including five convolution layers and three FC layers. The first and second convolution layers are followed by the normalization and max-pooling layers, while the last three convolution layers are linked to each other without any normalization and pooling intervention. In addition, there are some tricks that the authors used that allowed them to design a CNN model that is more efficient than previous ones. First, the activation function used is a ReLU, which contributes to significantly reduce the training time of the CNN compared to $tanh(x)$. Second, they have adopted local standardization, which facilitates generalization. Finally, they used an overlapping subsampling technique, which improves the invariance of the descriptor matrices to the various transformations. Traditionally, the subsampled neighboring neurons did not overlap; by adopting overlapping subsampling, they were able to reduce the learning error rate. Following the convolution layers, there are two FC layers called FC1 and FC2 with 4096 neurons. The last output layer, which is connected to the FC2 layer, contains 1000 neurons which is a prediction for each of the 1000 classes in the ImageNet database. In all, there are just over 60 million hyper-parameters on which the performance of the AlexNet depends.

For the application that concerns us, i.e. the discrimination between benign and malignant masses, we propose a three-tiered approach to decision making that allows radiologists to confirm or invalidate the diagnosis about the nature of the mass. In a first step, the AlexNet is applied on the basis of the ROIs, in order to extract in an unsupervised way the high-level descriptors that can be used to make a decision.

Then, the AlexNet is run on the images obtained after segmentation. Here, the role of the AlexNet is to extract different shape and contour descriptors related to the different masses contained in our learning database.

The architecture of a CNN is composed of several layers including the FCs which contain several neurons to provide the extracted descriptors. In our case, we used the second layer FC which we will call FC7, composed of 4096 neurons as a descriptor layer. Subsequently, these descriptors were used by a SVM classifier to build a learning model. This model was then used to classify and identify the masses as benign or malignant for decision support (Fig. 11.5).

Finally, as far as spicules are concerned, we propose to let the radiologists judge the relevance of taking them into account in their decision. Indeed, the random nature of the shapes, density of masses and types (length and number) of spicules in the case of breast cancer makes it unrealistic to propose any rule that would be generalizable. Nevertheless, we are certain through BI-RADS that spicules allow to classify almost perfectly benign masses from malignant ones or architectural distortions, because no benign mass could produce spicules.

In addition, the quantity and length of spicules could help to discriminate between spiculated masses and architectural distortions. However, only the expertise and experience of radiologists could help to achieve a high degree of identification. The help we propose through segmentation is to draw their attention to the real structures of interest by eliminating those that could cause confusion. Figure 11.6 shows the two types of information taken into account in decision support.

Once the descriptors are extracted, they are used in a SVM classifier to obtain a learning model that will help make a decision. This model is the first step in our breast cancer identification and recognition system, shown in Fig. 11.5.

11.4.2.2 Breast Cancer Database

In order to set up the breast mass classifier, we used a data set, the same as the one used by [KDM12] in order to carry out an objective comparison which we divided into three different partitions which are: DB_{train}, DB_{valid}, DB_{test}. DB_{train} represents the image database on which the model is trained in order to adjust its parameters. DB_{valid} is the database on which the selection of the hyper-parameters

Fig. 11.5 The different stages of the breast mass identification system

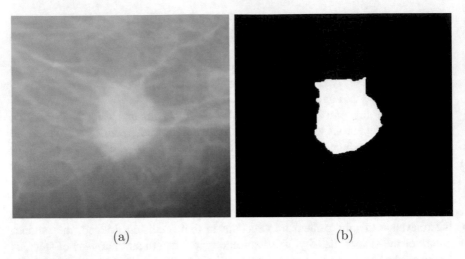

(a)   (b)

Fig. 11.6 Illustrations of the images involved in the decision support process: (**a**) ROI, (**b**) binary image obtained after segmentation

Table 11.1 Composition of the databases

	DDSM 248 images	
Databases	Benin	Malignes
DB$_{train}$	57	57
DB$_{valid}$	56	58
DB$_{test}$	14	6

and the evaluation of the performance of the model are made. As for DB$_{test}$, it represents the dataset used to assess overall generalization performance.

For this purpose, images from the DDSM containing only the masses are selected to be used in the experimental part. The selected database is shown in Table 11.1. Given the architecture of our CNN (five convolution layers and three FC layers) and considering the number of parameters (more than 60 million) to be learned, we used an NVIDIA GEFORCE GTX 1070 graphics card. The MatConvNet and/or Caffe libraries were used for the computing. Finally, we scan 100 times each of the images in our database per iteration to extract both DB$_{train}$ and DB$_{valid}$ descriptors.

11.4.3 Identification Results

In order to verify the relevance of the model learned on our dataset for the quality of the classification of benign or malignant masses, we decided to make a comparative study with the most efficient methods of the state of the art. The features used for the comparison are those proposed by [KDM12], namely SEP, PS and SMD. This choice is justified on the one hand by the fact that the authors have shown that the performance of these shape and contour descriptors surpasses both standard

Table 11.2 Performances obtained by the different descriptors

Descriptors	ROC (%)
CNN	99
SMD	97
PS	93
SEP	92

and specific descriptors. On the other hand, because we have exactly the same data set, i.e. the same regions and images as the one used by [KDM12] to evaluate the performance of its descriptors. In order to rigorously compare the performance of the descriptors extracted by the AlexNet with the SEP, the PS and the SMD, the evaluation will be made in the following using ROC curves.

In order to compare our results objectively to these different contributions, we used the ROC curve as it is the metric used in the cited works. Both the evaluation of the average performance in terms of classification of our recognition system through DB_{valid}, and the evaluation of the generalization process (DB_{test}). For each of the images contained in either DB_{valid} or DB_{test}, AlexNet extracts a set of 4096 descriptors using FC7. Once the descriptors are obtained, the learning model built from the DB_{train} database is used to predict the nature of the mass, whether it is benign or malignant. Finally, the average precision of the system is calculated using the results obtained by the ROC curves. Table 11.2, presents the accuracy obtained by the different descriptors. The best accuracy is obtained by the architecture using AlexNet.

11.5   Conclusion

Deep learning-aided medical image analysis poses as many challenges as promises for better pathology diagnosis. We started this chapter with a strong focus on trust as a mandatory rule for AI adoption in healthcare. Yet AI adoption does not necessarily mean replacing human decisions with machine decisions. Data-driven prediction models can be used to infer details about complex texture patterns typically found in microscopic, MRI, PET or ultrasound images, generate new hypotheses and maybe turn them into clinical, human-detectable symptoms through experimental testing. Improving trust requires quantifiable measures, some of which are yet to be defined for diagnostic medicine. Obviously, finding the best balance between human and artificial intelligence is not about what *can* to be automated, but what *should* be automated. As mentioned earlier, deep learning algorithms have many advantages compared to humans, and humans have many advantages compared to deep learning algorithms. Throughout these pages we have presented works in neuroscience and interpretable AI-related literature that we hope can show where the lines between human and machine expertise can be drawn when it comes to healthcare. We are convinced that because of their complex nature and increasing numbers, medical images need new paradigms for bridging the gap between standard analysis based on tissue visualization and ML-based data representations such as the ones presented in Sect. 11.2.2. A concrete example was given in Sect. 11.3, where the visualization of MRS data was augmented with CAM curves to better understand the effects of Alzheimer's Disease on metabolites. We also put focus on machine learning approaches developed for breast cancer diagnosis, and discussed the advances and emerging challenges in this particular domain. Both studies show that although deep learning has shown strong performance in a multitude of ML challenges, its direct use for practical healthcare is still an open issue with many unmet needs that warrant further investigations.

References

[AB18]   Amina Adadi and Mohammed Berrada. Peeking inside the black-box: A survey on explainable artificial intelligence (XAI). *IEEE Access*, 6:52138–52160, 2018.

[AFG+20]   O. B. Ahmed, S. Fezzani, C. Guillevin, L. Fezai, M. Naudin, B. Gianelli, and C. Fernandez-Maloigne. Deepmrs: An end-to-end deep neural network for dementia disease detection using mrs data. In *2020 IEEE 17th International Symposium on Biomedical Imaging (ISBI)*, pages 1459–1463, 2020.

[BBM+15]   Sebastian Bach, Alexander Binder, Grégoire Montavon, Frederick Klauschen, Klaus-Robert Müller, and Wojciech Samek. On pixel-wise explanations for non-linear classifier decisions by layer-wise relevance propagation. *PloS one*, 10(7), 2015.

[BLB90]   Dragana Brzakovic, Xiao Mei Luo, and P Brzakovic. An approach to automated detection of tumors in mammograms. *IEEE Transactions on Medical Imaging*, 9(3):233–241, 1990.

[BM17]   Erik Brynjolfsson and ANDREW Mcafee. The business of artificial intelligence. *Harvard Business Review*, pages 1–20, 2017.

[BMKV18]   Charlynne Bolton, Veronika Machová, Maria Kovacova, and Katarina Valaskova. The power of human–machine collaboration: Artificial intelligence, business automation, and the smart economy. *Economics, Management, and Financial Markets*, 13(4):51–56, 2018.

[CHH+17]   Min Chen, Yixue Hao, Kai Hwang, Lu Wang, and Lin Wang. Disease prediction by machine learning over big data from healthcare communities. *Ieee Access*, 5:8869–8879, 2017.

[CKP+16]   Radoslaw Martin Cichy, Aditya Khosla, Dimitrios Pantazis, Antonio Torralba, and Aude Oliva. Comparison of deep neural networks to spatio-temporal cortical dynamics of human visual object recognition reveals hierarchical correspondence. *Scientific reports*, 6:27755, 2016.

[CLG+15]   Rich Caruana, Yin Lou, Johannes Gehrke, Paul Koch, Marc Sturm, and Noemie Elhadad. Intelligible models for healthcare: Predicting pneumonia risk and hospital 30-day readmission. In *Proceedings of the 21th ACM SIGKDD international conference on knowledge discovery and data mining*, pages 1721–1730, 2015.

[CPC19]   Diogo V Carvalho, Eduardo M Pereira, and Jaime S Cardoso. Machine learning interpretability: A survey on methods and metrics. *Electronics*, 8(8):832, 2019.

[CSVZ14]   Ken Chatfield, Karen Simonyan, Andrea Vedaldi, and Andrew Zisserman. Return of the devil in the details: Delving deep into convolutional nets. *arXiv preprint arXiv:1405.3531*, 2014.

[DCSM17] Dhritiman Das, Eduardo Coello, Rolf F Schulte, and Bjoern H Menze. Quantification of metabolites in magnetic resonance spectroscopic imaging using machine learning. In *International Conference on Medical Image Computing and Computer-Assisted Intervention*, pages 462–470. Springer, 2017.

[Dir15] Cüneyt Dirican.   The impacts of robotics, artificial intelligence on business and economics. *Procedia-Social and Behavioral Sciences*, 195:564–573, 2015.

[DT18] Laura Delponte and G Tamburrini. *European Artificial Intelligence (AI) leadership, the path for an integrated vision*. European Parliament, 2018.

[DVKB+17] Finale Doshi-Velez, Mason Kortz, Ryan Budish, Chris Bavitz, Sam Gershman, David O'Brien, Stuart Schieber, James Waldo, David Weinberger, and Alexandra Wood. Accountability of ai under the law: The role of explanation. *arXiv preprint arXiv:1711.01134*, 2017.

[DY14] Li Deng and Dong Yu. Deep learning: methods and applications. *Foundations and trends in signal processing*, 7(3–4):197–387, 2014.

[EBCV09] Dumitru Erhan, Yoshua Bengio, Aaron Courville, and Pascal Vincent. Visualizing higher-layer features of a deep network. *University of Montreal*, 1341(3):1, 2009.

[EGVT17] Michael Eickenberg, Alexandre Gramfort, Gaël Varoquaux, and Bertrand Thirion. Seeing it all: Convolutional network layers map the function of the human visual system. *NeuroImage*, 152:184–194, 2017.

[EKAK17] Bradley J Erickson, Panagiotis Korfiatis, Zeynettin Akkus, and Timothy L Kline. Machine learning for medical imaging. *Radiographics*, 37(2):505–515, 2017.

[Eur18] European Commission (EC). Artificial intelligence for Europe. COM(2018) 237 final, April 2018. Communication from the Commission to the European parliament, the European Council, the Council, the European Economic and Social Committee and the Committee of the Regions.

[FBI+19] Samuel G Finlayson, John D Bowers, Joichi Ito, Jonathan L Zittrain, Andrew L Beam, and Isaac S Kohane.   Adversarial attacks on medical machine learning. *Science*, 363(6433):1287–1289, 2019.

[Fel16] Ed Felten. Preparing for the future of artificial intelligence. *Washington DC: The White House, May*, 3, 2016.

[FGSO+05] Alberto Fernández, Juan M García-Segura, Tomás Ortiz, Julia Inés Escobar Montoya, Fernando Maestú, Pedro Gil-Gregorio, Pablo Campo, and Juan Carlos Viano. Proton magnetic resonance spectroscopy and magnetoencephalographic estimation of delta dipole density: a combination of techniques that may contribute to the diagnosis of alzheimer's disease. *Dementia and geriatric cognitive disorders*, 20(2–3):169–77, 2005.

[FH17] Ethan Fast and Eric Horvitz. Long-term trends in the public perception of artificial intelligence. In *Thirty-First AAAI Conference on Artificial Intelligence*, 2017.

[FSR+19] Jean-Marc Fellous, Guillermo Sapiro, Andrew Rossi, Helen S Mayberg, and Michele Ferrante. Explainable artificial intelligence for neuroscience: Behavioral neurostimulation. *Frontiers in Neuroscience*, 13:1346, 2019.

[FYL+18] Xinyang Feng, Jie Yang, Zachary C Lipton, Scott A Small, Frank A Provenzano, Alzheimer's Disease Neuroimaging Initiative, et al. Deep learning on MRI affirms the prominence of the hippocampal formation in alzheimer's disease classification. *bioRxiv*, page 456277, 2018.

[FYLA18] Xinyang Feng, Jie Yang, Andrew F Laine, and Elsa D Angelini. Discriminative analysis of the human cortex using spherical cnns-a study on alzheimer's disease diagnosis. *arXiv preprint arXiv:1812.07749*, 2018.

[GB14] F Gao and Peter B Barker. Various mrs application tools for alzheimer disease and mild cognitive impairment. *American Journal of Neuroradiology*, 35(6 suppl):S4–S11, 2014.

[GBC16] Ian Goodfellow, Yoshua Bengio, and Aaron Courville. *Deep learning*. MIT Press, 2016.

[GRCS07] Denise Guliato, Rangaraj M Rangayyan, Juliano D Carvalho, and Sérgio A Santiago. Polygonal modeling of contours of breast tumors with the preservation of spicules. *IEEE Transactions on Biomedical Engineering*, 55(1):14–20, 2007.

[Gun17] David Gunning. Explainable artificial intelligence (xai). *Defense Advanced Research Projects Agency (DARPA), nd Web*, 2, 2017.

[GvG15] Umut Güçlü and Marcel AJ van Gerven. Deep neural networks reveal a gradient in the complexity of neural representations across the ventral stream. *Journal of Neuroscience*, 35(27):10005–10014, 2015.

[HDWF+17] Mohammad Havaei, Axel Davy, David Warde-Farley, Antoine Biard, Aaron Courville, Yoshua Bengio, Chris Pal, Pierre-Marc Jodoin, and Hugo Larochelle. Brain tumor segmentation with deep neural networks. *Medical image analysis*, 35:18–31, 2017.

[Hig19] High-Level Expert Group on Artificial Intelligence (AI HLEG). Ethics guidelines for trustworthy AI. COM(2018) 237 final, April 2019. Published by the European Commission.

[HMK+17] Andreas Holzinger, Bernd Malle, Peter Kieseberg, Peter M Roth, Heimo Müller, Robert Reihs, and Kurt Zatloukal. Towards the augmented pathologist: Challenges of explainable-ai in digital pathology. *arXiv preprint arXiv:1712.06657*, 2017.

[HNP09] Alon Halevy, Peter Norvig, and Fernando Pereira. The unreasonable effectiveness of data. *IEEE Intelligent Systems*, 24(2):8–12, 2009.

[Hob52] Eric J Hobsbawm. The machine breakers. *Past & Present*, (1):57–70, 1952.

[Hon18] Milo Honegger. Shedding light on black box machine learning algorithms: Development of an axiomatic framework to assess the quality of methods that explain individual predictions. *arXiv preprint arXiv:1808.05054*, 2018.

[HZRS16] Kaiming He, Xiangyu Zhang, Shaoqing Ren, and Jian Sun. Deep residual learning for image recognition. In *Proceedings of the IEEE conference on computer vision and pattern recognition*, pages 770–778, 2016.

[KAH19] Naimul Mefraz Khan, Nabila Abraham, and Marcia Hon. Transfer learning with intelligent training data selection for prediction of alzheimer's disease. *IEEE Access*, 7:72726–72735, 2019.

[KDK18] Sreenath P Kyathanahally, André Döring, and Roland Kreis. Deep learning approaches for detection and removal of ghosting artifacts in mr spectroscopy. *Magnetic resonance in medicine*, 80(3):851–863, 2018.

[KDM12] Imene Cheikhrouhou Kachouri, Khalifa Djemal, and Hichem Maaref. Characterisation of mammographic masses using a new spiculated mass descriptor in computer aided diagnosis systems. *International Journal of Signal and Imaging Systems Engineering*, 5(2):132–142, 2012.

[KHA+19] Pieter-Jan Kindermans, Sara Hooker, Julius Adebayo, Maximilian Alber, Kristof T Schütt, Sven Dähne, Dumitru Erhan, and Been Kim. The (un) reliability of saliency methods. In *Explainable AI: Interpreting, Explaining and Visualizing Deep Learning*, pages 267–280. Springer, 2019.

[KRA19] Incheol Kim, Sivaramakrishnan Rajaraman, and Sameer Antani. Visual interpretation of convolutional neural network predictions in classifying medical image modalities. *Diagnostics*, 9(2), 2019.

[KSH12] Alex Krizhevsky, Ilya Sutskever, and Geoffrey E Hinton. Imagenet classification with deep convolutional neural networks. In *Advances in neural information processing systems*, pages 1097–1105, 2012.

[KWG+17] Been Kim, Martin Wattenberg, Justin Gilmer, Carrie Cai, James Wexler, Fernanda Viegas, and Rory Sayres. Interpretability beyond feature attribution: Quantitative testing with concept activation vectors (tcav). *arXiv preprint arXiv:1711.11279*, 2017.

[LBH15] Yann LeCun, Yoshua Bengio, and Geoffrey Hinton. Deep learning. *nature*, 521(7553):436–444, 2015.

[LCG12] Yin Lou, Rich Caruana, and Johannes Gehrke. Intelligible models for classification and regression. In *Proceedings of the 18th ACM SIGKDD international conference on Knowledge discovery and data mining*, pages 150–158, 2012.

[Lip18] Zachary C Lipton. The mythos of model interpretability. *Queue*, 16(3):31–57, 2018.

[LL17] Scott M Lundberg and Su-In Lee. A unified approach to interpreting model predictions. In *Advances in neural information processing systems*, pages 4765–4774, 2017.

[LLWS19] Chunfeng Lian, Mingxia Liu, Li Wang, and Dinggang Shen. End-to-end dementia status prediction from brain MRI using multi-task weakly-supervised attention network. In *International Conference on Medical Image Computing and Computer-Assisted Intervention*, pages 158–167. Springer, 2019.

[LSL+16] Mengchen Liu, Jiaxin Shi, Zhen Li, Chongxuan Li, Jun Zhu, and Shixia Liu. Towards better analysis of deep convolutional neural networks. *IEEE transactions on visualization and computer graphics*, 23(1):91–100, 2016.

[LWB+19] Sebastian Lapuschkin, Stephan Wäldchen, Alexander Binder, Grégoire Montavon, Wojciech Samek, and Klaus-Robert Müller. Unmasking clever hans predictors and assessing what machines really learn. *Nature communications*, 10(1):1–8, 2019.

[LY17] Choong Ho Lee and Hyung-Jin Yoon. Medical big data: promise and challenges. *Kidney research and clinical practice*, 36(1):3, 2017.

[MD13] Travis B Murdoch and Allan S Detsky. The inevitable application of big data to health care. *Jama*, 309(13):1351–1352, 2013.

[MDFF16] Seyed-Mohsen Moosavi-Dezfooli, Alhussein Fawzi, and Pascal Frossard. Deepfool: a simple and accurate method to fool deep neural networks. In *Proceedings of the IEEE conference on computer vision and pattern recognition*, pages 2574–2582, 2016.

[Mil19] Tim Miller. Explanation in artificial intelligence: Insights from the social sciences. *Artificial Intelligence*, 267:1–38, 2019.

[MLB+17] Grégoire Montavon, Sebastian Lapuschkin, Alexander Binder, Wojciech Samek, and Klaus-Robert Müller. Explaining nonlinear classification decisions with deep Taylor decomposition. *Pattern Recognition*, 65:211–222, 2017.

[MNG+20] Xingjun Ma, Yuhao Niu, Lin Gu, Yisen Wang, Yitian Zhao, James Bailey, and Feng Lu. Understanding adversarial attacks on deep learning based medical image analysis systems. *Pattern Recognition*, page 107332, 2020.

[Mol20] Christoph Molnar. *Interpretable machine learning*. Lulu. com, 2020.

[Mos86] Jorge G Moser. Integration of artificial intelligence and simulation in a comprehensive decision-support system. *Simulation*, 47(6):223–229, 1986.

[MSM18] Grégoire Montavon, Wojciech Samek, and Klaus-Robert Müller. Methods for interpreting and understanding deep neural networks. *Digital Signal Processing*, 73:1–15, 2018.

[N+02] Alessandro Nuvolari et al. The "machine breakers" and the industrial revolution. *Journal of European Economic History*, 31(2):393–426, 2002.

[NR12] Fabián Narváez and Eduardo Romero. Breast mass classification using orthogonal moments. In *International Workshop on Digital Mammography*, pages 64–71. Springer, 2012.

[NYC16] Anh Nguyen, Jason Yosinski, and Jeff Clune. Multifaceted feature visualization: Uncovering the different types of features learned by each neuron in deep neural networks. *arXiv preprint arXiv:1602.03616*, 2016.

[O'n16] Cathy O'neil. *Weapons of math destruction: How big data increases inequality and threatens democracy*. Broadway Books, 2016.

[Rah17] A Rahimi. Machine learning has become alchemy. In *Thirsty-first Conference on Neural Information Processing Systems*, 2017.

[RDS+15] Olga Russakovsky, Jia Deng, Hao Su, Jonathan Krause, Sanjeev Satheesh, Sean Ma, Zhiheng Huang, Andrej Karpathy, Aditya Khosla, Michael Bernstein, et al. Imagenet large scale visual recognition challenge. *International journal of computer vision*, 115(3):211–252, 2015.

[REFDA97] Rangaraj M Rangayyan, Nema M El-Faramawy, JE Leo Desautels, and Onsy Abdel Alim. Measures of acutance and shape for classification of breast tumors. *IEEE Transactions on medical imaging*, 16(6):799–810, 1997.

[Ren19] Andrea Renda. Artificial intelligence: Ethics, governance and policy challenges. *CEPS Task Force Report*, 2019.

[RHDV17] Andrew Slavin Ross, Michael C Hughes, and Finale Doshi-Velez. Right for the right reasons: Training differentiable models by constraining their explanations. *arXiv preprint arXiv:1703.03717*, 2017.

[RR17] Ali Rahimi and Ben Recht. Reflections on random kitchen sinks, 2017.

[RSG16] Marco Tulio Ribeiro, Sameer Singh, and Carlos Guestrin. "why should I trust you?" explaining the predictions of any classifier. In *Proceedings of the 22nd ACM SIGKDD international conference on knowledge discovery and data mining*, pages 1135–1144, 2016.

[Rud19] Cynthia Rudin. Stop explaining black box machine learning models for high stakes decisions and use interpretable models instead. *Nature Machine Intelligence*, 1(5):206–215, 2019.

[RWD+16] Daniele Ravì, Charence Wong, Fani Deligianni, Melissa Berthelot, Javier Andreu-Perez, Benny Lo, and Guang-Zhong Yang. Deep learning for health informatics. *IEEE journal of biomedical and health informatics*, 21(1):4–21, 2016.

[SA19] Deniz Susar and Vincenzo Aquaro. Artificial intelligence: Opportunities and challenges for the public sector. In *Proceedings of the 12th International Conference on Theory and Practice of Electronic Governance*, pages 418–426, 2019.

[SCD+17] Ramprasaath R Selvaraju, Michael Cogswell, Abhishek Das, Ramakrishna Vedantam, Devi Parikh, and Dhruv Batra. Grad-cam: Visual explanations from deep networks via gradient-based localization. In *Proceedings of the IEEE international conference on computer vision*, pages 618–626, 2017.

[Sch15] Jürgen Schmidhuber. Deep learning in neural networks: An overview. *Neural networks*, 61:85–117, 2015.

[SDBR14] Jost Tobias Springenberg, Alexey Dosovitskiy, Thomas Brox, and Martin Riedmiller. Striving for simplicity: The all convolutional net. *arXiv preprint arXiv:1412.6806*, 2014.

[Smu19] Nathalie A Smuha. The eu approach to ethics guidelines for trustworthy artificial intelligence. *CRi-Computer Law Review International*, 2019.

[SR13] Jimeng Sun and Chandan K Reddy. Big data analytics for healthcare. In *Proceedings of the 19th ACM SIGKDD international conference on Knowledge discovery and data mining*, pages 1525–1525, 2013.

[SSWR18] David Sculley, Jasper Snoek, Alex Wiltschko, and Ali Rahimi. Winner's curse? on pace, progress, and empirical rigor. 2018.

[STB+18] Tyler C Steed, Jeffrey M Treiber, Michael G Brandel, Kunal S Patel, Anders M Dale, Bob S Carter, and Clark C Chen. Quantification of glioblastoma mass effect by lateral ventricle displacement. *Scientific reports*, 8(1):1–8, 2018.

[STE13] Christian Szegedy, Alexander Toshev, and Dumitru Erhan. Deep neural networks for object detection. In *Advances in neural information processing systems*, pages 2553–2561, 2013.

[SVS19] Jiawei Su, Danilo Vasconcellos Vargas, and Kouichi Sakurai. One pixel attack for fooling deep neural networks. *IEEE Transactions on Evolutionary Computation*, 23(5):828–841, 2019.

[SWM17] Wojciech Samek, Thomas Wiegand, and Klaus-Robert Müller. Explainable artificial intelligence: Understanding, visualizing and interpreting deep learning models. *arXiv preprint arXiv:1708.08296*, 2017.

[SZ14] Karen Simonyan and Andrew Zisserman. Very deep convolutional networks for large-scale image recognition. *arXiv preprint arXiv:1409.1556*, 2014.

[SZ16] Siuly Siuly and Yanchun Zhang. Medical big data: neurological diseases diagnosis through medical data analysis. *Data Science and Engineering*, 1(2):54–64, 2016.

[SZD+12] Xiaowei Song, Ningnannan Zhang, Ryan D'Arcy, Steven Beyea, Robert Bartha, Denise Bernier, Sultan Darvesh, and Kenneth Rockwood. Increased creatine in the posterior cingulate cortex in early alzheimer's disease: A high-field magnetic resonance spectroscopy study. *Alzheimer's & Dementia*, 8(4, Supplement):P35, 2012. Alzheimer's Association International Conference 2012.

[SZS+13] Christian Szegedy, Wojciech Zaremba, Ilya Sutskever, Joan Bruna, Dumitru Erhan, Ian Goodfellow, and Rob Fergus. Intriguing properties of neural networks. *arXiv preprint arXiv:1312.6199*, 2013.

[TCD+19] Ziqi Tang, Kangway V Chuang, Charles DeCarli, Lee-Way Jin, Laurel Beckett, Michael J Keiser, and Brittany N Dugger. Interpretable classification of alzheimer's disease pathologies with a convolutional neural network pipeline. *Nature communications*, 10(1):1–14, 2019.

[TG19] Erico Tjoa and Cuntai Guan. A survey on explainable artificial intelligence (XAI): Towards medical XAI. *arXiv preprint arXiv:1907.07374*, 2019.

[TNM+19] Hidenori Tanaka, Aran Nayebi, Niru Maheswaranathan, Lane McIntosh, Stephen Baccus, and Surya Ganguli. From deep learning to mechanistic understanding in neuroscience: the structure of retinal prediction. In *Advances in Neural Information Processing Systems*, pages 8535–8545, 2019.

[TSG+16] Nima Tajbakhsh, Jae Y Shin, Suryakanth R Gurudu, R Todd Hurst, Christopher B Kendall, Michael B Gotway, and Jianming Liang. Convolutional neural networks for medical image analysis: Full training or fine tuning? *IEEE transactions on medical imaging*, 35(5):1299–1312, 2016.

[VLFM04] Michael Van Lent, William Fisher, and Michael Mancuso. An explainable artificial intelligence system for small-unit tactical behavior. In *Proceedings of the national conference on artificial intelligence*, pages 900–907. Menlo Park, CA; Cambridge, MA; London; AAAI Press; MIT Press; 1999, 2004.

[Wan16] Ge Wang. A perspective on deep imaging. *Ieee Access*, 4:8914–8924, 2016.

[WS18] Scott A Wright and Ainslie E Schultz. The rising tide of artificial intelligence and business automation: Developing an ethical framework. *Business Horizons*, 61(6):823–832, 2018.

[WTW+15] Hui Wang, Lan Tan, Hui-Fu Wang, Ying Liu, Rui-Hua Yin, Wen-Ying Wang, Xiao-Long Chang, Teng Jiang, and Jin-Tai Yu. Magnetic resonance spectroscopy in alzheimer's disease: Systematic review and meta-analysis. *Journal of Alzheimer's disease: JAD*, 46 4:1049–70, 2015.

[WYB+10] Miles N Wernick, Yongyi Yang, Jovan G Brankov, Grigori Yourganov, and Stephen C Strother. Machine learning in medical imaging. *IEEE signal processing magazine*, 27(4):25–38, 2010.

[YHZL19] Xiaoyong Yuan, Pan He, Qile Zhu, and Xiaolin Li. Adversarial examples: Attacks and defenses for deep learning. *IEEE transactions on neural networks and learning systems*, 30(9):2805–2824, 2019.

[YPBI20] Evangeline Yee, Karteek Popuri, Mirza Faisal Beg, and Alzheimer's Disease Neuroimaging Initiative. Quantifying brain metabolism from FDG-PET images into a probability of alzheimer's dementia score. *Human brain mapping*, 41(1):5–16, 2020.

[YPBtADNI20] Evangeline Yee, Karteek Popuri, Mirza Faisal Beg, and the Alzheimer's Disease Neuroimaging Initiative. Quantifying brain metabolism from FDG-PET images into a probability of alzheimer's dementia score. *Human Brain Mapping*, 41(1):5–16, 2020.

[YRR18] Chengliang Yang, Anand Rangarajan, and Sanjay Ranka. Visual explanations from deep 3d convolutional neural networks for alzheimer's disease classification. In *AMIA Annual Symposium Proceedings*, volume 2018, page 1571. American Medical Informatics Association, 2018.

[ZF14] Matthew D Zeiler and Rob Fergus. Visualizing and understanding convolutional networks. In *European conference on computer vision*, pages 818–833. Springer, 2014.

[ZKL+16] Bolei Zhou, Aditya Khosla, Agata Lapedriza, Aude Oliva, and Antonio Torralba. Learning deep features for discriminative localization. In *Proceedings of the IEEE conference on computer vision and pattern recognition*, pages 2921–2929, 2016.

Chapter 12
Improving Video Quality with Generative Adversarial Networks

Leonardo Galteri, Lorenzo Seidenari, Tiberio Uricchio, Marco Bertini, and Alberto del Bimbo

12.1 Introduction

Video streaming has become a basic technology for everyday life: entertainment, education and communication are increasingly relying on it. Stimulated in the recent years by developments in networking like the 5G and modern video codecs; streaming has seen a dramatic increase with the global spread of COVID-19. People constrained at home by the emergency have adopted videoconferencing as main media of communication, for business meetings, lectures and to talk with friends and family members. This network traffic has been added to all the video streaming services that provide access to TV shows and movies. As a result, global networks have been profoundly impacted with an excessive traffic that they were not prepared to receive, resulting in a reduced user experience. To transmit or store a raw video, it must be compressed to reduce bandwidth and storage requirements. This happens at the cost of the perceived quality which strongly depends on the amount of available bandwidth and the compression algorithm. A model of the human visual system is commonly used by video coding algorithms to maintain the highest possible perceptual quality loss. They do not take into account the semantic of the video, ignoring the cues on which information is the most important to a viewer. The amount of bitrate allocated by modern video codecs may favor the wrong portion of a frame, limiting the quality of what a viewer may want to observe (e.g. the face of a speaker in a video call). In fact, it is a standard feature of commercial video call systems to blur the background [sky] or even replace it with a more pleasant image [zoo]. In contrast, computer vision techniques are recently able to correctly locate

L. Galteri · L. Seidenari (✉) · T. Uricchio · M. Bertini · A. del Bimbo
University of Florence, Firenze, Italy
e-mail: leonardo.galteri@unifi.it; lorenzo.seidenari@unifi.it; tiberio.uricchio@unifi.it; marco.bertini@unifi.it; alberto.delbimbo@unifi.it

© Springer Nature Switzerland AG 2021
J. Benois-Pineau, A. Zemmari (eds.), *Multi-faceted Deep Learning*,
https://doi.org/10.1007/978-3-030-74478-6_12

objects, assess the semantic and thus hint codecs to reserve more bitrate on such locations, with little or no effect on the user experience [WCRR+19].

A recent line of works in computer vision address the problem of recovering the reduced quality of compressed video. The core idea is that a video can be highly compressed and sub-sampled to reduce the required bandwidth, especially in the areas that are less important. Then, using approaches based on generative adversarial networks (GANs) [GSB+19], compressed areas may be reconstructed to provide the user experience of an equivalent video compressed with a higher bitrate. In a real time application such video calls, the solution should run in real time even on tablets and smartphones [GSBDB19, GSBD19].

Before describing the GANs based approaches that can enable reconstruction of reduced quality media, we need to focus on the different directions that researchers are currently considering. Restoration and compression are intertwined approaches which share similarities on the basic networks but are directed to different use cases. Restoration aims at recovering the lost details after compression, while compression aims at preserving as much details as possible according to the saliency of frames.

The chapter is organized as follows. We will first describe the restoration approaches, the computer vision assisted compression algorithms and hybrid approaches. Then, we will discuss how to measure the human semantic perceptual quality of videos. The GAN based approaches are the main focus, including a discussion on the network architectures, the latest loss functions and the recent NoGAN training. Finally, we will go beyond recovering the missing details from a fixed compressed video. We will discuss an approach that exploits semantics in the encoding stage that, working in tandem with the recovery stage, obtains superior performance.

12.2 Related Works

We can classify techniques to improve video quality in two main types: (1) those that restore a compressed image or frame to obtain a high quality one, eliminating compression artefacts and recreating lost details, and (2) those that improve the compression algorithm, so to preserve as much as possible details of relevant parts and details of the image that may be lost due to lossy coding.

12.2.1 Video and Image Restoration

Improving visual quality is a topic that has a large attention from the scientific community in the past, considering compression artifact removal and also other disturbs like blur and noise. Many of the past approaches are based on image processing techniques [DBEG16, JLJ+17, LHQ+17, LGTB14, ZXZ+16, ZXF+13]. More recently, have been studied learning-based methods for visual quality enhancement, such as [KHZ+15, DDCLT15, MSY16, SHBZ16, WLC+16, GSBD17, GSBD19, CHB17, YLK18, MNDS18]. These approaches use deep convolutional architectures to transform images corrupted by compression artifacts into high-quality ones.

The first work employing CNNs for compression artifact removal is [DDCLT15]. Their network design is specialized for JPEG compression, as in the more recent work of Fu et al. [FZW+19], while more recent works [SHBZ16, CHB17, YLL+21] employ general purpose architectures; these works share some common features such as residual learning and skip connections, bringing the benefit of allowing several layers of representation and propagating directly information from earlier layers to the final reconstruction; this latter effect can be obtained using also the U-Net architecture [MBGDB21]. It has to be noted that more perceptually satisfying results are obtained using Generative Adversarial Networks [GSBD19, MBGDB21]. In [GSBD17, GSBD19] Galteri et al. show that GAN based image restoration can be performed on various encoders even in a setting where no information about the encoding is available, by predicting coding parameters (Fig. 12.1).

12.2.2 Video and Image Compression

Semantic video coding techniques can be used to address bandwidth requirements. The basic idea is to recognize parts of the frame which are more relevant for

Fig. 12.1 Comparison of image quality comparing a GAN based restoration pipeline [GSBD19] with the deep learning based codec by wave.one. We show an uncompressed (RAW) frame and the source frame (x.264) restored by a GAN

the viewer, either perceptually or because they contain an object of interest, and improve their appearance allocating adaptively more bits for their encoding. We can frame two main lines of research: visual saliency [AT18, LEGV17] and object [WCRR+19, GBSDB18] based video coding. In the former approach some function of the image is computed pixel-wise, irrespective of the semantic content of the image, to compute the perceptual relevance of frame regions. Object based video instead assumes some form of semantic segmentation has been applied to obtain masks of relevant objects. This approach requires using robust object segmentation [HGDG17, BZXL19] which are nowadays implemented with high computational efficiency. Semantic video coding approaches have been used in very different domains, such as airplane cockpits [MFC+19], sport videos [BDBPC06], drone videos [WCC16], vehicles [ASBA+19], and surveillance videos [BBDBS11].

A few recent approaches have been proposed to directly perform video and image coding using neural networks [RB17, RNL+19], also using GAN approaches [ATM+19]. However, these approaches are currently not deployable with satisfying visual results in real-world applications due to their very high computational requirements. Moreover, fully learned compression requires the standardisation and diffusion of a novel technology, a fact that raises a high market barrier to entry. This can be mitigated if the decoding end of the pipeline continues to follow a standard. As partially reviewed in [RBB+16] there are two main strategies to improve video quality while still relying on standardized encoding solutions: pre-processing based and post-processing based. Talebi et al. [TKL+20] is an example of a pre-processing based hybrid approach to improve the quality of compressed images. Instead of relying on a deep network for encoding and decoding the authors learn a deep network for pre-processing images before standard JPEG compression. The training objective is to jointly minimize entropy and image distortion for a given JPEG quality factor.

12.2.3 Hybrid Approaches

The idea of combining methods belonging to the two families of approaches described above has been less studied, so far.

A system that combines learning based image enhancement methods, applying it on videos which have been semantically encoded, has been recently proposed in [GBS+20a], aiming at improving the quality of video conferences, using semantically coded faces. The system is able to provide comparable quality for videos talking humans with a third of the bandwidth. Mentzer et al. [MTTA20] combine GANs and learned compression to obtain a state-of-the-art generative lossy compression system that is able to obtain a visual quality comparable to other competing approaches with half the bandwidth.

12.2.4   Quality Metrics

Different visual quality metrics can be used to evaluate the performance of the encoding-decoding pipeline. These metrics can be classified as (1) full-reference, i.e. a reference image is available allowing to compare it to the restored image to assess image quality; (2) no-reference, i.e. only features and characteristics of the restored image are used. The recent work from Blau and Michaeli [BM19] has shown that there is a rate-distortion-perception trade-off showing that optimizing the statistical similarity of source and decoded images will increase the signal distortion rate. This is in line with the copious amount of results that shows how images ranked higher by humans obtains a lower score according to full-reference SSIM (Structural SIMilarity) and PSNR metrics. Also in [KLCB20] it has been shown that many existing image quality algorithms like SSIM are unreliable when used to assess GAN generated content, since images generated by such generative approaches may match poorly with a reference image when considering metrics based on pixel comparisons, although they may be realistic and perceptually similar to an original image.

For these reasons, metrics based on some evaluation of "naturalness" are more suitable, and a lot of work has been dedicated to obtain more reliable metrics for image quality assessment [KLCB20, ZIE+18, MSB13, MSB16]. In the following we will report results using modern LPIPS metric as a full-reference evaluation, and BRISQUE [MMB12] and NIQE [MSB13] for a no-reference image quality assessment. These latter metrics evaluate the naturalness of an image.

12.3   Generative Adversarial Networks vs Standard Enhancement CNNs

The goal of compression artifact removal is to obtain a reconstructed output image I^R from a compressed input image I^C. In this scenario, $I^C = A(I)$ is the output image of a compression algorithm A and I is an uncompressed input image. Different A algorithms will produce different I^C images, with different compression artifacts. Many image and video compression algorithms (e.g. JPEG, JPEG2000, WebP, H.264/AVC, H.265/HEVC) work in the YCrCb color space, separating luminance from chrominance information. This allows a better de-correlation of color components leading to a more efficient compression; it also permits a first step of lossy compression sub-sampling chrominance, considering the reduced sensitivity of the human visual system to its variations.

We represent images I^R, I^C and I as real valued tensors with dimensions $W \times H \times C$, where W and H are width and height, respectively, and C is the number of color channels.

The compression of an uncompressed image $I \in [0, 255]^{W \times H \times C}$ is performed according to:

$$I^C = A\,(I,\,QF) \in [0,\,255]^{W \times H \times C} \tag{12.1}$$

using a function A, representing some compression algorithm, which is parametrized by some quality factor QF. The problem of compression artifacts removal can be seen as to compute an inverse function $G \approx A_{QF}^{-1}$ that reconstructs I from I^C:

$$G\left(I^C\right) = I^R \approx I \tag{12.2}$$

Each generator can in principle be trained with images obtained from different qualities. In practice we show, in Sect. 12.3.3, that single quality generators perform better and can be driven by a quality predictor.

To this end, we train a convolutional neural network $G\left(I^C; \theta_g\right)$ where $\theta_g = \{W_{1:K}; b_{1:K}\}$ are the parameters representing weights and biases of the K layers of the network. Given N training images we optimize a custom loss function l_{AR} by solving:

$$\hat{\theta}_g = \arg\min_{\theta_g} \frac{1}{N} \sum_{n=1}^{N} L_{AR}\left(I, G\left(I^C, \theta_g\right)\right) \tag{12.3}$$

12.3.1 Network Architectures

The elimination of compression artifacts is a task that belongs to the class of image transformation problems, that comprises other tasks such as super-resolution and style-transfer. This category of tasks is conveniently addressed using generative approaches, i.e. learning a fully convolutional neural network (FCN) [LSD15] that given a certain input image is able to output an improved version of it. A reason to use FCN architectures in image processing is that they are extremely convenient to perform local non-linear image transformations, and can process images of any size. Interestingly, we take advantage of such property to speed up the training. Indeed, the artifacts we are interested in removing appear at scales close to the block size. For this reason we can learn models on smaller patches using larger batches.

In this chapter, we cover three architectures for the image enhancement network which we will interchangeably refer to also as the generator. A fully convolutional architecture that can be either optimized with direct supervision or combined in a generative adversarial framework with a novel discriminator. We will analyze how inference can be make efficient by replacing some of the most computational expensive blocks. Finally we will show how to speed-up training by using the so called NoGAN approach with a fast to converge U-Net like architecture.

12.3.1.1 Fully convolutional Generator

In [GSBD19] has been proposed to use a deep residual generative network, composed only by blocks of convolutional layers with non-linear LeakyReLU activations. This generator is inspired by [HZRS16]. We use layers with 64 convolution kernels with a 3×3 support, followed by LeakyReLU activations. After a first convolutional layer, we apply a layer with stride two to half the size of feature maps. Then we apply 15 residual blocks using a 1 pixel padding after every convolution with replication strategy to mitigate border effects. A nearest-neighbour upsampling layer is used to obtain feature maps at the original size [ODO16]. Considering that upsampling may lead to artifacts we apply another stride one convolutional layer. Finally, to generate the image we use single kernel convolutional layer with a *tanh* activation. This produces output tensors with values in $[-1, 1]$, which are therefore comparable to the rescaled image input. Adding batch normalization helps training of the GAN, resulting in a moderately improved performance.

This network can be coupled with a discriminator and used as generator to obtain a generative adversarial framework. The recent approach of adversarial training [GPAM+14] has shown remarkable performances in the generation of photo-realistic images and in super-resolution tasks [LTH+16].

In this approach, the generator network G is encouraged to produce solutions that lay on the manifold of the real data by learning how to fool a discriminative network D. On the other hand, the discriminator is trained to distinguish reconstructed patches I^R from the real ones I. In particular, we use a conditional generative approach, i.e. we provide as input to the generative network both positive examples $I|I^C$ and negative examples $I^R|I^C$, where $\cdot|\cdot$ indicates channel-wise concatenation. For samples of size $N \times N \times C$ we discriminate samples of size $N \times N \times 2C$.

12.3.1.2 Improving the Efficiency of Enhancement Architectures

Recent state-of-the-art neural networks for image restoration have very high demanding requirements that exceed mobile devices capabilities and embedded systems. We can improve the efficiency of such networks by manipulating the components of the generator in order to reduce the total number of the operations needed while maintaining high quality reconstruction outputs. The architecture of our generator is based on MobileNetV2 [SHZ+18], which is a very efficient network designed for mobile devices to perform classification tasks. Differently from [GSBD19], we replace standard residual blocks with bottleneck depth-separable convolutions blocks, as shown in Table 12.1, to reduce the overall amount of parameters. We set the expansion factor t to 6 for all the experiments.

After a first standard convolutional layer, feature maps are halved twice with strided convolutions and then we apply a chain of B bottleneck residual blocks. The number of convolution filters doubles each time the feature map dimensions are halved. We use two combinations of nearest-neighbour up-sampling and standard convolution layer to restore the original dimensions of feature maps. Finally, we

Table 12.1 Bottleneck
residual block used in our
generator network

Layer	Output
Conv2d 1×1, ReLU6	$m \times n \times t * c$
Dw Conv2d 3×3, ReLU6	$m \times n \times t * c$
Conv2d 1×1	$m \times n \times c$

Table 12.2 Parameters of the different GANs used. Compared to the previous work [GSBD19], our new "Fast" and "Very Fast" networks have much smaller number of parameters, resulting in improved computation time

Model	# Filters	# Blocks	# Params
Galteri et al. [TMM'19]	64	16	5.1M
Our fast	32	12	1.8M
Our very fast	8	16	145k

generate the RGB image with a 1×1 convolution followed by a *tanh* activation, to keep the output values between the $[-1, 1]$ range. In all our trained models we employed Batch Normalization to stabilize the training process. Table 12.2 reports the number of filters, blocks and weights of the GAN used in a previous work [GSBD19], and two variations of the proposed network, called "Fast" and "Very Fast" since they are designed to attain real-time performance. It can be observed that the new GAN architectures have much smaller number of parameters, resulting in reduced computational costs, that allow to reach the required real-time performance.

When using other approaches, such as U-Net, it is worth considering different backbones. In [MBGDB21], for example, a U-Net architecture, with a ResNet and a MobileNet backbone and self-attention mechanism has been used, following its success in performing image and video colorization of archive monochrome media. Substituting the original deep ResNet backbones (Resnet101), that has $\sim 40M$ parameters with smaller and faster backbones, like MobileNetV3 [HSC+19] that has only $\sim 5.5M$ parameters. This allows to use reduce computational costs, allowing to process large video archives. As reported in [MBGDB21], this technique allows to halve processing time, considering a NVIDIA TitanX, from 0.27 s to 0.15 and 0.12 s, depending on the use of the standard MobileNetV3 or of the Small MobileNetV3 variant.

12.3.1.3 Discriminative Network

When performing GAN training we need to define a suitable Discriminator network to couple with the generator. The architecture of the discriminator is typically based on a series of convolutional layers without padding and with single-pixel stride followed by LeakyReLU activations. The number of filters is doubled every two layers, with the exception of the last one. There are no fully connected layers. The size of the feature map is decreased solely because of the effect of convolutions reaching unitary dimension in the last layer, in which the activation function used is a sigmoid.

The set of weights ψ of the D network are learned by minimizing:

$$L_d = -\log\left(D_\psi\left(I|I^C\right)\right) - \log\left(1 - D_\psi\left(I^R|I^C\right)\right)$$

Discrimination is performed at the sub-patch level; this is motivated by the fact that compression algorithms decompose images into patches and thus artifacts are typically created within them. To encourage the generation of images with realistic patches, I and I^R are partitioned into P patches of size 16×16, that are then fed into the discriminator network. This approach has a beneficial effect in the reduction of mosquito noise and ringing artifacts.

12.3.2 Loss Functions

Now, we discuss losses employed to learn an enhancement network with direct supervision and using an adversarial approach.

We refer to direct supervision when the loss is computed as a function of the reconstructed image I^R and of the original uncompressed input image I. In this situation classical backpropagation is used to update the network weights. Several losses can be applied in this case, in the following we review the most effective ones. Note that even in a generative adversarial setting these losses will be still used to ensure the final image does not deviate from the original appearance.

12.3.2.1 Pixel-Wise MSE Loss

The baseline loss often employed is the Mean Squared Error loss (MSE):

$$L_{MSE} = \frac{1}{WH} \sum_{x=1}^{W} \sum_{y=1}^{H} \left(I_{x,y} - I_{x,y}^R\right)^2 . \tag{12.4}$$

This loss is commonly used in image reconstruction and restoration tasks [DDCLT15, SHBZ16, MSY16]. It has been shown that l_{MSE} is effective to recover the low frequency details from a compressed image, but the drawback is that high frequency details are suppressed. From a practical point of view this loss will end up recover good quality image with a blurry features.

12.3.2.2 SSIM Loss

The Structural Similarity (SSIM) [WBSS04] has been successfully proposed as an alternative to MSE and Peak Signal-to-Noise Ratio (PSNR) image similarity measures, because both these measures have shown to be inconsistent with the human visual perception of image similarity.

The formula to compute the SSIM of the uncompressed image I and the reconstructed image I^R is:

$$SSIM\left(I, I^R\right) = \frac{\left(2\mu_I \mu_{I^R} + C_1\right)\left(2\sigma_{I I^R} + C_2\right)}{\left(\mu_I^2 + \mu_{I^R}^2 + C_1\right)\left(\sigma_I^2 + \sigma_{I^R}^2 + C_2\right)} \tag{12.5}$$

Considering that the SSIM function is fully differentiable a loss can be defined as:

$$L_{SSIM} = -\frac{1}{WH}\sum_{x=1}^{W}\sum_{y=1}^{H} SSIM\left(I_{x,y}, I_{x,y}^R\right) \tag{12.6}$$

The network can then be trained minimizing Eq. 12.6, which means maximizing the structural similarity score computed on uncompressed and reconstructed image pairs.

12.3.2.3 Perceptual Loss

Several work proposed to measure the similarity of images in a space which preserves semantic features instead of the low-level pixel-wise similarity. Such losses have been proposed in Dosovitskiy and Brox [DB16], Johnson et al. [JAFF16], Bruna et al. [BSL15] and Gatys et al. [GEB15] and are referred as perceptual losses. The distance between images is computed after projecting I and I^R on a feature space by some differentiable function ϕ and taking the Euclidean distance between the two feature representations:

$$L_P = \frac{1}{W_f H_f}\sum_{x=1}^{W_f}\sum_{y=1}^{H_f}\left(\phi\left(I\right)_{x,y} - \phi\left(I^R\right)_{x,y}\right)^2 \tag{12.7}$$

where W_f and H_f are respectively the width and the height of the feature maps. The images reconstructed by the model trained with the perceptual loss are not necessarily accurate according to the pixel-wise distance measure, but on the other hand the output will be more similar from the point of view of feature representation. A possible choice would be to compute $\phi(I)$ by extracting the feature maps from a pre-trained VGG-19 model [SZ15], using the second convolution layer before the last max-pooling layer of the network.

The VGG-based perceptual loss has been used also for colorization in DeOldify [AHM19]. In [MBGDB21], the perceptual loss has been modified to use LPIPS visual quality metric. A motivation for this choice is that LPIPS metric has been shown to be strongly related to human visual perception and thus can be considered as real perceptual loss, while VGG-based loss can be considered more as a feature loss. Another benefit is that this metric has been designed to operate on small patches so it suits perfectly approaches that use a patch-based training, like [MBGDB21].

12.3.2.4   Adversarial Patch Loss

When the image enhancement network is trained in an adversarial setting we need to add a loss component to the generator loss. Such loss penalizes reconstructions that are not good enough to *fool* the discriminator network. So while the discriminator is trained with a standard binary cross-entropy loss the generator is trained with an adversarial loss which increase when the discriminator loss decreases.

Finally, as previously introduced the generator is trained by combining the perceptual loss with the adversarial loss thus obtaining:

$$L_{AR} = L_P + \lambda L_{adv}. \tag{12.8}$$

Where l_{adv} is the standard adversarial loss:

$$L_{adv} = -\log\left(D_\psi\left(I^R | I^C.\right)\right) \tag{12.9}$$

12.3.2.5   Relativistic GAN

In [WYW+18] the authors based their super-resolution network on the Relativistic GAN [JM18] instead of the standard GAN setup to get better reconstruction outputs. Here, the key idea is to drive the discriminator to estimate the probability that a ground truth image I is relatively more realistic than a generated one I^R. We define $D(I, I^R) = \sigma(C(I) - \mathbb{E}_{I^R}[C(I^R)])$ as the output of the relativistic discriminator, where σ, $C(.)$ and $\mathbb{E}_{I^R}[.]$ stand for the sigmoid activation, the dense layer output of the discriminator and the average for all reconstructed images in the mini-batch, respectively. In this case the discriminator loss is defined as:

$$L_D = -\mathbb{E}_I[\log(D(I, I^R))]$$
$$-\mathbb{E}_{I^R}[1 - \log(D(I^R, I))] \tag{12.10}$$

and the adversarial loss for the generator as:

$$L_{Adv} = -\mathbb{E}_I[1 - \log(D(I, I^R))]$$
$$-\mathbb{E}_{I^R}[\log(D(I^R, I))] \tag{12.11}$$

In case we have more control on the encoding/decoding process which is especially true when we can assume content semantics to be limited, we can devise specialised loss functions to train the image enhancement network. We pick as an example the work we presented in [GBS+20b], where we define a more effective perceptual loss by realizing that our data to be reconstructed is not homogeneous. As a matter of fact, the semantic encoding partitions the image in two regions that differs from content (background/faces) and quality (low-quality/high-quality). Therefore, the network needs to learn the reconstruction of different parts according to separate objectives using the semantic masks computed by the face parser. We define M as the foreground binary mask and $\overline{M}$ as the background mask that is computed by the logical negation of M.

In this situation, a perceptual loss based on VGG-19 for the background could be employed, limiting its computation to the parts of the image where $\overline{M}$ values are not zero. Naming such perceptual loss based for the background as L_B:

$$L_B = \mathbb{E}_{(I,I^R)} \left[||VGG(I \odot \overline{M}) - VGG(I^R \odot \overline{M})|| \right] \qquad (12.12)$$

where $\odot$ stands for element by element multiplication.

Considering as an example a case in which the foreground is composed of human faces, a different extractor must be used to handle this specific category of features. The logical choice is to extract such features from a pre-trained network that has processed millions of face images, such as VGG-Face [PVZ15]. As VGG-Face is based on VGG-16 backbone, we extract the output taken from the third convolutional layer of the fifth block before the ReLU activation for the loss computation. Under these assumptions, we define the perceptual loss constrained to the foreground as:

$$L_F = \mathbb{E}_{(I,I^R)} \left[||VGGFace(I \odot M) - VGGFace(I^R \odot M)|| \right] \qquad (12.13)$$

The total loss for the generator is:

$$L_G = L_B + L_F + \lambda L_{Adv} \qquad (12.14)$$

where λ is a fixed coefficient to balance the contribution of the adversarial loss.

12.3.3 Quality Agnostic Artifact Removal

The quality of an image can not be known in advance. To apply a model in real-world scenarios we can not depend on the prior knowledge of such information. The trivial approach of training a single GAN fed with all QFs is not viable unfortunately; in fact, we observe mode collapse towards higher compression rates. We believe that this effect is due to the fact that images with lower QFs contain more artifacts and generate more signal, thus overcoming the learning of subtle pattern removal that is needed at better qualities.

To cope with this problem, our full solution comprises two modules. The first module predicts, via regression the true QF of an image. This is possible with an extremely high precision. The compression quality estimator is used to drive the image signal to one of the fixed QF trained GANs of our ensemble. A schema of the system is shown in Fig. 12.2.

Our compression quality predictor consists of a stack of convolutional layers, each one followed by a non-linearity and Batch Normalization, and two Fully Connected layers in the last part. The architecture is shown in detail in Table 12.3.

The training set is selected from the DIV2k dataset [AT17], that contains 800 high definition high resolution raw images. During the training process, we compress the images to a random QF in a 5–95 range and we extract 128×128 patches. For the optimization, we used a standard MSE loss, computed over predicted and ground truth QF. We train the model as a regressor rather than a classifier since the wrong predictions that are close to the ground truth should not be penalized too much, as the corresponding reconstructions still result acceptable. On the other hand, predictions that are far from the ground truth lead to bad reconstructions, therefore we should penalize them accordingly in the training process.

In the inference phase, we extract eight random crops of 128×128 from a compressed image, we feed them into the QF predictor and we average the

Fig. 12.2 Example of masking for loss defined in Eq. 12.13, 12.12 and 12.14

Table 12.3 Network architecture of the proposed QF predictor

Layer	KernelSize/Stride	OutputSize
Conv11	$3 \times 3/1$	$128 \times 128 \times 64$
Conv12	$3 \times 3/2$	$64 \times 64 \times 64$
Conv21	$3 \times 3/1$	$64 \times 64 \times 128$
Conv22	$3 \times 3/2$	$32 \times 32 \times 128$
Conv31	$3 \times 3/1$	$32 \times 32 \times 256$
Conv32	$3 \times 3/2$	$16 \times 16 \times 256$
Conv41	$3 \times 3/1$	$16 \times 16 \times 512$
Conv42	$3 \times 3/2$	$8 \times 8 \times 512$
FC5	–	1024
FC6	–	1

prediction results. We use this prediction to reconstruct the corrupted image with the appropriate model for the input image quality. For this reason, we have trained six different generators, each one with fixed QF training images (5,10,20,30,40,60). Depending on the prediction, we give in input the corrupted image to the fixed QF reconstruction network closer to the QF predictor output.

In Fig. 12.3 show the performance of single quality models inside the ensemble by applying them to images of different qualities. Figure 12.3 measures mAP of a pretrained object detector for JPEG images compressed at various QFs and reconstructed with QF specific models. It can be seen that when models for similar QFs mAP varies smoothly. Interestingly, for images with lower QFs such as 5 and 10 we see improvements for every model applied. Higher quality images must be restored with models trained with higher QFs, otherwise performance can even degrade.

12.3.4 NoGAN Training

NoGAN is a method to train a GAN architecture that aims at obtaining better results, stabilizing both training and generation of images; it has been originally proposed

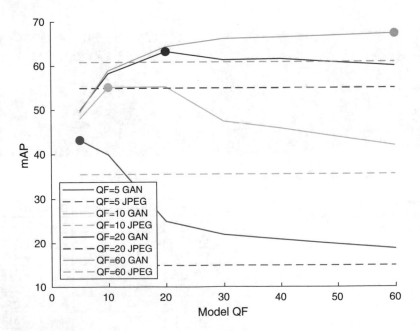

Fig. 12.3 Mean Average Precision on PASCAL VOC2007 varying image quality (QF) as well as the model used for restoration. Original mAP reported as dashed lines for every QF. Circular markers indicate the maximum mAP, obtained with the correct model

in the context of image and video colorization [AHM19]. The main idea of this approach is to pre-train separately the generator and the discriminator networks, performing then a standard final adversarial training step for a short period. In this setting the generator is initially trained using some perceptual loss (i.e. feature loss like the one reported in Sect. 12.3.2.3), then the generated images are used to train the discriminator as a binary classifier. The separate training of generator and discriminator is carried on for the vast majority of time, while the standard GAN training may take only 5–10% of the time used for the separate training, e.g. alternating five iterations of the discriminator for every two iterations of the generator as in [MBGDB21].

Table 12.4 compares the results obtained using the NoGAN approach (post-GAN) with direct training of the generator (pre-GAN), and compares the performance of different perceptual/features losses. Using LPIPS instead of other signal based losses provides better results, although it may be hard to decide a winner between the final GAN training step vs. direct training. However manual inspection of results shows that the final GAN training helps to reduce high frequency artefacts near the borders of the objects (Fig. 12.4).

These results have been obtained using the Div2K dataset [IT+19], using the standard setup of 800 images for training, 100 for validation and 100 for testing. Images have been compressed as JPEG with Quality Factor set to 20 (QF20).

Table 12.4 Quality metrics for different losses and training steps. A higher SSIM score is better, lower LPIPS, BRISQUE and NIQE scores are better. Best results are highlighted in bold. "small" indicates the use of MobileNetV3 Small backbone

Loss	SSIM ↑	LPIPS ↓	BRISQUE ↓	NIQE ↓
LPIPS (pre-GAN)	0.6933	**0.1243**	85.09	**16.76**
LPIPS (post-GAN)	**0.7374**	0.1526	89.31	17.57
LPIPS (post-GAN small)	0.7301	0.1502	**84.14**	17.67
MSE	0.7293	0.1661	88.91	17.82
SSIM	0.7354	0.1830	86.87	17.57
Target	*n.a.*	*n.a.*	85,32	15,69

Fig. 12.4 *Left* LPIPS loss, pre-GAN; *Right* LPIPS loss, post-GAN; adding a final GAN training reduces blocky artifacts near the borders of the objects

12.4   Exploiting Transmitter and Receiver for Improvement

In [GBS+20a] we presented an approach based on the idea that a compression artefact removal method can restore videos with a better perceived visual quality by exploiting semantically encoded videos. We first describe how the semantic video encoding is performed on the transmitter. Then, we report the GAN-based video restoration approach to improve the perceptual visual quality on the receiver party.

12.4.1   Semantic Video Encoding

The main idea of semantic video encoding is to allocate more bits to the regions that depict semantic content of interest for the viewer, to the detriment of background. Ideally, the amount of bits should be enough to maximize the perceptual quality of the objects of interest for a viewer. Semantic video encoding is related to saliency based video encoding [LEGV17, LEPV19] as they both consider regions that should be stored with a higher amount of data. Nonetheless they aim at different targets. Semantic encoding aims at transferring the high level semantic content that is of most interest of the viewer, regardless of any other element of background. The saliency based encoding, instead, has no specific knowledge of objects of the scene. It aims at transferring the content which is most probably observed by the eyes of a viewer, regardless of its importance.

To perform the encoding, we construct a semantic mask for each frame where we label each pixel as foreground (i.e. pixels of regions that are allotted more bits) or background. Depending on the domain, the foreground may be different. In our considered context of a video conference application, the foreground is the speaking person, more specifically its face. Hence, we employ the popular BiSeNet [YWP+18] image segmentation method, trained on CelebAMask-HQ [LLWL20] to perform face parsing. We label each pixel detected as a part of face and neck as foreground, the remainder as background.

The final video is encoded using a h.264 encoder which has been modified to allot a predetermined P percentage of a given bitrate to an input mask. We employ the implementation of [LEGV17] which uses a non-trivial estimation of macroblock sizes with respect to the quantization of parameter of h.264 constant quantizer.

12.4.2   Video Restoration

Most restoration approaches based on deep learning tackle the artifact removal problem trying to minimize the squared pixel-wise Euclidean distance between a reference raw frame I and the generated output I^R from a compressed input I^C. However, this kind of training strategy leads to feeble restored images as they appear often blurry and lacking important details. Besides, the h.264 encoder typically contains a strong loop de-blocking filter at the end of the compression pipeline, which leads to producing blurry frames, so that using an MSE based neural network to restore the images is even less effective.

Generative Adversarial Networks have been broadly used for both restoration and enhancement tasks to solve the aforementioned issues. The GAN framework tries to estimate a model distribution that approximate a target distribution, and it comprises two distinct entities, a generator and a discriminator. In this setup, the aim of the generator is to produce the model distribution given some noisy input and the role of the discriminator is to discern the model distribution from the target one. The two networks are trained one after another while gradually the distance between the model distribution and the generator decreases.

Since we do not want to generate completely novel images from the model distribution, but we rather want to restore some distorted data, we need to condition the training procedure of the GAN accordingly. Therefore, we feed the discriminator with real samples $I|I^C$ and fake samples $I^R|I^C$ where the operator $\cdot|\cdot$ defines the channel-wise concatenation of the inputs.

In Fig. 12.5 it is clear the advantage of controlling both ends, indeed just by using the saliency coding there is a gain in term of quality. Applying a GAN trained to reconstruct saliency encoded videos we obtain further increase in image quality.

12.5   Conclusion

Media streaming and storage have become a pillar of modern society, enabling widespread diffusion of information and allowing remote interaction both in social and work contexts. In this chapter we discussed a set of deep learning based tools allowing the enhancement of compressed media. We devised a setting in which

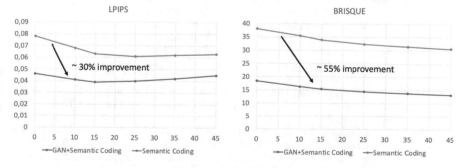

Fig. 12.5 Performance of Semantic Coding combined with GAN using BRISQUE and LPIPS (both lower are better). Increasing the percentage of bitrate allocated to the salient part there is a gain in performance with and without the GAN artifact removal. Applying GAN leads to a 30% and 55% improvement on both metrics respectively

the degradation process of media is known and replicable, allowing the generation of high quality large training sets at will. Leveraging existing deep learning tools and knowledge we have shown how this technology can be made efficient enough to run on mobile devices. Combining restoration networks with standard computer vision tools enable a set of vertical applications where the control of the transmitter and receiver nodes ensure higher quality for much less bandwidth. Our view of the approaches presented in this chapter is that more and more deep learning will be used to enhance the quality of consumed media, either in real-time on devices or in batch processes to improve existing lower quality digital archives. While in this chapter we mostly covered images and video we expect this enhancement approach to be extended to other media such as audio and 3D meshes and point clouds.

References

[AHM19]   Jason Antic, Jeremy Howard, and Uri Manor. Decrappification, deoldification, and super resolution, 2019.
[ASBA+19] Noor Al-Shakarji, Filiz Bunyak, Hadi Aliakbarpour, Guna Seetharaman, and Kannappan Palaniappan. Multi-cue vehicle detection for semantic video compression in georegistered aerial videos. In *Proc. of (CVPR) Workshops*, June 2019.
[AT17]   Eirikur Agustsson and Radu Timofte. Ntire 2017 challenge on single image super-resolution: Dataset and study. In *Proc. of IEEE CVPR Workshops*, 2017.
[AT18]   A Diana Andrushia and R Thangarjan. Saliency-based image compression using Walsh–Hadamard transform (WHT). In *Biologically rationalized computing techniques for image processing applications*, pages 21–42. Springer, 2018.
[ATM+19]   Eirikur Agustsson, Michael Tschannen, Fabian Mentzer, Radu Timofte, and Luc Van Gool. Generative adversarial networks for extreme learned image compression. In *Proceedings of the IEEE/CVF International Conference on Computer Vision (ICCV)*, October 2019.
[BBDBS11] Andrew D. Bagdanov, Marco Bertini, Alberto Del Bimbo, and Lorenzo Seidenari. Adaptive video compression for video surveillance applications. In *Proc. of International Symposium on Multimedia*, 2011.
[BDBPC06] Marco Bertini, Alberto Del Bimbo, Andrea Prati, and Rita Cucchiara. Semantic adaptation of sport videos with user-centred performance analysis. *IEEE Transactions on Multimedia*, 8(3):433–443, Jun 2006.
[BM19]   Yochai Blau and Tomer Michaeli. Rethinking lossy compression: The rate-distortion-perception tradeoff. In *Proc. of ICML*, 2019.
[BSL15]   Joan Bruna, Pablo Sprechmann, and Yann LeCun. Super-resolution with deep convolutional sufficient statistics. *CoRR*, abs/1511.05666, 2015.
[BZXL19]   Daniel Bolya, Chong Zhou, Fanyi Xiao, and Yong Jae Lee. Yolact: real-time instance segmentation. In *Proc. of International Conference on Computer Vision*, pages 9157–9166, 2019.
[CHB17]   Lukas Cavigelli, Pascal Hager, and Luca Benini. CAS-CNN: A deep convolutional neural network for image compression artifact suppression. In *Proc. of IJCNN*, 2017.
[DB16]   A. Dosovitskiy and T. Brox. Generating images with perceptual similarity metrics based on deep networks. In *Proc. of NIPS*, 2016.
[DBEG16]   Y. Dar, A. M. Bruckstein, M. Elad, and R. Giryes. Postprocessing of compressed images via sequential denoising. *IEEE Transactions on Image Processing*, 25(7):3044–3058, July 2016.

[DDCLT15]   Chao Dong, Yubin Deng, Chen Change Loy, and Xiaoou Tang. Compression artifacts reduction by a deep convolutional network. In *Proc. of International Conference on Computer Vision*, 2015.

[FZW+19]   Xueyang Fu, Zheng-Jun Zha, Feng Wu, Xinghao Ding, and John Paisley. Jpeg artifacts reduction via deep convolutional sparse coding. In *Proceedings of the IEEE/CVF International Conference on Computer Vision (ICCV)*, October 2019.

[GBS+20a]   Leonardo Galteri, Marco Bertini, Lorenzo Seidenari, Tiberio Uricchio, and Alberto Del Bimbo. Increasing video perceptual quality with gans and semantic coding. In *Proc. of ACM International Conference on Multimedia (ACM MM)*, MM '20, pages 862–870, New York, NY, USA, 2020. Association for Computing Machinery.

[GBS+20b]   Leonardo Galteri, Marco Bertini, Lorenzo Seidenari, Tiberio Uricchio, and Alberto Del Bimbo. Increasing video perceptual quality with gans and semantic coding. In *Proceedings of the 28th ACM International Conference on Multimedia*, pages 862–870, 2020.

[GBSDB18]   Leonardo Galteri, Marco Bertini, Lorenzo Seidenari, and Alberto Del Bimbo. Video compression for object detection algorithms. In *Proc. of International Conference on Pattern Recognition*, pages 3007–3012. IEEE, 2018.

[GEB15]   Leon A. Gatys, Alexander S. Ecker, and Matthias Bethge. Texture synthesis and the controlled generation of natural stimuli using convolutional neural networks. *CoRR*, abs/1505.07376, 2015.

[GPAM+14]   Ian Goodfellow, Jean Pouget-Abadie, Mehdi Mirza, Bing Xu, David Warde-Farley, Sherjil Ozair, Aaron Courville, and Yoshua Bengio. Generative adversarial nets. In *Proc. of NIPS*, 2014.

[GSB+19]   Leonardo Galteri, Lorenzo Seidenari, Marco Bertini, Tiberio Uricchio, and Alberto Del Bimbo. Fast video quality enhancement using gans. In *Proc. of ACM Multimedia*, MM '19, pages 1065–1067, New York, NY, USA, 2019. Association for Computing Machinery.

[GSBD17]   Leonardo Galteri, Lorenzo Seidenari, Marco Bertini, and Alberto Del Bimbo. Deep generative adversarial compression artifact removal. In *Proc. of International Conference on Computer Vision*, 2017.

[GSBD19]   L. Galteri, L. Seidenari, M. Bertini, and A. Del Bimbo. Deep universal generative adversarial compression artifact removal. *IEEE Transactions on Multimedia*, pages 1–1, 2019.

[GSBDB19]   Leonardo Galteri, Lorenzo Seidenari, Marco Bertini, and Alberto Del Bimbo. Towards real-time image enhancement gans. In *Proc. of International Conference on Analysis of Images and Patterns (CAIP)*. IAPR, 2019.

[HGDG17]   Kaiming He, Georgia Gkioxari, Piotr Dollár, and Ross Girshick. Mask r-cnn. In *Proc. of International Conference on Computer Vision*, pages 2961–2969, 2017.

[HSC+19]   Andrew Howard, Mark Sandler, Grace Chu, Liang-Chieh Chen, Bo Chen, Mingxing Tan, Weijun Wang, Yukun Zhu, Ruoming Pang, Vijay Vasudevan, et al. Searching for mobilenetv3. In *Proceedings of the IEEE International Conference on Computer Vision*, pages 1314–1324, 2019.

[HZRS16]   Kaiming He, Xiangyu Zhang, Shaoqing Ren, and Jian Sun. Deep residual learning for image recognition. In *Proc. of IEEE Computer Vision and Pattern Recognition*, 2016.

[IT+19]   Andrey Ignatov, Radu Timofte, et al. Pirm challenge on perceptual image enhancement on smartphones: report. In *European Conference on Computer Vision (ECCV) Workshops*, January 2019.

[JAFF16]   Justin Johnson, Alexandre Alahi, and Li Fei-Fei. Perceptual losses for real-time style transfer and super-resolution. In *Proc. of European Conference on Computer Vision*, 2016.

[JLJ+17]   V. Jakhetiya, W. Lin, S. P. Jaiswal, S. C. Guntuku, and O. C. Au. Maximum a posteriori and perceptually motivated reconstruction algorithm: A generic framework. *IEEE Transactions on Multimedia*, 19(1):93–106, 2017.

[JM18] Alexia Jolicoeur-Martineau. The relativistic discriminator: a key element missing from standard gan. *arXiv preprint arXiv:1807.00734*, 2018.

[KHZ+15] L. W. Kang, C. C. Hsu, B. Zhuang, C. W. Lin, and C. H. Yeh. Learning-based joint super-resolution and deblocking for a highly compressed image. *IEEE Transactions on Multimedia*, 17(7):921–934, 2015.

[KLCB20] H. Ko, D. Y. Lee, S. Cho, and A. C. Bovik. Quality prediction on deep generative images. *IEEE Transactions on Image Processing*, 29:5964–5979, 2020.

[LEGV17] Vitaliy Lyudvichenko, Mikhail Erofeev, Yury Gitman, and Dmitriy Vatolin. A semiautomatic saliency model and its application to video compression. In *Proc. of IEEE International Conference on Intelligent Computer Communication and Processing (ICCP)*, 2017.

[LEPV19] Vitaliy Lyudvichenko, Mikhail Erofeev, Alexander Ploshkin, and Dmitriy Vatolin. Improving video compression with deep visual-attention models. In *Proc. of International Conference on Intelligent Medicine and Image Processing*, IMIP '19, pages 88–94, New York, NY, USA, 2019. Association for Computing Machinery.

[LGTB14] Yu Li, Fangfang Guo, Robby T. Tan, and Michael S. Brown. A contrast enhancement framework with JPEG artifacts suppression. In *Proc. of European Conference on Computer Vision*, 2014.

[LHQ+17] Tao Li, Xiaohai He, Linbo Qing, Qizhi Teng, and Honggang Chen. An iterative framework of cascaded deblocking and super-resolution for compressed images. *IEEE Transactions on Multimedia*, 2017.

[LLWL20] Cheng-Han Lee, Ziwei Liu, Lingyun Wu, and Ping Luo. Maskgan: Towards diverse and interactive facial image manipulation. In *Proc. of CVPR*, 2020.

[LSD15] Jonathan Long, Evan Shelhamer, and Trevor Darrell. Fully convolutional networks for semantic segmentation. In *Proc. of IEEE Computer Vision and Pattern Recognition*, 2015.

[LTH+16] Christian Ledig, Lucas Theis, Ferenc Huszar, Jose Caballero, Andrew P. Aitken, Alykhan Tejani, Johannes Totz, Zehan Wang, and Wenzhe Shi. Photo-realistic single image super-resolution using a generative adversarial network. *CoRR*, abs/1609.04802, 2016.

[MBGDB21] Filippo Mameli, Marco Bertini, Leonardo Galteri, and Alberto Del Bimbo. A NoGAN approach for image and video restoration and compression artifact removal. In *Proc. of International Conference on Pattern Recognition (ICPR)*, 2021.

[MFC+19] I. Mitrica, A. Fiandrotti, M. Cagnazzo, E. Mercier, and C. Ruellan. Cockpit video coding with temporal prediction. In *Proc. of EUVIP*, pages 28–33, 2019.

[MMB12] A. Mittal, A. K. Moorthy, and A. C. Bovik. No-reference image quality assessment in the spatial domain. *IEEE Transactions on Image Processing*, 21(12):4695–4708, Dec 2012.

[MNDS18] Danial Maleki, Soheila Nadalian, Mohammad Mahdi Derakhshani, and Mohammad Amin Sadeghi. Blockcnn: A deep network for artifact removal and image compression. In *CVPR Workshops*, pages 2555–2558, 2018.

[MSB13] Anish Mittal, Rajiv Soundararajan, and Alan C Bovik. Making a "completely blind" image quality analyzer. *IEEE Signal Processing Letters*, 20(3):209–212, 2013.

[MSB16] Anish Mittal, Michele A Saad, and Alan C Bovik. A completely blind video integrity oracle. *IEEE Transactions on Image Processing*, 25(1):289–300, 2016.

[MSY16] Xiaojiao Mao, Chunhua Shen, and Yu-Bin Yang. Image restoration using very deep convolutional encoder-decoder networks with symmetric skip connections. In *Proc. of NIPS*, 2016.

[MTTA20] Fabian Mentzer, George Toderici, Michael Tschannen, and Eirikur Agustsson. High-fidelity generative image compression. 2020.

[ODO16] Augustus Odena, Vincent Dumoulin, and Chris Olah. Deconvolution and checkerboard artifacts. *Distill*, 2016. http://distill.pub/2016/deconv-checkerboard.

[PVZ15] Omkar M Parkhi, Andrea Vedaldi, and Andrew Zisserman. Deep face recognition. 2015.

[RB17] Oren Rippel and Lubomir Bourdev. Real-time adaptive image compression. In *Proc. of ICML*, 2017.

[RBB+16] Alessandro Redondi, Luca Baroffio, Lucio Bianchi, Matteo Cesana, and Marco Tagliasacchi. Compress-then-analyze versus analyze-then-compress: What is best in visual sensor networks? *IEEE Transactions on Mobile Computing*, 15(12):3000–3013, 2016.

[RNL+19] Oren Rippel, Sanjay Nair, Carissa Lew, Steve Branson, Alexander G Anderson, and Lubomir Bourdev. Learned video compression. In *Proc. of ICCV*, pages 3454–3463, 2019.

[SHBZ16] Pavel Svoboda, Michal Hradis, David Barina, and Pavel Zemcik. Compression artifacts removal using convolutional neural networks. *arXiv preprint arXiv:1605.00366*, 2016.

[SHZ+18] Mark Sandler, Andrew Howard, Menglong Zhu, Andrey Zhmoginov, and Liang-Chieh Chen. MobileNetV2: Inverted residuals and linear bottlenecks. In *Proc. of IEEE Computer Vision and Pattern Recognition*, June 2018.

[sky] Skype video conferencing application. http://www.skype.com.

[SZ15] Karen Simonyan and Andrew Zisserman. Very deep convolutional networks for large-scale image recognition. In *Proc. of ICLR*, 2015.

[TKL+20] Hossein Talebi, Damien Kelly, Xiyang Luo, Ignacio Garcia Dorado, Feng Yang, Peyman Milanfar, and Michael Elad. Better compression with deep pre-editing. *arXiv preprint arXiv:2002.00113*, 2020.

[TMM'19] Galteri, L., Seidenari, L., Bertini, M., and Del Bimbo, A. (2019). Deep universal generative adversarial compression artifact removal. *IEEE Transactions on Multimedia*, 21(8):2131–2145.

[WBSS04] Zhou Wang, A. C. Bovik, H. R. Sheikh, and E. P. Simoncelli. Image quality assessment: from error visibility to structural similarity. *IEEE Transactions on Image Processing*, 13(4):600–612, April 2004.

[WCC16] Xiaoli Wang, Aakanksha Chowdhery, and Mung Chiang. Skyeyes: Adaptive video streaming from uavs. In *Proc. of Workshop on Hot Topics in Wireless*, HotWireless '16, pages 2–6, New York, NY, USA, 2016. Association for Computing Machinery.

[WCRR+19] Maarten Wijnants, Sven Coppers, Gustavo Rovelo Ruiz, Peter Quax, and Wim Lamotte. Talking video heads: Saving streaming bitrate by adaptively applying object-based video principles to interview-like footage. In *Proc. of ACM Multimedia*, MM '19, pages 2449–2458, New York, NY, USA, 2019. Association for Computing Machinery.

[WLC+16] Zhangyang Wang, Ding Liu, Shiyu Chang, Qing Ling, Yingzhen Yang, and Thomas S Huang. D3: Deep dual-domain based fast restoration of JPEG-compressed images. In *Proc. of IEEE Computer Vision and Pattern Recognition*, 2016.

[WYW+18] Xintao Wang, Ke Yu, Shixiang Wu, Jinjin Gu, Yihao Liu, Chao Dong, Yu Qiao, and Chen Change Loy. Esrgan: Enhanced super-resolution generative adversarial networks. In *Proceedings of the European Conference on Computer Vision (ECCV)*, pages 0–0, 2018.

[YLK18] Jaeyoung Yoo, Sang-ho Lee, and Nojun Kwak. Image restoration by estimating frequency distribution of local patches. In *Proc. of IEEE Computer Vision and Pattern Recognition*, 2018.

[YLL+21] Chia-Hung Yeh, Chu-Han Lin, Min-Hui Lin, Li-Wei Kang, Chih-Hsiang Huang, and Mei-Juan Chen. Deep learning-based compressed image artifacts reduction based on multi-scale image fusion. *Information Fusion*, 67:195–207, 2021.

[YWP+18] Changqian Yu, Jingbo Wang, Chao Peng, Changxin Gao, Gang Yu, and Nong Sang. Bisenet: Bilateral segmentation network for real-time semantic segmentation. In *Proc. of European Conference on Computer Vision*, pages 325–341, 2018.

[ZIE+18] Richard Zhang, Phillip Isola, Alexei A Efros, Eli Shechtman, and Oliver Wang. The unreasonable effectiveness of deep features as a perceptual metric. In *CVPR*, 2018.

[zoo] Zoom video conferencing application. http://www.zoom.us.

[ZXF+13] X. Zhang, R. Xiong, X. Fan, S. Ma, and W. Gao. Compression artifact reduction by overlapped-block transform coefficient estimation with block similarity. *IEEE Transactions on Image Processing*, 22(12):4613–4626, 2013.

[ZXZ+16] J. Zhang, R. Xiong, C. Zhao, Y. Zhang, S. Ma, and W. Gao. CONCOLOR: Constrained non-convex low-rank model for image deblocking. *IEEE Transactions on Image Processing*, 25(3):1246–1259, March 2016.

Chapter 13
Conclusion

Jenny Benois-Pineau and Akka Zemmari

In this book we were interested in a large set of problems the new artificial intelligence methods such as Deep learning are facing. We first reminded fundamentals of Deep Learning approach and then tackle the most interesting and open problems of machine learning in the framework of Deep Learning.

Despite proficiency of a large set of models we have presented, various questions the machine learning community faces still remain open. One of them is the availability of large annotated datasets for training Deep Neural Networks. In several chapters, the contributing authors have made an overview of available annotated corpora, such as sport video datasets. Furthermore, they show how to implement the zero-shot learning with Deep neural Network classifiers which require a large amount of training data. The lack of annotated training data naturally pushes the researchers to implement low supervision algorithms. Metric learning is a long term research but in the framework of Deep Learning approaches it gets freshness and originality.

Deep learning approaches do penetrate the whole chain of visual and multimodal content acquisition, mining, coding. For such a well developed area as video coding, we have seen two examples: prediction of coding modes and improvement of the quality of transmitted video.

In the book almost all kinds of Deep Neural Networks have been covered: simple CNNS, Twin architectures, GANs, FCN, Auto-Encoders, RNNs... As Deep Learning approaches imitate human cognition process: learning from examples, we state that a very active research is fulfilled in the introduction of (self) attention mechanisms into network training process thus imitating the selectivity of human attention. Such mechanisms accelerate convergence of network training, slightly,

J. Benois-Pineau (✉) · A. Zemmari
LaBRI UMR 5800, University of Bordeaux, Talence Cedex, France
e-mail: jenny.benois-pineau@u-bordeaux.fr; akka.zemmari@u-bordeaux.fr

© Springer Nature Switzerland AG 2021
J. Benois-Pineau, A. Zemmari (eds.), *Multi-faceted Deep Learning*,
https://doi.org/10.1007/978-3-030-74478-6_13

but increase classification scores, as we have seen e.g. for action recognition with 3D CNNs.

All contributed chapters were built on the same principle presenting proposed methodological innovations and illustrations with application examples on large variety of data we call "multimedia": images, video, sound, cross-modality and in different application domains such as general purpose content and medical images, audio and music, sport videos...

The rise of deep learning, has led to a considerable increase in the performance of AI systems, but has also raised the question of the reliability and explicability of their predictions for decision making. These shortcomings raise many ethical questions which are especially crucial in such a domain as analysis of medical images. We have also seen the first contributions to building of explainable and trustworthy models for medical image analysis in our book.

Despite the richness of models, methods, the variety of data for which they were developed as presented in this book, research in Deep Learning is far from been exhausted. A lot of effort has to be made not only to improve accuracy of decisions by Deep Neural Networks by designing adequate architectures, formulating new objective functions, improving training algorithms, but also in making them trustworthy,explaining them to humans and bringing their decisions as much as possible to human decision process.

Printed in the United States
by Baker & Taylor Publisher Services